AF389431

Conductive Polymers

Conductive Polymers: Materials and Applications

Special Issue Editor

César Quijada

MDPI • Basel • Beijing • Wuhan • Barcelona • Belgrade

Special Issue Editor
César Quijada
Universitat Politècnica de
València
Spain

Editorial Office
MDPI
St. Alban-Anlage 66
4052 Basel, Switzerland

This is a reprint of articles from the Special Issue published online in the open access journal *Materials* (ISSN 1996-1944) from 2018 to 2020 (available at: https://www.mdpi.com/journal/materials/special_issues/conductive_polymers_materials_applications).

For citation purposes, cite each article independently as indicated on the article page online and as indicated below:

LastName, A.A.; LastName, B.B.; LastName, C.C. Article Title. *Journal Name* **Year**, *Article Number*, Page Range.

ISBN 978-3-03936-497-8 (Hbk)
ISBN 978-3-03936-498-5 (PDF)

Contents

About the Special Issue Editor

César Quijada (Associate Professor in Physical Chemistry).   César Quijada received his Ph.D. degree (1997) from the University of Alicante (Spain) for his studies on the fundamental surface electrochemistry of SO_2.   He was then appointed a postdoctoral fellowship with the laboratory of Dr. L. Berlouis (University of Stratchclyde, U.K.).   In 1999, he obtained an Assistant Professor professor position with the Polytechnic University of Valencia (Spain), where he has been Associate Professor in Physical Chemistry since 2008.   His research interests span the electrochemical processing of textiles and the development of novel nanostructured metal oxide electrocatalysts and producing polymer-based hybrid materials for energy conversion, environmental, and photoluminescence applications.

Editorial

Special Issue: Conductive Polymers: Materials and Applications

César Quijada

Departamento de Ingeniería Textil y Papelera, Universitat Politècnica de València. Pza Ferrándiz y Carbonell, E-03801 Alcoy (Alicante), Spain; cquijada@txp.upv.es; Tel.: +34-966-528-419

Received: 12 May 2020; Accepted: 18 May 2020; Published: 20 May 2020

Abstract: Intrinsically conductive polymers (CPs) combine the inherent mechanical properties of organic polymers with charge transport, opto-electronic and redox properties that can be easily tuned up to those typical of semiconductors and metals. The control of the morphology at the nanoscale and the design of CP-based composite materials have expanded their multifunctional character even further. These virtues have been exploited to advantage in opto-electronic devices, energy-conversion and storage systems, sensors and actuators, and more recently in applications related to biomedical and separation science or adsorbents for pollutant removal. The special issue "Conductive Polymers: Materials and Applications" was compiled by gathering contributions that cover the latest advances in the field, with special emphasis upon emerging applications.

Keywords: polyacetylene; polyaniline; polypyrrole; PEDOT-PSS; copolymers; charge transport models; silica gel composite; carbon composite; antimicrobial; drug release; sensors; adsorption

Intrinsically conductive polymers (CPs) are a fascinating family of organic materials that can be easily synthesized with a large diversity of chemical structures and a wide variety of micro- and nano-morphologies in order to obtain tailored macroscopic physical and chemical properties [1]. The breakthrough discovery of polyacetylene, the first conducting polymer, by Heeger, MacDiarmid and Shirakawa in 1977 (jointly awarded the Nobel Prize in Chemistry for 2000), opened up a completely new field of research, tracking the boundaries between chemistry and solid-state physics. Since then, interest in this intriguing class of polymers has expanded incessantly to become a well-established area of highly dynamic, multidisciplinary research.

From a chemical viewpoint, CPs are π-conjugated organic polymers, that is to say, compounds with their skeletal carbon atoms linked by both σ-bonds and the extended overlap of π-electron orbitals. As a result, neutral CPs show a semiconductor electronic structure with a completely filled π-band (valence band) and an empty π*-band (conduction band), separated by an energy gap of the order of 1 eV. The intrinsic conductivity arises from *doping*, i.e., when electrons are withdrawn or injected onto the conjugated polymeric chain, while the overall electroneutrality is retained by the incorporation of counter ions, the so-called *dopants*. The doped state is achieved by simple oxidation (p-doping) or reduction (n-doping) reactions, leading to the formation of delocalized charged structural defects (polarons, bipolarons, solitons) that are energetically located within the energy gap and work as charge carriers. Thus, the electrical conductivity of these polymers can be tuned over the full range from insulation to metallic by facile and reversible chemical or electrochemical doping/de-doping.

Jointly with simple control of the doping level, the electronic structure of CPs can be "engineered" by designing novel chemical structures at the molecular level through judicious synthetic chemical or electrochemical strategies, thus enabling a set of desired and tunable optical, electronic and redox/electrochemical properties. This concept has evolved to produce a vast repertory of chemically diverse, lightweight and flexible tailor-made polymer structures, showing attractive properties for a broad spectrum of applications with high technological impact, many of them reaching the marketplace.

Some examples are their use as key materials in thin-film transistors, light-emitting diodes and solar cells, electrochromic displays, (bio)sensors and actuators, secondary batteries and supercapacitors or artificial muscles, just to cite a few [1]. Over the past few decades, some other new applications in biomedical science and tissue engineering, enhanced solid-phase extraction or as adsorbent/ion-exchange materials for environmental issues have emerged as promising growth areas.

In spite of the obvious advantages of CPs over their inorganic semiconducting and metallic counterparts in terms of chemical diversity, tunable conductivity, low density and cost and flexibility, much work is still to be done in order to overcome their inherent limitations regarding solubility/processability, conductivity and long-term stability. For this, the development of CP-based hybrid composites with enhanced properties and controlled shape and morphology has become a flourishing area of discovery within the field. Typical examples are CP nanocomposites with a variety of carbon nanomaterials, metal oxide nanoparticles, or hydrogel inorganic matrices.

The Special Issue "Conductive Polymers: Materials and Applications" was launched to cover the latest advances and developments in the synthesis, characterization, structure–properties relationship and applications of intrinsically conductive polymers, with particular attention given to novel functionalized polymer/copolymer structures or novel CP–inorganic composite materials and their use in emergent applications. The nine articles included in the issue touch different aspects of the goals pursued. A brief summary of their main achievements and conclusions is given below.

In undoped trans-polyacetylene, bond alternation has been widely recognized as the obvious consequence of the existence of two-equivalent degenerate ground states arising from the Peierls instability. Hudson [2] critically reviewed the available experimental studies (X-ray diffraction (XRD), nuclear magnetic resonance (NMR), Raman or infrared (IR) data) that support the existence of bond alternation and concluded that their results are compromised by the presence of finite chains or finite conjugation segments or other ambiguities. The author proposed a novel synthetic route with the aid of urea inclusion complexes to produce fully extended all s-trans polyacetylene, free of finite-chain polyene impurities.

In recent years, the application of conducting polymer-based materials to biomedical science and healthcare has been the subject of intense research activity. The papers by Robertson et al. [3] and Shah et al. [4] fall within this category. In the former, the authors investigated the activity of polyaniline (PANI) and poly(3-aminobenzoic acid) (P3ABA) as potential antimicrobial agents against *Escherichia coli* and *Staphylococcus aureus*. Both CPs showed high bactericidal action in suspension, but their efficacy was depressed when applied to agar (an adsorbent surface) and especially to styrene-ethylene-butyrene-styrene films (a non-adsorbent surface). This behavior was attributed to the decreasing contact occurring between bacterial cells and CPs on going from solution to adsorbent surfaces to non-adsorbent films. The remaining activity of P3ABA-containing surfaces was reported to be superior to that published for triclosan, a popular antimicrobial agent, which encouraged the use of this polymer in cost-effective antimicrobial surfaces to break pathogen transmission pathways in hospitals. On the other hand, Shah et al. [4] reported the use of polypyrrole (PPY)-based coatings loaded with clinically relevant drugs as systems for switchable drug delivery. The anti-inflammatory dexamethasone phosphate (DMP) and the antibiotic meropenem (MER) were loaded as the anionic dopants. Analysis by energy-dispersive X-ray (EDX) spectroscopy, Fourier transform infrared (FTIR) spectroscopy, ultraviolet–visible (UV–vis) spectroscopy, XRD and electrochemical techniques (cyclic voltammetry (CV) and electrochemical impedance spectroscopy (EIS)) confirmed the presence of the drugs in the films. The authors showed that the electrochemically-triggered release of drugs in vitro was enhanced relative to the passive delivery from non-stimulated samples. The higher improvement observed for MER was attributed to its higher hydrophilicity and polar character, which makes it more responsive to electrically stimuli within the PPY matrix.

The application of CPs and their composites to the detection of biologically active molecules of interest in medical diagnosis, food or plant analysis is the subject of another three articles. A feature paper by Dkhili et al. [5] reported the electrosynthesis of novel, chemically stable aniline-piperazine copolymer films that exhibited reversible hydroxyl-ketopiperazine redox transitions between the well-known redox processes of polyaniline. The new copolymer material showed reproducible linear response and high sensitivity for the analysis of dopamine and ascorbic acid, thus being potentially applicable to amperometric sensors. Hybrid materials consisting of silica gel matrices modified with CPs have found interesting applications in analytical science. Djelad et al. [6] described the synthesis of SWCNT-modified silica films by electro-assisted deposition from sol-gel precursors, and their further modification by reactive electrochemical insertion of PEDOT-PSS from the monomer and dopant constituents. The resulting SWCNT@SiO$_2$–PEDOT-PSS composite doubled the electroactive surface area and exhibited a threefold increase in the heterogeneous rate constant of ferrocene, a common electron shuttle in electrochemical biosensors. Sowa et al. [7] modified commercial silica gel particles with PANI by chemical polymerization. Confocal Raman microscopy revealed that PANI preferentially deposited inside the inner pores of the silica particles. The authors optimized an experimental procedure for the extraction and quantification of triterpenic acids from plant samples by using PANI-modified silica gel as the sorbent in dispersive and matrix solid-phase extraction (d-SPE and MSPD) coupled to diode array detection high-performance liquid chromatography (DAD-HPLC).

Finally, the special issue also illustrates the use of CP-based composites as adsorbent materials for the removal of environmentally hazardous chemicals. In a series of two papers, Muhammad et al. [8,9] examined the adsorption behavior of PANI-Fe$_3$O$_4$ composites for the removal of cationic Basic Blue 3 (BB3) [8] and anionic Acid Blue 40 (AB40) [9] dyes from aqueous solution. The composite materials were fully characterized by scanning electron microscopy (SEM), FTIR, EDX, UV and XRD [8]. The adsorption of BB3 and AB40 obeyed Langmuir and Freundlich isotherm models, respectively, while kinetics was of the pseudo-second order. PANI-Fe$_3$O$_4$ composites showed enhanced adsorption capability for BB3 when compared to the individual components and other typical low-cost adsorbents. This effect was associated with an increase in surface area and pore volume of the hybrid material. Instead, the composite material was a less efficient adsorbent for AB40 than pure PANI, probably because repulsive forces from Fe$_3$O$_4$ oxygen lone pairs offset the electrostatic attraction between oppositely charged sites of PANI and the dye. Quijada et al. [10] electro-synthesized PANI-activated carbon cloth (ACC) composites in a filter-press cell. Based on SEM, X-ray photoelectron spectroscopy (XPS), N$_2$ adsorption, thermal analysis, CV and direct current (DC) conductivity data, the authors suggested that thin polymer films containing a small amount of phenazine/phenoxazine segments formed within the micro- and mesopores of the carbon fibers, thus showing enhanced conductivity and pseudocapacitance. In heavily loaded PANI-ACC, a nanofibrous thick coating developed which caused strong pore blocking, diminished surface area, and loss of conductivity. Composites with moderate PANI loadings showed promoted pseudo-second order adsorption rate of Acid Red 27 from aqueous solution, which was related to the electrostatic interaction between dye-negative sites and positive N sites in acid-doped PANI.

Conflicts of Interest: The author declares no conflict of interest.

References

1. Le, T.-H.; Kim, Y.; Yoon, H. Electrical and electrochemical properties of conducting polymers. *Polymers* **2017**, *9*, 150. [CrossRef] [PubMed]
2. Hudson, B.S. Polyacetylene: Myth and Reality. *Materials* **2018**, *11*, 242. [CrossRef] [PubMed]
3. Robertson, J.; Gizdavic-Nikolaidis, M.; Swift, S. Investigation of polyaniline and a functionalised derivative as antimicrobial additives to create contamination resistant surfaces. *Materials* **2018**, *11*, 436. [CrossRef] [PubMed]
4. Shah, S.A.A.; Firlak, M.; Berrow, S.R.; Halcovitch, N.R.; Baldock, S.J.; Yousafzai, B.M.; Hathout, R.M.; Hardy, J.G. Electrochemically enhanced drug delivery using polypyrrole films. *Materials* **2018**, *11*, 1123. [CrossRef] [PubMed]

5. Dkhili, S.; López-Bernabeu, S.; Kedir, C.N.; Huerta, F.; Montilla, F.; Besbes-Hentati, S.; Morallón, E. An electrochemical study on the copolymer formed from piperazine and aniline monomers. *Materials* **2018**, *11*, 1012. [CrossRef] [PubMed]

6. Djelad, H.; Benyoucef, A.; Morallón, E.; Montilla, F. Reactive insertion of PEDOT-PSS in SWCNT@Silica composites and its electrochemical performance. *Materials* **2020**, *13*, 1200. [CrossRef] [PubMed]

7. Sowa, I.; Wójciak-Kosior, M.; Strzemski, M.; Sawicki, J.; Staniak, M.; Dresler, S.; Szwerc, W.; Modoch, J.; Latalski, M. Silica modified with polyaniline as a potential sorbent for matrix solid phase dispersion (MSPD) and dispersive solid phase extraction (d-SPE) of plant samples. *Materials* **2018**, *11*, 467. [CrossRef] [PubMed]

8. Muhammad, A.; Shah, A.A.; Bilal, S.; Rahman, G. Basic Blue dye adsorption from water using Polyaniline/Magnetite (Fe_3O_4) composites: Kinetic and thermodynamic aspects. *Materials* **2019**, *12*, 1764. [CrossRef] [PubMed]

9. Muhammad, A.; Shah, A.A.; Bilal, S. Comparative study of the adsorption of Acid Blue 40 on polyaniline, magnetic oxide and their composites: Synthesis, characterization and application. *Materials* **2019**, *12*, 2854. [CrossRef] [PubMed]

10. Quijada, C.; Leite-Rossa, L.; Berenguer, R.; Bou-Belda, E. Enhanced adsorptive properties and pseudocapacitance of flexible polyaniline-activated carbon cloth composites synthesized electrochemically in a filter-press cell. *Materials* **2019**, *12*, 2516. [CrossRef] [PubMed]

Review

Polyacetylene: Myth and Reality

Bruce S. Hudson

Department of Chemistry, Syracuse University, Syracuse, NY 13244-4100, USA; bshudson@syr.edu;
Tel.: +1-315-443-5805

Received: 31 December 2017; Accepted: 31 January 2018; Published: 6 February 2018

Abstract: Polyacetylene, the simplest and oldest of potentially conducting polymers, has never been made in a form that permits rigorous determination of its structure. *Trans* polyacetylene in its fully extended form will have a potential energy surface with two equivalent minima. It has been assumed that this results in bond length alternation. It is, rather, very likely that the zero-point energy is above the Peierls barrier. The experimental studies that purport to show bond alternation are reviewed and shown to be compromised by serious experimental inconsistencies or by the presence, for which there is considerable evidence, of finite chain polyenes. In this view, addition of dopants results in conductivity by facilitation of charge transport between finite polyenes. The double minimum potential that necessarily occurs for polyacetylene, if viewed as the result of elongation of finite chains, originates from admixture of the 1^1A_g ground electronic state with the 2^1A_g excited electronic singlet state. This excitation is diradical (two electron) in character. The polyacetylene limit is an equal admixture of these two ^1A_g states making theory intractable for long chains. A method is outlined for preparation of high molecular weight polyacetylene with fully extended chains that are prevented from reacting with neighboring chains.

Keywords: polyacetylene; double-minimum potential; Peierls barrier; zero-point level; cross-linking

---

1. Introduction/Background History

Polyacetylene is selected for review because of its relative simplicity; the small periodic repeat permits polyacetylene to be treated by sophisticated computational methods. The path from bond-alternate potential minimum to symmetric bond-equivalent maximum is along a single normal mode (the Peierls mode). The vibrational spectrum of polyacetylene is relatively simple. In particular, the Raman spectrum is quite sparse and the oligopolyenes that lead to polyacetylene show characteristic frequency shifts and relative intensity changes which increase in chain length. On the experimental side, however, polyacetylene is entirely insoluble, reactive with itself, and has not been obtained in crystalline form that yields to single crystal diffraction, a feature that it shares with most polymers. Another reason for this review is that there have been several recent experimental and theoretical studies of polyacetylene and finite chain oligopolyenes of known length that are relevant to studies of the structure of polyacetylene. Furthermore, the evidence for the existence of bond alternation in polyacetylene has never been critically reviewed. This exercise shows that the fully extended chain with an all *trans* geometry has never been made. We outline a new method for the preparation of polyacetylene with these properties.

The first publication on the electrical conductivity of doped polyacetylene in 1977 [1], and in a recent 2017 review [2], it is stated as known that *trans*-polyacetylene exhibits bond alternation as a consequence of Peierls instability. It should be noted that Peierls instability refers to a negative curvature in the potential energy for a one-dimensional lattice. The resulting "dimerization", if it occurs, would mean that the periodic repeat for the electron exchange integral is two CH units, and thus the material has filled bonding and empty antibonding π-orbital bands. Thus, *trans* polyacetylene in its fully extended periodic form would be a semiconductor if it has bond alternation and a conductor if it does not. A close reading of the literature suggests an alternative interpretation of the argument: "since polyacetylene is not a conductor, it must be a semiconductor, and thus it must exhibit bond alternation". The alternative to this argument is that "polyacetylene" is really a mixture of finite chains and cross-linked polymer segments.

In this first of many papers [1], it was noted that bond alternation can also be inferred in long linear conjugated polyenes because the allowed optical "band gap" transition converges to a constant value with increase in conjugation length. An alternative view based on strong electron correlation for this limit that does not require bond alternation [3] in order to exhibit a finite limiting electronic excitation energy was mentioned in this initial work [1]. It is argued below that this alternative Mott semiconductor (Hubbard Hamiltonian) picture is substantially correct for very long chains.

Bond alternation is the key ingredient in the interpretation of the observations for polyacetylene in terms of a semiconductor band gap and "doping" enhancement of electrical conduction. In the recent review [2] it is stated that this Peierls barrier is very high on the basis that polyacetylene cannot be made to have equal bond lengths at any reasonable temperature. This proposed double well potential is rarely shown with a quantitative energy scale. The Peierls effect in the case of a π-electron system with a stiff σ-framework might lead to a negligibly small barrier. It is only in classical mechanics that atomic positions are at the bottom of the potential.

The earliest reasonable estimation of the "dimerization energy" (Peierls barrier) is a 1983 work [4], where Hartree Fock and small basis set MP2 (Moller-Plesset second order) methods were used. The MP2 value for the difference in energy at the symmetry point and either minima of the potential is 2 kJ/mol (700 cm^{-1}) [4]. A more extensive 1997 treatment [5] used multiple linear conjugated polyenes of increasing length, for which optimized structures were compared with equal bond structures or with structures that had one of several variations of bond alternation with optimized bond lengths at the molecular ends changing to equal bonds in the middle of the chain. The resulting dimerization energy extrapolated to $1/N = 0$ at the MP2/6-31G* level was 0.4 ± 0.1 kJ/mol depending on the method of structure variation used. The low end of this range is 0.3 kJ/mol or 105 cm^{-1}. The point that has never before been considered in this context is that the harmonic vibrational motion at the bond alternation geometry has a frequency known since at least 1958 to be in the 1500–1600 cm^{-1} range corresponding to a double bond stretching motion [6–8]. The harmonic zero-point energy is higher than the estimated barrier height.

We have shown, as discussed below, that while polyacetylene must have a potential energy surface with two equivalent minima, it cannot exhibit bond alternation, i.e., a periodic alternation between short and long bonds. The band structure argument that leads to this conclusion, leads conversely to the conclusion that if polyacetylene did not have bond alternation, it would be metallic. It would appear, however, that polyacetylene is not metallic unless it is "doped" with electron donors or acceptors. The terminology is from elemental semi-conductors where the doping is elemental. Here the "dopant" is usually molecular. Thus, the experimental evidence appears to differ from a conclusion in regard to ground state structure of polyacetylene that seems elementary. The alternative point of view investigated in this paper is that this conflict between our conclusions regarding the necessary lack of bond alternation of polyacetylene and experimental results for so-called "polyacetylene" is that what is called "polyacetyene" is not in fact polyacetylene, but is, rather, a mixture of finite chain polyenes of various lengths. The addition of dopants permits electron transport from chain to chain. These oligopolyenes do exhibit bond alternation and give rise in spectroscopic studies this property but this is due to end effects.

The term "finite polyene" here means that the molecule exhibits bond alternation with the terminal carbon–carbon bond length being shorter than the average of the values near the center of the molecule by an amount of 0.003 Å or more; barely measurable but otherwise arbitrary. We do not know and cannot easily compute how long a polyene needs to be such that it has two minima in its bond alternation potential and further how long it must be to have two equal energy minima. Whatever that is, it defines "approaching infinite". Minima of exact energy equality energy define "infinity" for the chain length.

To reiterate, our conclusion is not that polyacetylene has a single minimum potential energy for the bond alternation atomic displacement mode, but rather that the zero-point level is above the barrier that separates two minima. It is the probability distribution of the zero-point barrier that determines the structure, not the minima of the potential. Both minima are equally populated in the zero-point probability distribution. Simulations with reasonable parameters predict, in fact, that the maximum of the probability distribution is at the symmetry point where the potential is a maximum.

When referring to studies of the preparations, we use the notation "polyacetylene". We restrict the terms *cis-* or *trans-*polyacetylene to hypothetical infinite chains in their fully extended conformation. We are primarily concerned here with *trans-*polyacetylene. An infinite transationally symmetric chain is the starting point for the standard treatment of polymer vibrations in general as developed by Born and von Karman. As in all periodic problems, the description of the nuclear motion is the product of a local function times a function with translational symmetry along the chain propagation direction. It is the vibrational levels of the periodic repeat unit that count in establishing the zero-point level energy and the "optical" vibrational excitations in infrared and Raman spectra.

The relevant internal degree of freedom in this case is the bond alternation or Peierls distortion mode, a mixture of double bond expansion and single bond contraction. The potential energy in which the nuclei move for this degree of freedom necessarily has two minima that are exactly equivalent in energy only for the infinite chain. The equivalence of the minima derives from the fact that the energy difference between the two patterns of bond alternation becomes negligibly small per repeat unit as the chain length becomes very long. This argument makes no statement as to the height of any barrier between the two minima, if there is one.

Assuming that there is a barrier, then the issue of how to treat the nuclear motion arises. In this regard, there are two relevant observations. (1) The harmonic vibrational frequency that corresponds to motion along the Peierls degree of freedom is computed to have a value in the harmonic approximation of ca. 1500 cm^{-1}. This corresponds to the strongest Raman active mode at a similar wavenumber. (2) The best estimate of the height of the Peierls barrier via extrapolations discussed above is 100–300 cm^{-1} [5]. The harmonic zero point level is thus 2–7 times larger than this barrier height. Use of the harmonic approximation is clearly not justified.

2. Summary of This Review

We first review in Section 3 the double minimum problem in general, for two molecular cases, and then in Section 4 the specific case of polyacetylene. This is followed in Section 5 by a survey of experimental observations on polyacetylene in the literature that are relevant to bond alternation. It is found that X-ray diffraction, solid state NMR (Nuclear Magnetic Resonance), and polarized IR studies are compromised by ambiguities internal to the studies or to the presence of the finite chains, or both. The electronic spectroscopy of finite conjugated polyenes is then discussed in Section 6. The conclusion of these spectroscopic studies is that a low-lying doubly excited "diradical" state with the same symmetry as the ground electronic state is the lowest energy electronic excitation. This conclusion for the best studied case of octatetraene has recently received theoretical treatment, whose results are in excellent agreement with experiment. Admixture of this 2^1A_g excited state with the ground 1^1A_g state is the origin of the double minimum barrier for polyacetylene in its ground electronic state. It is also the basis of the difficulty in dealing theoretically with the ground state of polyacetylene with current periodic quantum chemical computational methods, since it requires inclusion of at least all doubly excited configurations at a non-perturbative level. This is followed by a brief review in Section 7 of the experimental electronic and vibrational Raman spectra of finite linear polyenes facilitated by the recent availability of such materials in homologous series. In Section 8, it is shown how these Raman spectra are relevant to our ongoing experimental Raman and vibrational inelastic neutron scattering studies of a molecular crystal for which photochemical elimination polymerization has been demonstrated to occur that leads to polyacetylene constrained to be fully extended in parallel channels formed by an inert lattice that also prevents cross-linking reactions. The iodine atoms that are photochemically cleaved are able to leave the host crystal as iodine vapor. In Section 9, the salient features gleaned from the literature are reviewed and an outlook is presented.

3. Double Minimum Potential Vibrational Energy Levels: Ammonia and [18]-Annulene

The mathematical technology for determination of the vibrational energy levels of arbitrary one-dimensional potential is now straightforward. These methods were developed to treat numerous molecular potentials [9–15] that have two equivalent minima. The most famous of these is ammonia, where the tunneling splitting is ca. 0.8 cm^{-1}. A potential that fits the precise vibrational data is shown in Figure 1 [9–13]. A potential that has the form $V(x) = C_2 x^2 + C_4 x^4$ with $C_2 = -9000$ cm^{-1} A^{-2} and $C_4 = 10,000$ cm^{-1} A^{-4} and a reduced mass of 1.008 amu has a tunneling splitting of 0.45 cm^{-1} (vs. 0.79 cm^{-1} of Figure 1). The 0 to 1 transitions of 932.5 and 968.3 cm^{-1} are computed to be at 940.3 and 969.8 cm^{-1}. The barrier height of 2031 cm^{-1} is 2025 cm^{-1} in this simple treatment using the efficient FGH (Fourier Grid Hamiltonian) method [14,15]. The reduced mass for ammonia varies along the out-of-plane umbrella coordinate. For the equilibrium pyramidal geometry, the value is 1.18 amu, while at the trigonal D_{3h} maximum it is 1.20 amu. This increase relative to the mass of H reflects the small geometry-dependent contribution of the N atom to the inversion normal mode. The zero-point level tunneling splitting of ammonia corresponds to an inversion time for the pyramidal superposition state of about 11 ps. This follows from the tunneling splitting 0.45 cm^{-1} for NH$_3$ in the simplest model treatment. This same model gives 792 ps for the tunneling splitting for ND$_3$.

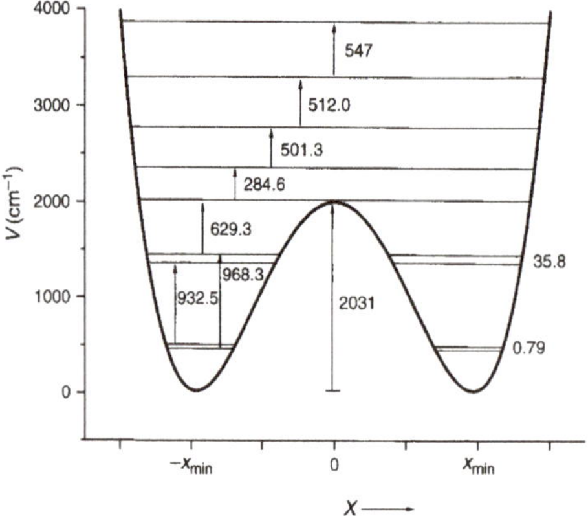

Figure 1. Umbrella mode potential for NH$_3$ with transition and level splittings indicated [9,10].

Figure 2. One of the D_{3h} Kekule structures of [18]-annulene.

Another molecular example of more relevance to polyacetylene is [18]-annulene, Figure 2 [16–18]. This simple cyclic C$_{18}$H$_{18}$ compound is the $4n + 2$ analog of benzene ($n = 1$) with $n = 4$, and is thus expected to be aromatic. To make a complicated story short, this conclusion is consistent with the observation of six-fold equivalent bonds in the X-ray diffraction structure but not with the computed NMR spectrum (for which the inside and outside protons are not shifted in opposite directions by the same amount as is the case for the D_{6h} symmetry). It has been proposed that [18]-annulene has a D_{3h} bond-alternate structure. A method of computation is found that results in a D_{3h} bond-alternate structure that results in agreement with the NMR spectrum [16]. This proposed geometry is either one of the structures corresponding to the minima of the potential in Figure 3. The zero-point level and probability distribution are shown. This proposed geometry is either one of the structures corresponding to the minima of the potential in Figure 3. The zero-point level and probability distribution are shown. A vibrational normal mode analysis at the symmetry point maximum and also at the minima gives in each case a reduced mass of 9.315 amu. The proton NMR spectrum computed for 200 points along bond order displacement coordinate weighted by the probability of Figure 3 gives a value in reasonable agreement with experiment. Other details of this density functional theory (DFT) and FGH treatment for [18]-annulene are in [17]. A classical MD (Molecular Dynamics) treatment for NMR averaging that includes this case is in [18]. An important factor for this case is that one of the normal modes of this molecule converts the structure from the maximum of the potential to either one of the minima and back. This example provides a demonstration that zero-point heavy atom averaging is expected in such cases because of the very stiff nature of the bonds prohibits localization into one of the minimum energy wells. The general point to keep in mind is that even with heavy atom motion, it is impossible to localize a carbon-based structure into a localized bond-alternate structure for a period of time that is significant on an experimental time scale. Benzene is the obvious example.

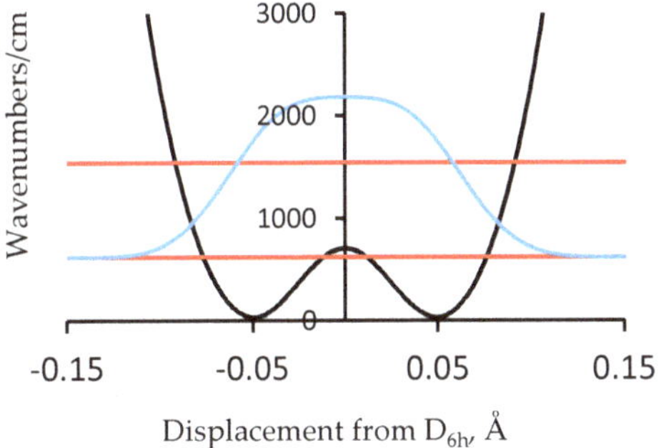

Figure 3. Computed potential energy as a function of displacement from 6-fold symmetry for [18]-annulene (black line) showing the two lowest vibrational energy levels (red) and the probability distribution for the ground state (blue) [17].

4. Double Minimum Potential Vibrational Energy Levels: Polyacetylene

For cases like ammonia, where the double minimum potential represents displacement of the three H atoms out of the molecular plane, and this case of a cyclic hydrocarbon, the potential must necessarily contain only even terms. The potential energy variation for polyacetylene must also necessarily be symmetric due to translational symmetry.

For the case of polyacetylene [19], for which periodic boundary conditions [20] apply, we have followed two independent paths of enquiry in Figures 4 and 5. In Figure 4, we compute the energy of the –CH–CH– periodic repeat using B3LYP/6-311G(2d,2p) with periodic boundary conditions evaluated at 240 points along the potential in one direction. This is then symmetrized by reflection. The barrier height computed by this DFT method is 110 cm^{-1}.

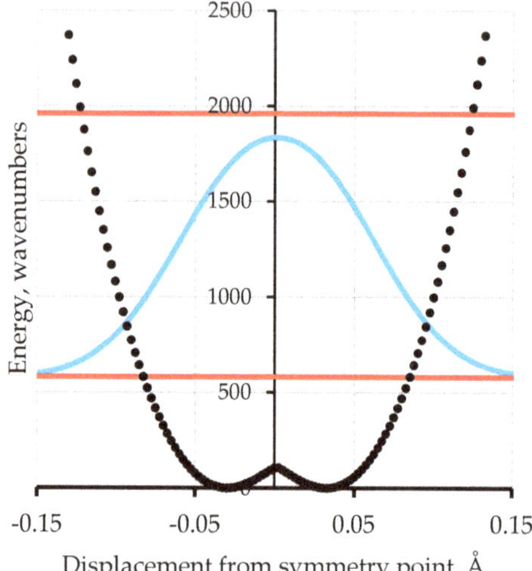

Displacement from symmetry point, Å

Figure 4. Computed potential energy of polyacetylene using periodic boundary conditions-density functional theory (PBC-DFT) with B3LYP 6-311G(2d,2p) at 240 points (black points) along one displacement direction with subsequent generation of the symmetric potential shown as blue dotted trace [19]. The horizontal red lines are the two lowest energy levels; the light blue line is the probability distribution.

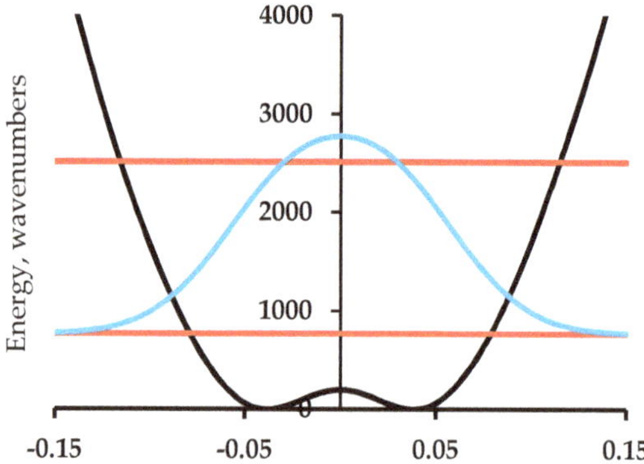

Figure 5. An analytical model potential energy (black curve) for the Peierls bond alternation mode of polyacetylene. The function is a harmonic oscillator plus a Gaussian barrier [19]. The lines are as in Figure 4.

Figure 5 shows the results of an empirical treatment of polyacetylene. The barrier height for polyacetylene has been established by the extrapolation procedure of Guo and Paldus [5] to be less than 200 cm^{-1}. In their treatment, a series of computational methods are applied to three structures for each member in a series of a finite polyene chains with an even number of carbon atoms. The structures are (1) the optimized bond-alternate structure, (2) the equal bond length (barrier) structure, and (3) the bond order reversed structure corresponding to the other minimum in the infinite chain case. These energies values relative to the optimized structure are plotted against $1/N$, where N is the number of C=C bonds. The values of this energy difference for the bond reversed and bond optimized cases must, of course, extrapolate to zero as $1/N$ goes to zero. The plot for the barrier height when extrapolated linearly gives a finite value of about 100 cm^{-1}. Figure 5 uses a barrier of 200 cm^{-1}. This value comes from the observation that for chain lengths that are sufficiently long, the computed values before extrapolation are below that value, so this value is an upper limit. The harmonic force constant for the model of Figure 5 is chosen to match the value of the force constant for C–C single bonds based on harmonic normal mode analyses for simple molecules like ethane. This is the lowest reasonable value. Higher values of this force constant parameter will result in a higher zero-point energy. The reduced mass for both cases is 4.33 amu. This is derived from a Gaussian computation for finite polyenes which uses the Wilson, Decius & Cross prescription [21]. This value depends on the C–C–C bond 120° C–C–C angle but is not crucially dependent on this value.

Calculations of the vibrational frequencies of long linear polyenes using MP2 wavefunction methods has been used to compute the force constants needed for harmonic treatment of polyacetylene in the periodic limit. It is found from the resulting vibrational eigenvalues that the vibrational motion that gives rise to the strongest feature in the Raman spectrum is the bond alternation or Peierls motion mode. Figures 4 and 5 show the computed first excited vibrational mode for these potentials are at 1379 cm^{-1} for the DFT computed potential and at 1460 cm^{-1} for the variable parameter treatment. At the time of this work, it was thought that the strongest Raman active mode of what was thought to be polyacetylene was at 1459 cm^{-1}. The 1460 cm^{-1} value was chosen as a target value in adjusting the barrier width of this analytic empirical model of Figure 5.

If this modelling procedure is followed with larger values of the barrier height including readjustment of the harmonic force constant so that the same value of the vibrational frequency is obtained, then with a barrier height of 2000 cm^{-1} the ground state zero-point level has a double maximum. Because of the symmetry, the probability of being in one well is the same as being in the other in this and every other state. If the barrier height is raised to 20,000 cm^{-1}, then the energy levels occur in pairs with a splitting for the lowest level of 20 cm^{-1} corresponding to femtosecond time scale tunneling. Bond-order alternation states will be exceedingly ephemeral.

5. Review of Experimental Observations on Polyacetylene with Emphasis on Bond Alternation

In the literature, especially in theoretical publications, bond alternation is often used to mean that there are two equivalent minima in the potential energy function. It is our argument above that this cannot give rise to populated bond-alternate structures with a periodic difference in bond lengths. If the zero-point energy is considered, then both structures are equally populated and, for what is considered to be at least a reasonable approximation to reality, the most probable geometry is at the symmetry point were the two bond lengths are equal. It might be argued that bond alternation in polyacetylene is known to occur on experimental grounds. This is, however, not the case when the experimental studies are viewed critically taking into account the likely presence of finite chains in the sample.

This section on experimental observations is divided into (1) X-ray diffraction, (2) infrared dichroism, (3) solid state NMR spectroscopy, (4) resonance Raman spectroscopy, and (5) a cautionary note on doping.

5.1. X-ray Diffraction

We begin with X-ray diffraction studies of polyacetylene. The initial 1982 work in this field [22] is discussed in a 1984 monograph [23], re-evaluated in a 1992 experimental study with new X-ray data [24], and then discussed in a 2010 computational paper [25]. Beginning with [22], it is noted that there are two possible monoclinic structures with two chains per unit cell P2$_1$/a and P2$_1$/n corresponding to the case of in-phase bond alternation or out-of-phase, respectively. For the former P2$_1$/a in-phase case, the (001) reflection is expected to be strong; for P2$_1$/n, it is forbidden. In this experiment [22], the (001) reflection is observed very weakly. The information in [22] as to bond alternation stems from an analysis of the shape of this (001) reflection with a two-parameter least squares fit of the alteration parameter u$_0$ and the monoclinic angle β to data with an intensity of unspecified physical origin, that depends on sample preparation, that clearly consists of more than one reflection and for which the statistics of the fit are not reported. The data presented in [22] are critically analyzed in [23] and in particular, the two parameter fit was repeated. No distinct minimum was observed for the least squared fit. The newer 1992 experimental study with new data of [24] took a global look at the full data set. It was decisively determined that the structure is $P2_1/n$ with expectation of a forbidden (001) but again, observed weakly. The authors note, "A possible explanation of the (001) intensity might be the coexistence of a small, variable fraction of bulk P21/a second phase. However, in our sample the (00L) intensities imply 20% $P2_1/a$ while the ratio of I(021)/I(011) gives an upper limit of only 4%. This inconsistency between on-axis and off-axis measures rule out the phase separation argument." They continue, "An alternative explanation is that local defects are responsible for the (001) intensity. For example, since the energy difference between the two structures is small, one might envision defects such as short chain segments which correlate in-phase with the surrounding long chains. Such defects would be a natural consequence of small molecular weight fragments, and could easily be quenched in from the polymerization. They would contribute to the (001) but not to general (HKL) intensities as is observed." Or they could be low molecular weight interstitial oligopolyenes. In the analysis of this data [24], the bond alternation parameter was not adjusted to fit the data. Instead, a value determined by NMR was used. This removes this X-ray study from having an impact on the bond-alternation issue. The validity of this NMR determination is discussed below.

5.2. Infrared Dichroism

Infrared (IR) absorption dichroism in the spectral region of the CH stretch [26] has been used to argue that polyacetylene exhibits bond-alternation. If polyacetylene has equal bond lengths for sequential bonds, then the 3013 cm^{-1} C–H stretch absorption will be, by symmetry, perpendicular to the chain (and thus perpendicular to the stretch direction for a stretched film). The absorption will be zero when the electric field lies along the orientation axis. It is found that this is not the case. The out-of-plane bending mode exhibits significant dichroism in the direction expected, indicating that the chain is well oriented. It was concluded that the local symmetry is C_s, not C_{2v}, permitting a dipole derivative with a component along the chain axis. However, a DFT calculation for the finite polyene chain $C_{60}H_{62}$ gives the dipole derivative for the IR-active CH mode at an angle of only about 25 degrees with the extended chain. This is presumably due to the very large axial polarizability. This axial component, however, vanishes in the postulated symmetric structure. It is expected that longer oligopolyene chains will have a larger axial/transverse ratio, and it is quite possible that the long-chain part of the distribution may dominate the IR spectrum, even if they are not the predominant species in the sample. The interpretation of this experiment depends on whether the signal is dominated by finite chains in the sample or by polyacetylene. This depends on both the amounts of these components and their relative intensities.

5.3. NMR Spectroscopy

The simplest indication of the nature of polyacetylene at the molecular level is the determination by CP MAS (Cross Polarization Magic Angle Spinning) ^{13}C NMR that $\approx$5% of the carbon atoms have sp^3 hybridization instead of the predominant sp^2 hybridization. The sp3 features "can probably be ascribed to chain terminations, cross-links, or hydrogenated double bonds" [27]. This corresponds to one atom in 20. If we ascribe the putative sp^3 carbons to cyclobutane ring formation then this corresponds to two cyclobutane defects in 200 carbons or 50 C=C double bonds from one to the next in two chains connected by four sp^3 carbons in cyclobutane rings at each end. This is, of course, only the average structure in a distribution.

Another NMR experiment aimed directly at detection of bond alternation is the determination of the ^{13}C–^{13}C dipolar coupling constant of polyacetylene prepared with low levels of acetylene–^{13}C$_2$ [28,29]. In this case, the observation of two coupling constants for *trans*-polyacetylene indicates two distinct bond lengths. This work has led to the prevailing bond alternation value and the individual C–C bond lengths for polyacetylene. For example, [30] compares a computed value to the bond lengths in [28]. There are, however, a few areas of concern in this work. The *cis*-polyacetylene isomer shows, as expected, only one coupling constant corresponding to a double bond length, 1.37 Å. Converting the sample of *cis*-polyacetyene to *trans*-polyacetylene is done by heating in vacuum at 160 °C for one hour. The resulting solid state nutation NMR spectra at 77 K show two coupling constants corresponding to 1.36 and 1.44 Å bond lengths. It is noted [28] that "The generation of approximately equal populations of singly and doubly bonded labeled carbon pairs in *trans*-$(CH)_x$ starting with only doubly bonded pairs in *cis*-$(CH)_x$ is intriguing." No explanation is given for this observation. The only obvious explanation is that half of the $^{13}C=^{13}C$ double bonds react with nearby predominantly $^{12}C=^{12}C$ double bonds to make $^{13}C-^{13}C$ single bonds from the original $^{13}C=^{13}C$ double bonds in cyclobutane rings. This is the kind of unambiguous experiment that is not done often enough. Presumably this also happens with half of the $^{12}C=^{12}C$ bonds. The result is best called *poly("ladderane")*. Performing a CP-MAS ^{13}C determination of random ^{13}C labeled polyacetylene treated in the same fashion. Both before and after thermal conversion to the *trans* form would be interesting.

The authors further investigated the temperature dependence of the $^{13}C-^{13}C$ splitting up to 300 K and did not see an expected coalescence of the features from defect migration along long chains. They speculated that the signals that they were observing came from chains that did not contain mobile defects necessary for thermal averaging, i.e., locked-in bond alternation due to cross-linking. In another work [29], it was concluded that "nuclear spin-lattice relaxation rates for ^1H and ^{13}C in polyacetylene cannot be adequately explained in terms of either nuclear spin diffusion to a static paramagnetic defect or rapid one-dimensional diffusion of the defect itself. A model in which only a small fraction of the molecular chains contain defects was proposed. Nuclei on these chains are rapidly relaxed, whereas the remainder achieve equilibrium by nuclear spin diffusion. The dependence of the measured relaxation rates upon frequency and isotopic concentration agreed with the predictions of the model." As indicated by the above, an alternative hypothesis for this signal is that it comes from finite polyene chains.

5.4. Resonance Raman Spectroscopy

Resonance Raman studies of "polyacetylene" [31–46] make use of the fact that Raman scattering under resonance conditions has a greatly enhanced contribution from species that have electronic excitation resonances near the Raman excitation frequency. If the sample is homogeneous, then the intensity of vibrational features in the resonance Raman spectrum will all increase or decrease together as the excitation frequency is changed. If, however, the sample is heterogeneous, with various components having different electronic excitation behavior and vibrational spectra, changing the excitation frequency will cause some vibrational features to increase and other decrease in intensity. In the case of "polyacetylene" containing multiple oligoene components, the electronic excitation spectrum moves to lower energy, and the strongly enhanced vibrational mode moves to lower frequency, as the chain length is increased. It is observed as expected that Raman excitation at longer wavelength results in lower frequency Raman active modes. The conclusion of most of these studies is that chain length heterogeneity is the most likely explanation for the experimental observations. Simple calculations reproduce the data, and alternative models were eliminated [43].

Finite linear conjugated polyenes have very strong Raman scattering. This is especially true when the Raman excitation is close to the strong electronic absorption bands of polyenes. The reason for the very strong Raman scattering of polyenes is that linear polyenes have a very strongly allowed electronic excitation that involves an appreciable geometry change in the excited state relative to the ground state. This geometry change is primarily along one normal mode of motion in the ground state near 1500 cm^{-1} with a smaller contribution from another mode around 1100 cm^{-1}. The details of this are discussed in Section 6 below. Here, we concentrate on "polyacetylene" in the context of its anticipated Raman scattering. This issue was noted in [44] in the context of periodic solids and in a way relevant to this review in [45,46]. In these publications, it is noted that the Condon mechanism for Raman scattering (also called A-term scattering) due to geometry changes in excited electronic states vanishes for periodic solids and for polyacetylene as we have defined it. This is because the one-electron excitation involved does not result in a significant geometry change for a system of effectively infinite size. The most argumentative statement of the issue is in [46], in which it is claimed that (a) since "polyacetylene" has a well-known Raman spectrum, it must be the case that something is missing that is beyond the Condon scattering mechanism; (b) a vibronic activity term is added (called B-term scattering), which (c) explains ("solving polyacetylene") by addition of the next term albeit with an unknown magnitude, and (d) that this refinement also removes the need to consider polyacetylene as a heterogeneous mixture of finite chains [31–42]. Specifically, quoting from [46], "In ref. 24 (of [46]), three samples of nearly monodisperse polyacetylene with lengths of about 200, 400, and 3800 unit cells were synthesized and their Raman spectra were obtained. The sidebands remained, and many of the earlier "polydisperse" explanations for the line shape quickly evaporated." The relevant ref. [24] is here ref. [35].

The authors of [46] have misrepresented [35] by claiming that the samples involved have very long conjugation length and ignoring the fact that in [35] itself these "monodisperse" samples are investigated as to the polydispersity of the conjugation lengths. Quoting from the abstract of [35] "After thermal isomerization, theoretical analysis of the Resonance Raman spectra using the Brivio, Mulazzi model indicate the ratio of long trans conjugated segments ($N \geq 30$) to short trans conjugated segments ($N \leq 30$) is significantly larger for 100,000 Dalton (3800 unit) polymer." It is the contention of this author that it is the $N = 15$–30 and perhaps a bit longer double bond species that have Raman spectra from the Condon mechanism, and that there is no evidence that anything that might be defined as polyacetylene makes any contribution to the Raman spectrum.

5.5. A Cautionary Note on Doping

Our argument, outlined above, is that it is not possible that the nuclear probability distribution will exhibit bond alternation in any experiment. The reported bond alternation in "polyacetylene" is evidence for the presence of finite chains. These may dominate the signals even if they are not the major component of the sample itself. The "semiconductor" properties attributed to "polyacetylene" are, in this interpretation, due to the lack of extended conducting chains. The observed effect of "doping" on conductivity, interpreted in terms of band theory, may rather be due to enhancement of conduction via electron transfer between finite chains rather than population of a conduction band, as assumed in the conventional model. In fact, iodine "doping" of polymers such as polybutadiene that do not contain any conjugated double bonds results in significant conductivity [47,48]. This is attributed [47] to the presence of ionic iodine chains such as I_3^- and I_5^- in the material. In [47], the conduction is claimed to be electronic.

6. Electronic Spectroscopy of Finite Linear Conjugated Polyenes

Interest in the electronic spectroscopy of linear conjugated polyenes began in the early days of the development of quantum mechanics. At that point, dealing with polyatomic molecules was based on qualitative molecular orbital and valence bond descriptions. Early treatments of butadiene with valence bond methods concluded that the lowest energy excitation would be a diradical with the same A_g symmetry in C_{2h} as the ground state. The observation of a strongly allowed electronic excitation contradicted this view. For linear conjugated polyenes with an even number of carbons in the C_{2h} point group, the symmetries of the non-degenerate molecular orbitals alternate in symmetry with increasing energy with a behavior similar to that of a particle in a box. Because of this, the HOMO-to-LUMO excitation is thus necessarily from an A_g ground state to an excited state of B_u electronic symmetry. This seemed to be in agreement with experiment in respect to a low-energy, strongly allowed transition that increased in intensity and decreased in energy as the polyene chain increased in length. There were, however, several aspects of linear conjugated polyene electronic spectroscopy and, especially, of the corresponding photophysics that attracted our attention that all was not right. The main things that we found of interest was the observation [49,50] that the intrinsic fluorescence (or radiative) lifetime of linear polyenes is much longer than expected on the basis of the integration of the absorption spectrum. The other items of interest were the series of spectroscopic papers on linear polyenes by Hauser and co-workers [51–56] and the discussion of that work by Mulliken [57]. In this discussion, Mulliken said:"A puzzling phenomenon reported by Kuhn, Hausser, and co-workers in their comparative study of absorption and fluorescence in polyene derivatives is the existence of a gap between the longest wave-length absorption band in the vibrational structure of N to V_1 and the shortest wave-length band the corresponding V_1 to N fluorescence spectrum. It is worth noting, however, that there seems to be still a small amount of absorption and emission at the middle of the gap. The width of the gap increases with increase in the number of conjugated double bonds. The absorption and fluorescence spectra are roughly mirror images of each other on a frequency scale. There seems to be no reason, especially in view of our theoretical analysis, to doubt that the fluorescence spectrum really is V_1 to N. According to the theory, no other excited singlet level should be below V_1." This is followed by an attempt to explain the observed spectral pattern in terms of an in-plane bending deformation of large amplitude. A similar explanation was posed for the long intrinsic lifetime [49,50].

The partially resolved vibronic spectra (Figure 6) of absorption and fluorescence of 1,3,5,7–octatetraene at 77 K have a gap between what appears to be the first absorption feature (the 0-0) near 32,000 cm^{-1} and the first fluorescence feature near 29,000 cm^{-1}. This is not anticipated if there is only one low-energy excited electronic state. On the other hand, if there is another low-lying state, this absorption should begin where the fluorescence begins near 29,000 cm^{-1}.

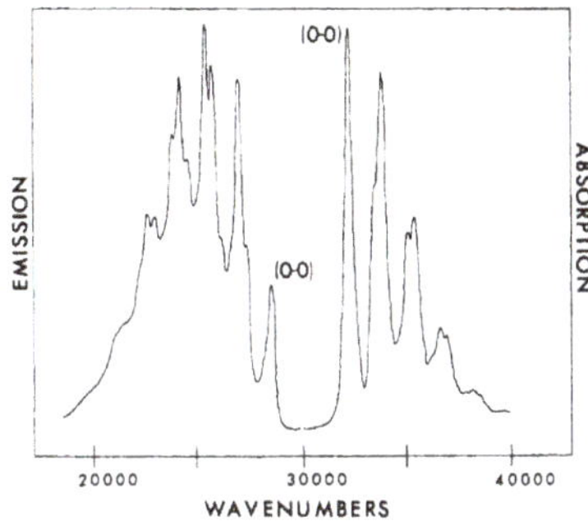

Figure 6. Fluorescence and absorption spectra of all *trans*-1,3,5,7-octatetraene in 3-methylpentane at 77K. Left, fluorescence on an arbitrary emission scale; right, absorption on an arbitrary absorbance scale [58].

In order to observe the hidden spectral features, it is necessary to use a solvent that at low temperature provides the same environment for all dissolved solute octatetraene molecules. This was found to be the case for n-octane, where apparently a centrosymmetric mixed crystal is formed. The spectra shown in Figures 7 and 8 are for polycrystalline matrices of n-octane with a very low concentration of octatetraene. The major features in these spectra are (a) the vibration-less origin (0,0) transition of the 1^1A_g to 1^1B_u electronic transition near 310 nm (32200 cm^{-1} in Figure 7), (b) the first absorption transition to the new low-energy, low-intensity transition near 348 nm (Figure 8) (28,730 cm^{-1} in Figure 7), and (c) the first emission transition from the as of yet unknown excited electronic state to the ground electronic state of 1^1A_g symmetry near 352 nm in Figure 8. The line shapes reflect phonon side-band structure.

Figure 8 is an expanded view of the origin region combining with addition in the upper trace the first one-photon feature of Figure 7. The two central features of the upper trace are "false origins" due to modes of b_u vibrational symmetry in the upper or the ground state. The lower trace is the two-photon excitation spectrum showing the true 0-0 at 350 nm plus phonon side bands and the lowest molecular a_g mode an in-plane bending vibration. These spectra present the classic pattern of Herzberg–Teller vibronic coupling, in which an electronic transition that is forbidden by symmetry is made allowed by a vibronic promoting mode that is non-totally symmetric and thus transiently reduces the molecular symmetry. The absence of the true origin transition in the one-photon excitation spectrum is a result of strict centrosymmetry in the n-octane crystal. The electronic symmetry of the ground electronic state is ^1A_g and so the upper level must also be ^1A_g. These are designated 1^1A_g and 2^1A_g respectively with the superscript 1 indicating single states. Use of n-nonane or n-heptane in place of n-octane induces significant intensity in the 0-0 transition due to the necessary loss of symmetry. The vibration-less origin transition is two-photon allowed as is observed. The strongly allowed one-photon transition at about 312 nm is ca. 10^5 times stronger than the first vibronic feature of the excitation spectrum in the 2^1A_g region. The vibronically active normal modes are expected to be of vibrational b_u symmetry because of the proximity of the strong 1^1B_u excitation. The two-photon high resolution spectral feature when compared to the high resolution fluorescence spectrum permits determination of the frequency the promoting mode as 86 cm^{-1}, which is an observed mode of this b_u symmetry. All of the above were the contribution of Bryan E. Kohler and his students [59–64] following the present author's initial contributions [65–68]. The above argument as to the presence of a low lying 2^1A_g state in octatetraene is airtight on experimental grounds. The development of the theoretical situation showed very early [69–72] that the low-lying 2^1A_g state is derived from doubly excited configurations that mix with singly excited configurations of the same symmetry to become the lowest excited state. It is now possible to compute the electronic excitations of octatetraene and get very good agreement with experiment using advanced *ab initio* methods [73].

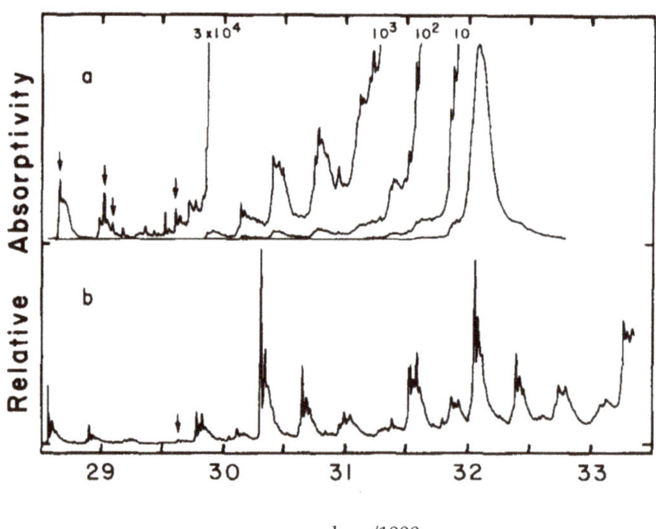

Figure 7. (**a**) The one-photon fluorescence excitation spectrum of octatetraene in n-octane matrix at 4.2 K. The arrows mark the vibronically induced transitions to the forbidden 2^1A_g excited state. The intense broad feature at 32,200 cm^{-1} is the vibration-less origin of the allowed electronic transition to the 1^1B_u excited state; (**b**) Two-photon fluorescence excitation spectrum of the same sample. All of the features are due to transitions to the 2^1A_g excited state. [59,67,68].

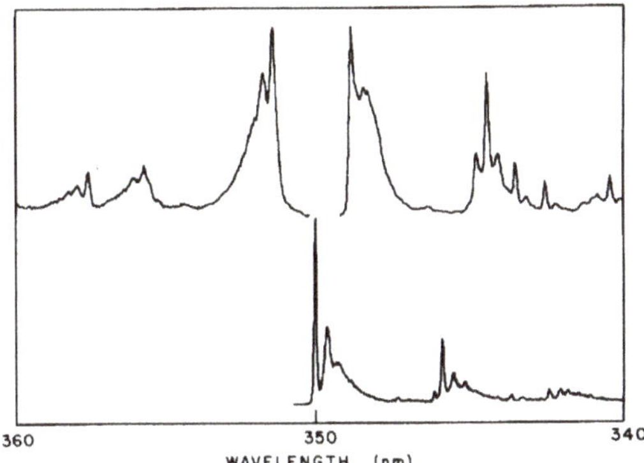

Figure 8. The upper trace left is the beginning of the fluorescence spectrum; the upper trace right is the beginning of the one-photon fluorescence excitation spectrum; the lower trace is the beginning of the two-photon fluorescence excitation spectrum. The two traces on the right are the same as the extreme left of Figure 7. [59,67,68].

Analysis of the pattern of the vibrational intensity of the strongly allowed absorption transition of spectra, like Figure 6, for a variety of polyenes shows that the features are due to the ca. 1500 and 1100 cm^{-1} C=C and C–C stretching modes. This is the case for both the transitions to or from the upper 2^1A_g electronic state and the allowed electronic absorption transition to the 1^1B_u state. This means that these excited states differ from the ground electronic 1^1A_g state by displacement along the total symmetric double bond and single bond contraction/expansion vibrational modes. The direction of the displacement is to upper states that have a reversal in their bond alternation pattern. The relative intensity of a vibration in an electronic transition is related to this displacement, which leads to finite overlap between the vibrational modes of the two states involved. These overlap integrals are called Franck–Condon factors. This displacement is the major mechanism that results in the vibrational structure and overall width of electronic spectra and in Raman activity of totally symmetric modes. This is called the A-term or Condon contribution to Raman scattering. This depends on the displacement of the potential energy minimum for the low-energy, strongly allowed electronic excitations. In a more general treatment of Raman scattering, there can also be cases where the mechanism of Raman intensity is due to the fact that some displacements of the atoms result in a change in the intensity of the electric dipole transition moment rather than a change in the energy of the potential energy surface that is required for non-zero Franck–Condon factors. This is especially important in electronic transitions that are between states and have electronic symmetries that cause the electric dipole transition moment to vanish at the equilibrium geometry, as is the case for the 1^1A_g to 2^1A_g electronic transitions of linear polyenes giving rise to the pattern of features shown above.

The relevance of the presence of a low-lying doubly excited 2^1A_g state in finite polyenes to properties of polyacetylene, as pointed out by Torii and Tasumi, is that polyacetylene in its ground electronic state must necessarily be an admixture of structures that have the standard pattern of bond alternation with another structure that has its bond order pattern reversed from that of the optimized structure [74]. The result is a polyacetylene with equal bond lengths. This has been investigated by Torii and Tasumi using the CASSCF (Complete Active Space Self Consistent Field) method. Their results for N = 6, dodecahexaene with an STO-3G basis set give an energy difference per CH group for the optimized geometry and that of the bond-reversed geometry of 2000 cm^{-1}. A slight inflection in the potential hints at the evolution toward a double minimum expected for longer chains.

There have been several recent synthetic efforts at preparation of oligopolyenes with defined lengths. One of these [75] is relevant to the location of the 2^1A_g state as a function of chain length and thus to the argument above concerning its ultimate admixture with the ground state. However, these are experimental studies and therefore the excitation energies of the 2^1A_g state from the 1^1A_g ground state already includes the mixing which will push the $2A_g$ state up as it is repelled by the receding 1^1A_g ground state. It is, in fact, already known that in finite polyenes the two lowest ^1A_g states are vibronically coupled [76–80].

7. Raman Vibrational Spectroscopy of Finite Polyenes

Another of these synthetic efforts concentrates on the Raman spectroscopy of finite conjugated polyenes [81]. This study is relevant to our investigation in progress of the preparation of polyacetylene *in situ* in a host–guest inclusion complex. In [81] the Raman spectra of a series of di-t-butyl polyenes with N = 3 to N = 12 C=C bonds were measured and discussed. The main observations for this series of compounds from [81] are:

- synthesis of these compounds is limited by solubility to N =12, i.e., N > 12 are insoluble;
- there are two strong Raman features near 1100 cm^{-1} and in the 1600–1500 cm^{-1} region;
- the lower C–C mode is not very sensitive to chain length;
- the higher C=C mode moves to lower frequency as the chain elongates;
- when plotted vs. 1/N, this C=C mode extrapolates to a value of ca. 1440 cm^{-1};
- the integrated intensity of the C–C mode increases relative the C=C mode as N increases.

All of the above are observed in the photochemical elimination polymerization reaction discussed below, which proceed for guest molecules in urea channel inclusion channels. In addition:

- there is a loss of mass corresponding quantitatively to the loss of iodine;
- loss of Raman intensity due to the decreasing effect of a one-electron excitation on a large chain.

8. In Situ Synthesis of Oriented Insulated Polyacetylene

Polyacetylene, whatever its limitations in terms of degree of polymerization, suffers from being entirely insoluble, conformationally disordered, subject to cross linking, and having a lack of macroscopic crystallinity. All of these factors make the characterization of this material problematic. We describe here the beginnings of a method for *in situ* generation of polyacetylene in a host inclusion complex. The objective of this is to force the *trans*-polyacetylene chain to be in its fully extended all s-*trans* configuration, to prevent close proximity of neighboring polyacetylene chains and to provide exclusion of oxygen. The initial approach to this objective is the use of urea inclusion compounds containing a reactive species. Urea inclusion compounds (UICs) are self-assembling crystalline structures formed from solution with inclusion of guest hydrophobic compounds with an extended structure. The most extensive example is the series of n-alkane urea inclusion complexes. These have a macroscopic hexagonal structure with the n-alkanes being ordered in two dimensions but disordered about their axis of rotation coincident with the hexagonal c-axis. An important aspect of urea inclusion crystals for our application is that the terminal atoms of the guest species are in contact in the complex. The urea host is not stable in the absence of the guest species; the urea lattice grows around the guest species. It is known that radical polymerization can be induced in UIC's, e.g., diene guests form very high molecular weight polymer poly(butadiene) [82–84]. The rigid urea tunnels allow guest rotations, translations, and lateral diffusion along the tunnel axis [82]. The guest molecules are generally more mobile than in single component molecular crystals, where reaction requires precise initial alignment.

The initial focus for our experimental effort aimed at the synthesis of long chain conjugated polyenes, in particular polyacetyelene in an all-*trans* extended conformation, begun with the preparation of a crystalline urea inclusion complex with E,E–1,4–diiodo–1,3–butadiene (DIBD) [85]. In Figure 9, we present the crystal structure of DIBD–urea as obtained by X-ray diffraction at 90 K viewed perpendicular and parallel to the channel axis [85]. Like all hexagonal urea inclusions, these crystals form as parallel channels in a host lattice densely packed with guest species [82–85]. The guest monomers, DIBD, are in end-to-end contact with each other and entrapped by the hydrogen bonded urea host. The internal diameter of the channels varies periodically along the channel from 5.5 to 5.8 Å. The separation of the parallel channels is 8.2 Å.

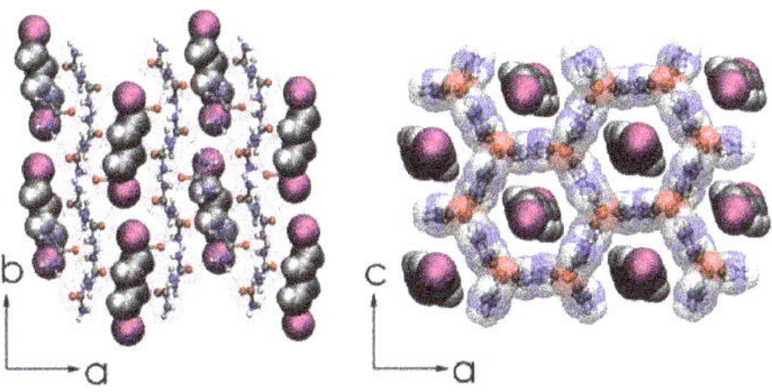

Figure 9. Representations of the commensurate, fully-ordered single-crystal DIBD–urea inclusion compound (UIC) complex as obtained by X-ray diffraction at 90 K viewed along the c (left panel) and b (right panel) crystal axes. Redrawn from structure Crystallographic Information File, cif of [85].

The DIBD–urea complex used in this study is unusual in that it is a commensurate structure, in contrast to most other UICs such as those formed by n-alkanes. DIBD–urea crystals have Raman features due to DIBD at 1600 and 1250 cm^{-1}. The strong tetragonal urea feature at 1010 cm^{-1} is shifted to 1022 cm^{-1}, the value observed for hexagonal urea. Irradiation with ultraviolet (UV) light results in new Raman modes near 1509 and 1125 cm^{-1} [86] (Figure 10a), nearly identical to spectra of *trans*-(CH)$_x$ prepared in solution [87]. The 254 nm radiation used in this experiment has very limited penetration into the DIBD–urea complex due to the high optical density. The change in composition of the urea channels is expected to be as shown in Figure 11.

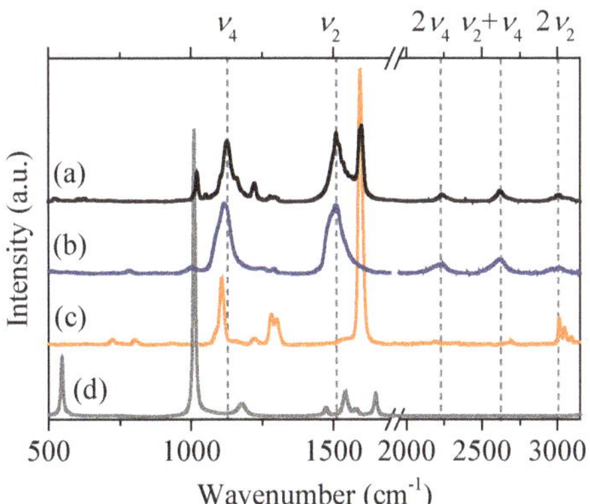

Figure 10. Raman spectra with 532 nm excitation of (**a**) DIBD–UIC after irradiation at 254 nm; (**b**) *trans* –(CH)$_x$; (**c**) crystalline DIBD; and (**d**) tetragonal urea [86]. The v_n values at the top are the mode frequencies for polyacetylene fundamental transitions and their overtones.

The overall progress of this irreversible second order sequential reaction is anticipated to be as illustrated in Figure 12. Continued progress requires that there be considerable axial diffusional motion of the chains in order to take up the space in the channel that has been vacated by the loss of iodine. This type of diffusional motion has been demonstrated with n-alkanes in UICs [88,89]. The overall progress of the reaction can be monitored with loss of mass due to release of iodine.

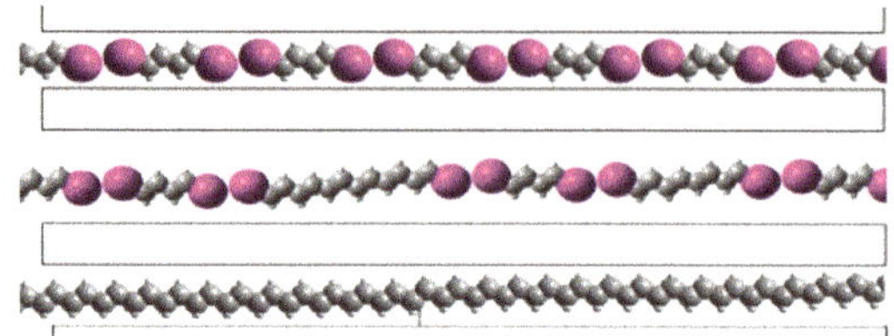

Figure 11. Schematic figure showing progress of the photochemical reaction from diiodobutadiene to polyacetylene with an intermediate stage showing a dimer and a trimer. The picture is to scale showing the large loss of channel filling with loss of iodine. There is a 2:1 ratio in the number of carbons in the bottom/top panels. Dimers and trimers have been shown by UV-vis of the extracted material. Longer chains have been shown by Raman.

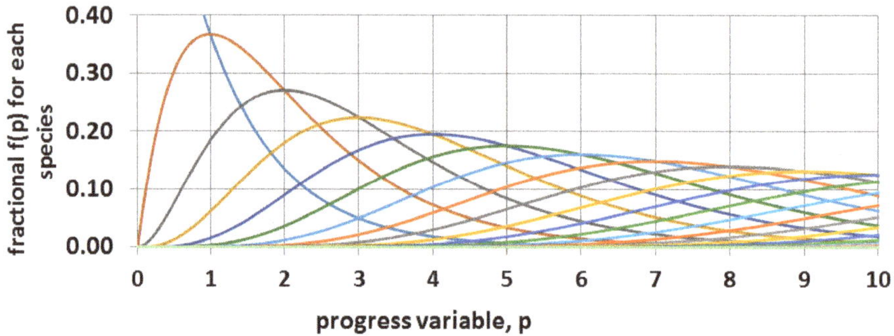

Figure 12. Irreversible sequential second order kinetics. The blue decreasing curve is for the monomer. The dimer peaks at $p = 1$ where $f_2 = f_1$, the trimer peaks at $p = 2$ where $f_3 = f_2$, etc. The number of carbons in the most frequent species is $C_N = 4(p + 1)$. The line colors differentiate the time dependence of the sequentially larger oligomers with their increasing delay.

9. Summary of Lessons from the Literature on Polyacetylene

The treatment of the vibrational level structure of polyacetylene unambiguously eliminates the possibility that bond alternation can be supported in the absence of terminal double bond end effects. Even if the barriers were much larger, the system would undergo tunneling. In reading the literature on this subject, the statement that the potential energy has a double minimum seems often to be the equivalent that the bond lengths will alternate. This is a fallacy. It is refreshing that all of the experimental methods used to establish bond alternation fail, in some cases in rather spectacular fashion.

In the case of X-ray diffraction the nature of the samples is essentially fibers and the periodic variation of internuclear separation is a relatively fine detail to pick out reliably from such data and that further blurred by disorder.

In the case of the NMR experiments, one wonders why the experiment was done. It could not be expected to succeed in its aim as designed, and it did not. It showed instead that polyacetylene undergoes cross-linking or some other form of conversion of double bonds to single bonds. This is also noted in the Raman studies of [35]. The better NMR experiment might be to use 50% random ^{13}C labeling.

It appears that the original interpretation of the extensive resonance Raman studies in terms of chain heterogeneity is sound with upper limits of chain length on the order of 40 double bonds.

It is our contention and that of several other workers that the Raman spectra that are observed in what is called "polyacetylene" is due to finite chains. This interpretation is consistent with observations that the most strongly enhanced feature in the Raman spectrum vary with the Raman excitation wavelength in an expected way with lower energy excitation resulting in stronger lower frequency vibrations due to preferential enhancement by longer chains. Examination of the available published data shows that there is no reliable evidence from diffraction, NMR, or IR data for bond alternation that can be demonstrated to be uncompromised by finite chains or finite conjugation segments or other ambiguities.

Our outlook is as follows. The methodology of *in situ* synthesis of oriented insulated polyacetylene has recently taken a step forward in the preparation of urea inclusion complexes with a different guest species that has the standard hexagonal lattice and morphology. This is expected to result in much faster reaction kinetics due to slightly looser packing, ease of orientation of the crystal channel axis, and the capability of making polyacetylene chains that are the length of the crystal. This is currently ca. 1.5 cm.

Acknowledgments: The author thanks Steluta A. Dinca, Damian G. Allis, Michael D. Moskowitz, Michael B. Sponsler, Mark Hollingsworth, and Luke Daemon for on-going access for discussions on specialist points.

Conflicts of Interest: The authors declare no conflict of interest.

References

1. Chiang, C.K.; Fincher, C.R.; Park, Y.W.; Heeger, A.J.; Shirakawa, H.; Louis, E.J.; Gau, S.C.; MacDiarmid, A.G. Electrical conductivity in doped polyacetylene. *Phys. Rev. Lett.* **1977**, *39*, 1098–1101. [CrossRef]

2. Swager, T.M. *50th Anniversary Perspective: Conducting/Semiconducting Conjugated Polymers*; A Personal Perspective on the Macromolecules; ACS Publications: Washington, DC, USA, 2017; Volume 50, pp. 4867–4886. [CrossRef]

3. Ovchinnikov, A.A.; Ukrainskii, I.I.; Kventsel, G.F. Theory of one-dimensional Mott semiconductors and the electronic structure of long molecules with conjugated bonds. *Uspekhi Fiziceskih Nauk* **1972**, *108*, 81–111. [CrossRef]

4. Suhai, S. Bond alternation in infinite polyene: Peierls distortion reduced by electron correlation. *Chem. Phys. Lett.* **1983**, *96*, 619–625. [CrossRef]

5. Guo, H.; Paldus, J. Estimates of the Structure and Dimerization Energy of Polyacetylene from Ab *Initio* Calculations on Finite Polyenes. *Int. J. Quant. Chem.* **1997**, *63*, 345–360. [CrossRef]

6. Lippincott, E.R.; White, C.E.; Siblia, J.P. Vibrational spectra and geometrical configuration of 1,3,5-hexatriene. *J. Am. Chem. Soc.* **1958**, *80*, 2926–2930. [CrossRef]

7. Lippincott, E.R.; Kenney, T.E. Vibrational spectra and geometric configuration of cis-1,3,5-hexatriene. *J. Am. Chem. Soc.* **1962**, *84*, 3641–3648. (includes updated data for *trans* form.) [CrossRef]

8. Guo, H.; Karplus, M. Ab initio studies of polyenes. I. 1,3-Butadiene. *J. Chem. Phys.* **1991**, *94*, 3679–3699. [CrossRef]

9. Laane, J. Eigenvalues of the Potential Function $V = z^4 \pm Bz^2$ and the Effect of Sixth Power Terms. *Appl. Spectrosc.* **1970**, *24*, 73–80. [CrossRef]

10. Bernath, P.F. *Spectra of Atoms and Molecules*; Oxford Univ. Press: New York, NY, USA, 1994; pp. 273–274.

11. Dennison, D.M.; Uhlenbeck, G.E. The two-minima problem and the ammonia molecule. *Phys. Rev.* **1932**, *41*, 313–321. [CrossRef]

12. Dennison, D.M. The infrared spectra of polyatomic molecules. II. *Rev. Mod. Phys.* **1940**, *12*, 175–214. [CrossRef]

13. Aquino, N.; Campoy, G.; Yee-Madeira, H. The inversion potential for NH_3 using a DFT approach. *Chem. Phys. Lett.* **1998**, *296*, 111–116. [CrossRef]

14. Marston, C.C.; Balin-Kurti, G.G. The Fourier grid Hamiltonian method for bound state eigenvalues and eigenfunctions. *J. Chem. Phys* **1989**, *91*, 3571–3576. [CrossRef]

15. Johnson III, R.D. Fourier Grid Hamiltonian (FGH) 1D Program. Available online: https://www.nist.gov/chemical-informatics-research-group/products-and-services/fourier-grid-hamiltonian-fgh-1d-program (accessed on 26 December 2017).

16. Wannere, C.S.; Sattelmeyer, K.W.; Schaefer III, H.F.; Schleyer, P.V.R. Aromaticity: The alternating C–C bond length structures of [14]-, [18]-, and [22]annulene. *Angew. Chem. Int. Ed.* **2004**, *43*, 4200–4206. [CrossRef] [PubMed]

17. Hudson, B.S.; Allis, D.G. The Structure of [18]-annulene: Computed Raman Spectra, Zero-point Level and Proton NMR Chemical Shifts. *J. Mol. Struct.* **2012**, *1023*, 212–215. [CrossRef]

18. Kwan, E.E.; Liu, R.Y. Enhancing NMR Prediction for Organic Compounds Using Molecular Dynamics. *J. Chem. Theory Comp.* **2015**, *11*, 5083–5089. [CrossRef] [PubMed]

19. Hudson, B.S.; Allis, D.G. Bond alternation in infinite periodic polyacetylene: Dynamical treatment of the anharmonic potential. *J. Mol. Struct.* **2013**, *1032*, 78–82. [CrossRef]

20. Born, M.; Kármán, T.V. Vibrations in Space-gratings (Molecular Frequencies). *Phys. Z.* **1912**, *13*, 297–309.

21. Wilson, E.B., Jr.; Decius, J.C.; Cross, P.C. *Molecular Vibrations: The Theory of Infrared and Raman Vibrational Spectra (Dover Books on Chemistry)*; Courier Corporation: North Chelmsford, MA, USA, 1980.

22. Fincher, C.R., Jr.; Chen, C.E.; Heeger, A.J.; MacDiarmid, A.G.; Hastings, J.B. Structural determination of the symmetry-breaking parameter in trans-polyacetylene (CH)x. *Phys. Rev. Lett.* **1982**, *48*, 100–104. [CrossRef]

23. Chien, J.C.W. *Polyacetylene: Chemistry, Physics, and Materials Science*; Academic Press: New York, NY, USA, 1984; pp. 112–117, ISBN 0-12-172460-3.

24. Zhu, Q.; Fischer, J.E.; Zuzok, R.; Roth, S. Crystal structure of polyacetylene revisited: An X-ray study. *Solid State Commun.* **1992**, *83*, 179–183. [CrossRef]

25. Zicovich-Wilson, C.M.; Kirtman, B.; Civalleri, B.; Ramirez-Solis, A. Periodic density functional theory calculations for 3-dimensional polyacetylene with empirical dispersion terms. *Phys. Chem. Chem. Phys.* **2010**, *1*, 3289–3293. [CrossRef] [PubMed]

26. Castiglioni, C.; Zerbi, G.; Gussoni, M. Peierls distortion in trans-polyacetylene—Evidence from infrared intensities. *Solid State Commun.* **1985**, *56*, 863–866. [CrossRef]

27. Maricq, M.M.; Waugh, J.S.; MacDiarmid, A.G.; Shirakawa, H.; Heeger, A.J. Carbon-13 nuclear magnetic resonance of cis- and trans-polyacetylenes. *J. Am. Chem. Soc.* **1978**, *100*, 7729–7730. [CrossRef]

28. Yannoni, C.S.; Clarke, T.C. Molecular Geometry of *cis*- and *trans*-Polyacetylene by Nutation NMR Spectroscopy. *Phys. Rev. Lett.* **1983**, *51*, 1191–1193. [CrossRef]

29. Scott, J.C.; Clarke, T.C. Nuclear magnetic relaxation in polyacetylene. *J. Phys. Colloq.* **1983**, *44*, 365–368. [CrossRef]

30. Izmaylov, A.F.; Scuseria, G.E. Efficient evaluation of analytic vibrational frequencies in Hartree-Fock and density functional theory for periodic nonconducting systems. *J. Chem. Phys.* **2007**, *127*, 144106/1–144106/9. [CrossRef] [PubMed]

31. Lefrant, S.; Lichtmann, L.S.; Temkin, H.; Fitchen, D.B.; Miller, D.C.; Whitewell II, G.E.; Burlitch, J.M. Raman scattering in (polyacetylene) and (polyacetylene) treated with bromine and iodine. *Solid State Commun.* **1979**, *29*, 191–196. [CrossRef]

32. Brivio, P.; Mulazzi, E. Absorption and resonant Raman scattering from trans-polyacetylene. *Chem. Phys. Lett.* **1983**, *95*, 555–560. [CrossRef]

33. Mulazzi, E.; Brivio, P.; Falques, E.; Lefrant, S. Experimental and theoretical Raman results in transpolyacetylene. *Solid State Commun.* **1983**, *46*, 851–855. [CrossRef]

34. Brivio, P.; Mulazzi, E. Theoretical analysis of absorption and resonant Raman scattering spectra of *trans*-polyacetylene ((CH)x). *Phys. Rev. B* **1984**, *30*, 876–882. [CrossRef]

35. Schen, M.A.; Chien, J.C.W.; Perrin, E.; Lefrant, S.; Mulazzi, E. Resonant Raman scattering of controlled molecular weight polyacetylene. *J. Chem. Phys.* **1988**, *89*, 7615–7620. [CrossRef]

36. Masetti, G.; Campani, E.; Gorini, G.; Piseri, L.; Tubino, R.; Piaggio, R.P.; Dellepiane, G. Resonance Raman-spectra of highly oriented *trans*-polyacetylene. *Solid State Commun.* **1985**, *55*, 737–742. [CrossRef]

37. Kuzmany, H. Resonance Raman-Scattering from Neutral and Doped Polyacetylene. *Phys. Status Soldi* **1980**, *97*, 521–531. [CrossRef]

38. Schuegerl, F.B.; Kuzmany, H. Optical modes of trans-polyacetylene. *J. Chem. Phys.* **1981**, *74*, 953–958. [CrossRef]

39. Kuzmany, H.; Imhoff, E.A.; Fitchen, D.B.; Sarhangi, A. Franck-Condon approach for optical absorption and resonance Raman scattering in trans-polyacetylene. *Phys. Rev. B* **1982**, *26*, 7109–7112. [CrossRef]

40. Kuzmany, H. The particle in the box model for resonance Raman scattering in polyacetylene. *Pure Appl. Chem.* **1985**, *57*, 235–246. [CrossRef]

41. Kuzmany, H.; Knoll, P. Application of the particle in the box model for resonance Raman scattering to recent experimental results of polyacetylene. *Mol. Cryst. Liq. Cryst.* **1985**, *117*, 385–392. [CrossRef]

42. Kuzmany, H.; Knoll, P. The Dispersion effect of Resonance Raman Lines in Trans-Polyacetylene, Springer Series in Solid-State Sciences. *Electron. Prop. Polym. Relat. Compd.* **1985**, *63*, 114–121.

43. Eckhardt, H.; Steinhauser, S.W.; Chance, R.R.; Schott, M.; Silbey, R. Anti-Stokes Raman-scattering in *trans* polyacetylene. *Solid State Commun.* **1985**, *55*, 1075–1079. [CrossRef]

44. Martin, R.M.; Falicov, L.M. *Resonance Raman Scattering in Light Scattering in Solids*; Cardona, M., Guntherodt, G., Eds.; Springer: New York, NY, USA, 1975; pp. 79–145, ISBN 03034216.

45. Jin, B.; Silbey, R. Theory of resonance Raman scattering for finite and infinite polyenes. *J. Chem. Phys.* **1995**, *102*, 4251–4260. [CrossRef]

46. Heller, E.J.; Yang, Y.; Kocia, L. Raman Scattering in Carbon Nanosystems: Solving Polyacetylene. *ACS Cent. Sci.* **2015**, *1*, 40–49. [CrossRef] [PubMed]

47. Thakur, M. A Class of Conducting Polymers Having Nonconjugated Backbones. *Macromolecules* **1988**, *21*, 661–664. [CrossRef]

48. Shang, Q.; Pramanick, S.; Hudson, B. Chemical nature of conduction in iodine-doped trans-1,4-poly(buta-1,3-diene) and some of its derivatives: The presence of I_3^- and the effect of double-bond configuration. *Macromolecules* **1990**, *23*, 1886–1889. [CrossRef]

49. Birks, J.B.; Dyson, D.J. The relations between the fluorescence and absorption properties of organic molecules. *Proc. R. Soc. Lond. Ser. A* **1963**, *275*, 135–148. [CrossRef]

50. Birks, J.B.; Birch, D.J.S. Fluorescence of diphenyl- and retinolpolyenes. *Chem. Phys. Lett.* **1975**, *31*, 608–610. [CrossRef]

51. Hausser, K.W.; Kuhn, R.; Smakula, A.; Kreuchen, K.H. Absorption of light and double bonds. I. Problem and methods. *Z. Phys. Chem.* **1935**, *B29*, 363–370.

52. Hausser, K.W.; Kuhn, R.; Smakula, A.; Hoffer, M. Absorption of light and double bonds. II. Polyene aldehydes and polyene carboxylic acids. *Z. Phys. Chem.* **1935**, *B29*, 371–377.

53. Hausser, K.W.; Kuhn, R.; Smakula, A.; Deutsch, A. Absorption of light and double bonds. III. Investigation in the furane series. *Z. Phys. Chem.* **1935**, *B29*, 378–383.

54. Hausser, K.W.; Kuhn, R.; Smakula, A. Absorption of light and double bonds. IV. Diphenylpolyenes. *Z. Phys. Chem.* **1935**, *B29*, 384–389.

55. Hausser, K.W.; Kuhn, R. Seitz, Absorption of light and double bonds. V. The absorption of compounds with conjugate double bonds of carbon at low temperature. *Z. Phys. Chem.* **1935**, *B29*, 391–416.

56. Hausser, K.W.; Kuhn, R.; Kuhn, E. Absorption of light and double bonds. VI. The fluorescence of diphenylpolyenes. *Z. Phys. Chem.* **1935**, *B29*, 417–454.

57. Mulliken, R.S. Intensities of electronic transitions in molecular spectra. VII. Conjugated polyenes and carotenoids. *J. Chem. Phys.* **1939**, *7*, 364–373. [CrossRef]

58. D'Amico, K.L.; Manos, C.; Christensen, R.L. Electronic energy levels in a homologous series of unsubstituted linear polyenes. *J. Am. Chem. Soc.* **1980**, *102*, 1777–1782. [CrossRef]

59. Granville, M.F.; Holtom, G.R.; Kohler, B.E.; Christensen, R.L.; D'Amico, K.L. Experimental confirmation of the dipole forbidden character of the lowest excited singlet state in 1,3,5,7-octatetraene. *J. Chem. Phys.* **1979**, *70*, 593–594. [CrossRef]

60. Granville, M.F.; Holtom, G.R.; Kohler, B.E. High-resolution one and two photon excitation spectra of *trans, trans*-1,3,5,7-octatetraene. *J. Chem. Phys.* **1980**, *72*, 4671–4675. [CrossRef]

61. Hudson, B.S.; Kohler, B.E. Polyene spectroscopy. Lowest energy excited singlet state of diphenyloctatetraene and other linear polyenes. *J. Chem. Phys.* **1973**, *59*, 4984–5002. [CrossRef]

62. Adamson, G.; Gradl, G.; Kohler, B.E. Photochemical hole burning for 1,3,5,7-octatetraene in n-hexane. *J. Chem. Phys.* **1989**, *90*, 3038–3042. [CrossRef]

63. Kohler, B.E. *Electronic Properties of Linear Polyenes*; Conjugated Polymers; Brédas, J.L., Silbey, R.J., Eds.; Kluwer Academic Publishers: Dordrecht, The Netherlands, 1991; pp. 405–434.

64. Kohler, B.E. Octatetraene photoisomerization. *Chem. Rev.* **1993**, *93*, 41–54. [CrossRef]

65. Hudson, B.S.; Kohler, B.E. Low-lying weak transition in the polyene α,ω-diphenyl-octatetraene. *Chem. Phys. Lett.* **1972**, *14*, 299–304. [CrossRef]

66. Hudson, B.; Kohler, B. Linear polyene electronic structure and spectroscopy. *Annu. Rev. Phys. Chem.* **1974**, *25*, 437–460. [CrossRef]

67. Hudson, B.S.; Kohler, B.E.; Schulten, K. Linear polyene electronic structure and potential surfaces. *Excit. States* **1982**, *6*, 1–95.

68. Hudson, B.; Kohler, B. Electronic structure and spectra of finite linear polyenes. *Synth. Met.* **1984**, *9*, 241–253. [CrossRef]

69. Schulten, K.; Karplus, M. Origin of a low-lying forbidden transition in polyenes and related molecules. *Chem. Phys. Lett.* **1972**, *14*, 305–309. [CrossRef]

70. Schulten, K.; Ohmine, I.; Karplus, M. Correlation effects in the spectra of polyenes. *J. Chem. Phys.* **1976**, *64*, 4422. [CrossRef]

71. Tavan, P.; Schulten, K. The low-lying electronic excitations in long polyenes: A PPP-MRD-CI study. *J. Chem. Phys.* **1986**, *85*, 6602–6609. [CrossRef]

72. Tavan, P.; Schulten, K. Electronic excitations in finite and infinite polyenes. *Phys. Rev. B* **1987**, *36*, 4337–4358. [CrossRef]

73. Angeli, C.; Pastore, M. The lowest singlet states of octatetraene revisited. *J. Chem. Phys.* **2011**, *134*, 184302. [CrossRef] [PubMed]

74. Torii, H.; Tasumi, M. Changes in the electronic structures of trans-polyenes in the 1^1A_g and 2^1A_g states induced by molecular vibrations. *Chem. Phys. Lett.* **1996**, *260*, 195–200. [CrossRef]

75. Christensen, R.L.; Enriquez, M.M.; Wagner, N.L.; Peacock-Villada, A.Y.; Scriban, C.; Schrock, R.R.; Polivka, T.; Frank, H.A.; Birge, R.R. Energetics and Dynamics of the Low-Lying Electronic States of Constrained Polyenes: Implications for Infinite Polyenes. *J. Phys. Chem. A* **2013**, *117*, 1449–1465. [CrossRef] [PubMed]

76. Orlandi, G.; Zerbetto, F. Vibronic coupling in polyenes: The frequency increase of the active C = C a_g stretching mode in the absorption spectra. *Chem. Phys.* **1986**, *108*, 187–195. [CrossRef]

77. Zerbetto, F.; Zgierski, M.Z.; Orlandi, G. Correlation between the frequency of the Franck-Condon active carbon:carbon a_g stretch vibration and the excitation energy of the $1B_u$ electronic state in polyenes. *Chem. Phys. Lett.* **1987**, *141*, 138–142. [CrossRef]

78. Buma, W.J.; Zerbetto, F. The large $1^1A_g^- $-$2^1A_g^-$ C=C and C–C stretch vibronic interaction in all-trans polyenes. *Chem. Phys. Lett.* **1998**, *289*, 118–124. [CrossRef]

79. Buma, W.J.; Zerbetto, F. Modeling the Spectroscopy of the Lowest Excited Singlet State of *cis,trans*-1,3,5,7-Octatetraene: The Role of Symmetry Breaking and Vibronic Interactions. *J. Phys. Chem. A* **1999**, *103*, 2220–2226. [CrossRef]

80. Fuss, W.; Haas, Y.; Zilberg, S. Twin states and conical intersections in linear polyenes. *Chem. Phys.* **2000**, *259*, 273–295. [CrossRef]

81. Schaffer, H.E.; Chance, R.R.; Silbey, R.J.; Knoll, K.; Schrock, R.R. Conjugation length dependence of Raman scattering in a series of linear polyenes: Implications for polyacetylene. *J. Chem. Phys.* **1991**, *94*, 4161–4170. [CrossRef]

82. Hollingsworth, M.D.; Harris, K.D.M. Urea, Thiourea, and Selenourea. In *Comprehensive Supramolecular Chemistry*; Solid State Supramolecular Chemistry: Crystal Engineering; Atwood, J.L., Davies, J.E.D., MacNicol, D.D., Vogtle, F., Eds.; Elsevier: Oxford, UK, 1996; Volume 6, pp. 177–237.

83. Harris, K.D.M. Fundamental and applied aspects of urea and thiourea inclusion compounds. *Supramol. Chem.* **2007**, *19*, 47–53. [CrossRef]

84. Harris, K.D.M.; Palmer, B.A.; Edwards-Gau, G.R. Reactions in Solid-State Inclusion Compounds. In *Supramolecular Chemistry: From Molecules to Nanomaterials*; Gale, P., Steed, J., Eds.; John Wiley & Sons, Ltd.: Chichester, UK, 2012; Volume 4, pp. 1589–1612, ISBN 978-0-470-74640-0.

85. Lashua, A.F.; Smith, T.M.; Hu, H.; Wei, L.; Allis, D.G.; Sponsler, M.B.; Hudson, B.S. Commensurate urea inclusion crystals with the guest (E,E)-1,4-diiodo-1,3-butadiene. *Cryst. Growth Des.* **2013**, *13*, 3852–3855. [CrossRef]

86. Dincă, S.A.; Allis, D.G.; Lashua, A.F.; Sponsler, M.B.; Hudson, B.S. Insulated polyacetylene chains in an inclusion compound by photopolymerization. *MRS Online Proc. Libr. Arch.* **2015**, *1799*, 1–6. [CrossRef]

87. Schuehler, D.E.; Williams, J.E.; Sponsler, M.B. Polymerization of acetylene with a ruthenium olefin metathesis catalyst. *Macromolecules* **2004**, *37*, 6255–6257. [CrossRef]

88. Marti-Rujas, J.; Desmedt, A.; Harris, K.D.M.; Guillaume, F. Kinetics of molecular transport in a nanoporous crystal studied by confocal raman microspectrometry: Single-file diffusion in a densely filled tunnel. *J. Phys. Chem. B* **2007**, *111*, 12339–12344. [CrossRef] [PubMed]
89. Marti-Rujas, J.; Desmedt, A.; Harris, K.D.M.; Guillaume, F. Bidirectional transport of guest molecules through the nanoporous tunnel structure of a solid inclusion compound. *J. Phys. Chem. C* **2009**, *113*, 736–743. [CrossRef]

Article

Investigation of Polyaniline and a Functionalised Derivative as Antimicrobial Additives to Create Contamination Resistant Surfaces

Julia Robertson [1], Marija Gizdavic-Nikolaidis [2] and Simon Swift [1,*

[1] Department of Molecular Medicine and Pathology, The University of Auckland, Auckland 1023, New Zealand; julia.robertson@auckland.ac.nz
[2] School of Chemical Sciences, The University of Auckland, Auckland 1142, New Zealand; m.gizdavic@auckland.ac.nz
* Correspondence: s.swift@auckland.ac.nz; Tel.: +64-9-373-7599 (ext. 86273)

Received: 2 February 2018; Accepted: 13 March 2018; Published: 16 March 2018

Abstract: Antimicrobial surfaces can be applied to break transmission pathways in hospitals. Polyaniline (PANI) and poly(3-aminobenzoic acid) (P3ABA) are novel antimicrobial agents with potential as non-leaching additives to provide contamination resistant surfaces. The activity of PANI and P3ABA were investigated in suspension and as part of absorbent and non-absorbent surfaces. The effect of inoculum size and the presence of organic matter on surface activity was determined. PANI and P3ABA both demonstrated bactericidal activity against *Escherichia coli* and *Staphylococcus aureus* in suspension and as part of an absorbent surface. Only P3ABA showed antimicrobial activity in non-absorbent films. The results that are presented in this work support the use of P3ABA to create contamination resistant surfaces.

Keywords: antimicrobial; surfaces; infection control; polyaniline; *Escherichia coli*; *Staphylococcus aureus*

---

1. Introduction

Microbial resistance to antimicrobial agents is increasing worldwide and it represents a major threat to the successful treatment of infectious diseases [1]. Development of antimicrobial resistance is an inescapable consequence of natural selection and is associated with exposure to antimicrobial agents [1]. Efforts need to be made to decrease the unnecessary exposure of bacteria to antibiotics to reduce the selective pressure driving the development of resistance so that existing antibiotics retain their efficacy for as long as possible [1]. In part, this can be achieved by controlling the spread of pathogenic bacteria and therefore reducing the number of infections that require antibiotic treatment [2].

Healthcare-associated infections are a major contributor to patient morbidity and mortality, and occur in part due to bacterial contamination of hospital surfaces [3]. Surfaces in hospitals that come into contact with hands are regularly contaminated with nosocomial pathogens [4,5]. Infected patients shed pathogenic bacteria, including methicillin-resistant *Staphylococcus aureus* and vancomycin-resistant *Enterococcus* spp., into their immediate environment [5–10]. Surfaces near shedding patients, such as walls, door handles, bed frames, and light switches, tend to be touched frequently and therefore are more likely to be contaminated [3,10,11]. Once a surface is contaminated, a single hand contact event is sufficient to transmit bacteria from the surface to a person [7,8,12].

Bacteria that have been transferred to a surface can persist for a period of time or actively colonise to form a biofilm. Bacterial persistence on a surface is influenced by dynamic environmental conditions, including organic soiling, humidity, and temperature [4,13,14]. Biofilms are highly recalcitrant to antimicrobial treatments and facilitate the persistence of bacteria on surfaces resulting in surface associated pathogen reservoirs, which increase the risk of transmission [15–17]. The level of bacterial transfer that occurs between a contaminated surface and a hand following contact has been demonstrated to occur at a comparable level to direct contact with an infectious patient, which is a well-established transmission route [4,18,19]. Hand washing can help to control the spread of infection in hospitals; however, without the decontamination of surfaces, the reservoirs of pathogens will seed further spread. Nosocomial pathogens isolated from hospital surfaces are typically in the range of 100–10,000 colony forming units (CFU)/cm^2 [5,10]. For a microbial burden exceeding 250 CFU/100 cm^2, transmission from the surfaces to health care workers and/or patients increases [10,20,21]. Therefore, despite the relatively low inocula present, any contamination of a hospital surface by a pathogen should be considered to be a transmission risk [10,20,21].

The involvement of contaminated surfaces in pathogen transmission pathways in hospitals necessitates the improved control of surface microbiology. Reduction of microbial contamination on hospital surfaces could disrupt transmission pathways and potentially reduce infectious disease incidence rates and the associated antibiotic usage [22]. Utilisation of antibacterial surfaces is a promising means of reducing microbial surface load as well as preventing formation of biofilms and surface associated pathogen reservoirs [2]. An ideal antimicrobial surface would be active against relevant bacteria at appropriate bacterial loads and active in environmental conditions relating to potential applications in terms of temperature, relative humidity, pH, exposure to cleaning products, and contaminating organic matter [23–25]. The time that is required for decontamination would need to be sufficiently short to be effective in breaking transmission pathways. Activity needs to be retained for sufficiently long periods of time, and after repeated bacterial challenges, to be cost effective [26,27]. Activity overtime is informed by whether the antimicrobial agent is immobilised on the surface or if it has to be released to elicit an effect [28,29]. The release of an antimicrobial agent over time means that the surface concentration of the agent will fall below the threshold needed to exert antimicrobial activity [27]. An ideal antimicrobial surface would also need to be cheap and easy to make, suitable for large-scale production, and have regulatory approval for the intended use [26,30].

To create an antimicrobial surface, we can take one of two basic approaches. First, a coating may be applied to a material or a modification of the surface chemistry of the material made to provide an antimicrobial surface [31]. Alternatively, the material may be fabricated by incorporating an antimicrobial into the material, which can be challenging as manufacturing procedures can involve extreme environmental conditions, including high temperatures and shear forces, which can negatively impact on bactericidal activity [26,28,32]. Covalent attachment of an antimicrobial agent to a surface may cause side reactions that result in conformational changes in the agent, ultimately causing a loss of activity [33]. Therefore, the method of antimicrobial surface production may affect the resulting surface activity.

The activity of an antimicrobial surface is also influenced by the nature of the surface. Surfaces can be absorbent allowing water droplets to move into the surface or they can be non-absorbent, in which water droplets sit on top of the surface [34,35]. These surface properties may affect the antimicrobial activity as a bacterium in a water droplet would have more contact with the antimicrobial agent if it has absorbed into the surface. Non-absorbent surfaces in hospitals are frequently contaminated with pathogens, and include walls, door handles, and bed frames. Much of the focus of development of antimicrobial surfaces in the published literature is on model non-absorbent surfaces, such as metal coupons and plastic films [36–38]. Many absorbent surfaces in hospitals are fabric-based, such as apparel worn by healthcare workers and patient privacy curtains [36,39,40]. Privacy curtains are high-touch areas that are contacted by the hands of the healthcare worker before, during, and after patient care, and are infrequently changed [40,41]. It has been demonstrated that more than 90% of privacy curtains can become contaminated within a week of use [41]. Contaminated absorbent surfaces in hospitals may be involved in pathogen transmission [36,39]. Absorbent surfaces are harder to clean or disinfect than non-absorbent surfaces, while the latter facilitates a greater transfer of bacteria [8,41,42]. Therefore, the development of both absorbent and non-absorbent antimicrobial surfaces would help to curtail the spread of infection in hospitals [1].

Antimicrobial polymers are good candidates for immobilised biocides. These polymers can be either polymeric biocides (the repeating unit is a biocide) or biocidal polymers (the active principle is embodied by the whole macromolecule) [28,29]. In this article, we investigate the antimicrobial activity of polyaniline (PANI) and a functionalised derivative (fPANI), homopolymer poly(3-aminobenzoic acid) (P3ABA), as surface-immobilised biocidal polymers. Utilisation of PANI for potential applications is restricted because of its insolubility in common solvents, which renders it difficult to process [43,44]. fPANIs are easily and inexpensively synthesised using substituted aniline monomers, which improves the solubility, and thus the processability, of the resulting polymer [43,44]. PANI and P3ABA are good candidates for incorporation into surfaces because they have thermal stability up to 300 °C, environmental stability in the conducting form, simple and inexpensive synthetic procedures [45–48], and have been demonstrated to be biocompatible with mammalian cells [49–52], all of which increases their commercial viability. Surfaces containing PANI and P3ABA are non-leaching [45,53,54], which promotes activity over a longer period of time and reduces both personal and environmental safety concerns [27].

In this study, we investigated the potential of PANI and P3ABA as surface antimicrobial agents. Initial testing involved the challenge of target organisms in suspension, mirroring standard antimicrobial susceptibility testing methods [55,56]. The target organisms selected were the antimicrobial susceptibility testing strains, *Escherichia coli* 25922 and *S. aureus* 6538, which reflect bacteria that are commonly isolated from surfaces in hospitals [10,57]. Susceptibility to antimicrobial activity can be influenced by media composition through its effects on bacterial cell physiology [58]. Therefore, *E. coli* was challenged in Lennox Broth (LB)—a rich media on which cells grow at high rates—and in minimal A salts with 0.4% succinate as the carbon source, which only contains nutrients that are essential for growth [59–61]. The slow growth of bacteria in a minimal media environment is similar to what may occur on surfaces in nature [62].

Following confirmation of activity in suspension, PANI and P3ABA were incorporated into absorbent and non-absorbent surfaces. The effect of incorporation on antimicrobial activity was determined in 96 well plate based assays, which allowed for the the testing of many concentrations and treatment times against one inoculum [27]. Absorbent surfaces were modelled using agar mixed with varying amounts of PANI or P3ABA [63]. Drops of liquid containing bacteria absorbed into the solidified agar test surfaces [64]. Non-absorbent surfaces were established in the form of compression moulded Styrene Ethylene Butylene Styrene (SEBS) films [63]. The activity of non-absorbent surfaces containing PANI and P3ABA were then characterised in relevant environmental conditions, including challenging with a range of inocula and in the presence of organic matter [62]. The experimental strategy is summarised in Figure 1, and, taken together, the results presented demonstrate the activity of PANI and P3ABA in suspension and in surfaces, in application relevant settings. The efficacious activity of P3ABA supports the utilisation of this polymer to create contamination resistant surfaces.

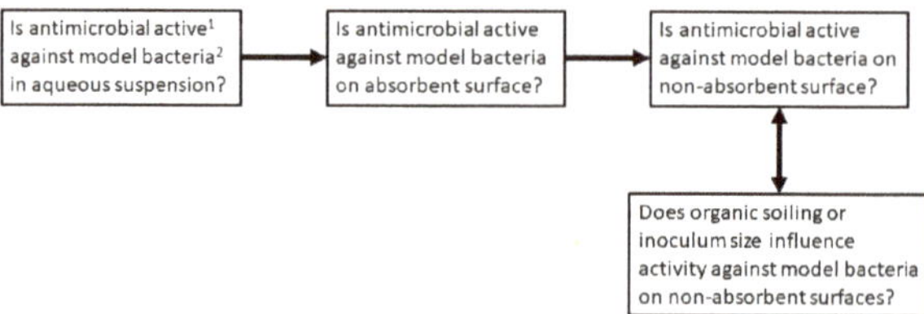

Figure 1. Experimental strategy. Polyaniline (PANI) and functionalised derivative (fPANI) are tested according to the scheme presented. [1] Activity is measured as reduction in the number of viable cells recovered from surfaces. [2] The Gram positive bacterium *S. aureus*, and the Gram-negative bacterium *E. coli* were tested as model species.

2. Results

2.1. Activity of PANI and P3ABA Suspensions against E. coli and S. aureus

To examine the activity of PANI and P3ABA, cell viability assays were performed on *E. coli* 25922 *lux* and *S. aureus* 6538 challenged with 0.5% (*w/v*) suspensions. Activity was determined in rich media, LB broth, for *E. coli* and *S. aureus*, as well as a minimal media, minimal A salts with succinate as the carbon source, for *E. coli*. At 0.5 h, 1 h, 2 h, and 4 h time points the treated cells were enumerated using the drop count method [64].

E. coli and *S. aureus* treated with PANI suspension were present at similar cell numbers at the earlier, 0.5 h and 1 h, time points (Figure 2A). For the later, 2 h and 4 h, time points *E. coli* was knocked down by 1 to 2 logs (measured as the difference between inoculum and the median number of viable cells remaining), while *S. aureus* cell numbers decreased by only ~0.5 log (Figure 2A). The overall difference in sensitivity between *E. coli* and *S. aureus* to 0.5% PANI suspension was statistically significant (linear regression analysis, intercepts are different, *p* value: less than 0.05).

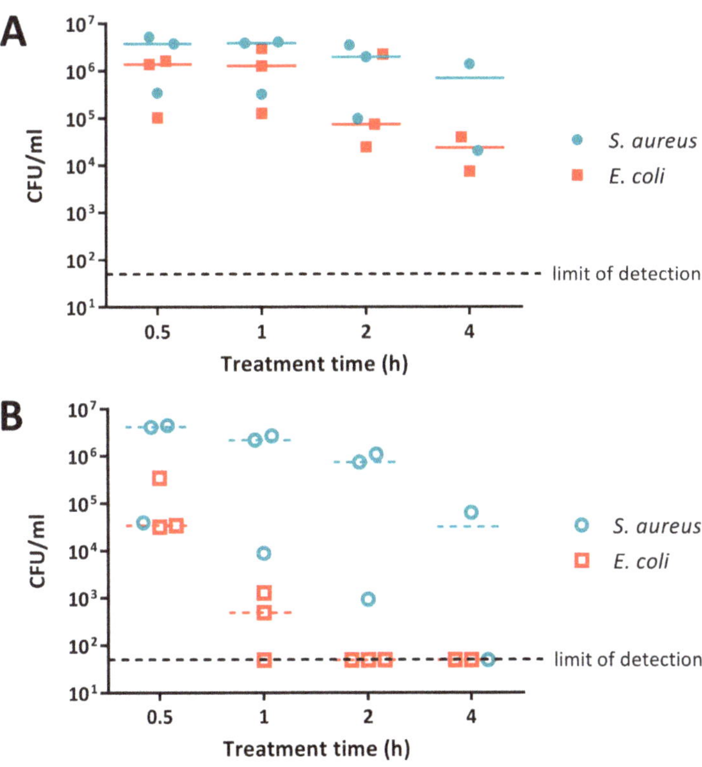

Figure 2. *Cont.*

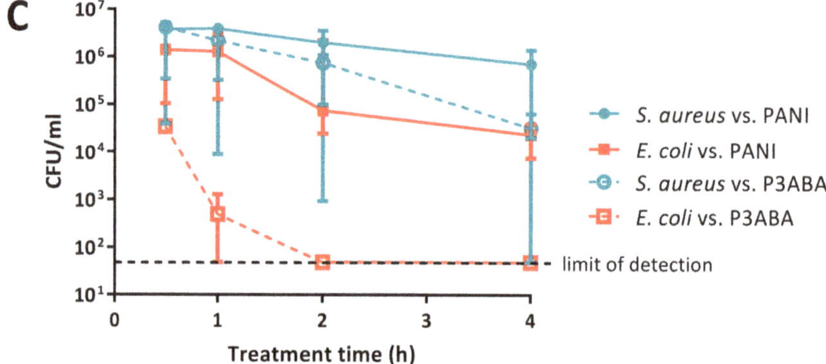

Figure 2. Sensitivity of *E. coli* 25922 *lux* and *S. aureus* 6538 to PANI and poly(3-aminobenzoic acid) (P3ABA) suspensions. Cell viability assays of ~10^6 CFU/mL *E. coli* and *S. aureus* treated with 0.5% PANI suspension (**A**) and 0.5% P3ABA suspension (**B**) in Lennox broth (LB), with each symbol representing the median of three technical replicates and the bar representing the median of each biological replicate. The data in A and B is replotted together in (**C**) where each point represents the median of biological replicates and the error bars are the range Viable cell counts (colony forming units (CFU)/mL) were obtained for each strain at 0.5 h, 1 h, 2 h, and 4 h time points. The limit of detection is 50 CFU/mL.

Both *E. coli* and *S. aureus* were more susceptible to P3ABA suspension when compared to PANI suspension (Figure 2). P3ABA suspension reduced *E. coli* viable cell numbers to below the limit of detection within 2 h, while *S. aureus* was knocked down by ~2 log following a 4 h exposure (Figure 2B; i.e., the difference between inoculum and the median number of viable cells remaining was about 2 logs, or 100-fold). As observed for PANI suspension, P3ABA suspension was more active against *E. coli* as compared to *S. aureus* (Figure 2B,C). The difference in susceptibility between *E. coli* and *S. aureus* to 0.5% P3ABA suspension was statistically significant (linear regression analysis, intercepts are different, *p* value: less than 0.05). These results confirm that PANI and P3ABA in suspension are active against the model Gram-negative and Gram-positive organisms tested and support investigation of their surface activity.

The antimicrobial activity of 0.5% PANI and P3ABA suspensions in rich and minimal media was determined against *E. coli* 25922 *lux*. PANI suspension mediated a greater reduction in the viable cell count in minimal media when compared to rich media (Figure 3A). *E. coli* in minimal media reached the limit of detection (~4 log reduction) after a 2 h challenge, while a decrease of ~1–2 log was observed in rich media after 4 h (Figure 3A). The greater activity of PANI suspensions against *E. coli* in minimal media when compared to LB broth was statistically significant (linear regression analysis; intercepts are different, *p* value: less than 0.05).

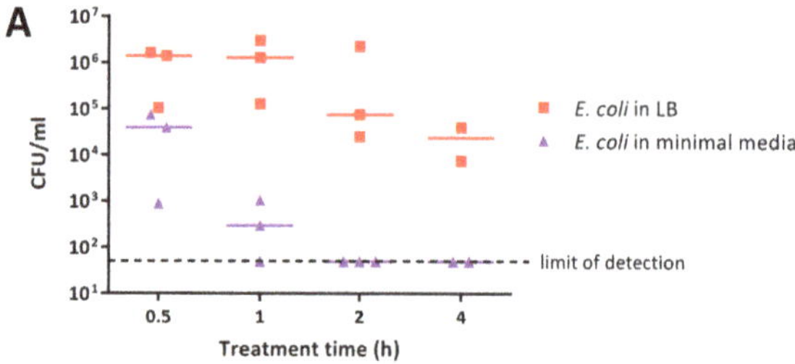

Figure 3. *Cont.*

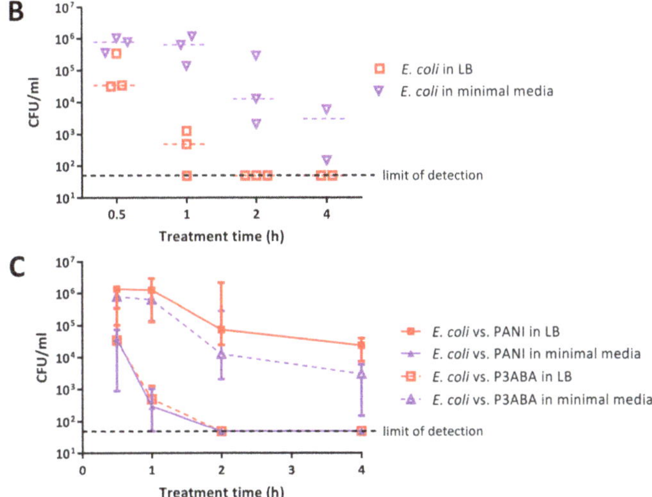

Figure 3. Sensitivity of *E. coli* 25922 *lux* to PANI and P3ABA suspensions in rich and minimal media. Cell viability assays of ~10^6 CFU/mL *E. coli* treated with 0.5% PANI suspension (**A**) and 0.5% P3ABA suspension (**B**) in Lennox broth (LB) and minimal A salts, with each symbol representing the median of three technical replicates and the bar representing the median of each biological replicate. The data in A and B is replotted together in (**C**) where each point represents the median of biological replicates and the error bars are the range. Viable cell counts (CFU/mL) were obtained for experimental sample at 0.5 h, 1 h, 2 h, and 4 h time points. The limit of detection is 50 CFU/mL.

P3ABA suspension was more active against *E. coli* in rich media as compared to minimal media (Figure 3B). *E. coli* in LB broth was knocked down ~3 log after 1 h and reached the limit of detection by 2 h (Figure 3B). In comparison, the levels of viable *E. coli* in minimal media were stable at the 0.5 h and 1 h time points (Figure 3B). After 4 h of treatment, P3ABA knocked down *E. coli* in minimal media by ~2 log (Figure 3B). The difference in activity of P3ABA against *E. coli* in LB broth and *E. coli* in minimal media was statistically significant (linear regression analysis, slopes are different, *p* value: less than 0.05). These results confirm that PANI and P3ABA in suspension are active against *E. coli* 25922 in rich and minimal media, and support the investigation of their surface activity.

2.2. Activity of Absorbent Surfaces Containing PANI and P3ABA against E. coli and S. aureus

Following the demonstration of activity for PANI and P3ABA against *E. coli* and *S. aureus*, we investigated the antibacterial activity of absorbent surfaces containing PANI and P3ABA to simulate surfaces that absorb water droplets, such as fabrics [65]. LB agar as used as a model of an absorbent surface. Agar containing 1% and 2% concentrations of PANI and P3ABA in agar were tested along with a higher concentration of PANI (8% PANI), because PANI was less active than P3ABA in suspension (Figure 2).

To investigate antimicrobial activity, *E. coli* 25922 *lux* or *S. aureus* 6538 were inoculated onto the experimental agar, with cells being rescued at various time points in fresh media. Survival of the PANI or P3ABA in agar challenge was based on growth of rescued cells. A *lux*-tagged version of *E. coli* 25922 was used in this work as bioluminescence is a practical alternative to enumeration by plate counts [66–68] (Appendix A, Figures A1 and A2). Growth of *E. coli* 25922 *lux* was determined by measuring bioluminescence using the VICTOR X Multilabel Plate Reader. Growth of *S. aureus* 6538 was determined by examining optical density at 600 nm (OD_{600}) using the μQuant™ Microplate Spectrophotometer as a *lux*-tagged *S. aureus* 6538 strain was not available for testing.

Agar containing 8% PANI mediated a decrease in bioluminescence levels of *E. coli* 25922 *lux* to background levels after 1 h of treatment, while 2% PANI agar required 2 h to reduce surface bacterial load (Figure 4A). The limit of detection for this assay is ~20 CFU (Figure A3). Agar containing 1% PANI did not exhibit antimicrobial activity against *E. coli* (Figure 4A). P3ABA agar had greater surface antimicrobial activity than PANI agar (Figure 4). Agar containing 1% P3ABA and 2% P3ABA reduced bioluminescence levels from *E. coli* 25922 *lux* to background levels within 1 h and 15 min, respectively (Figure 4B).

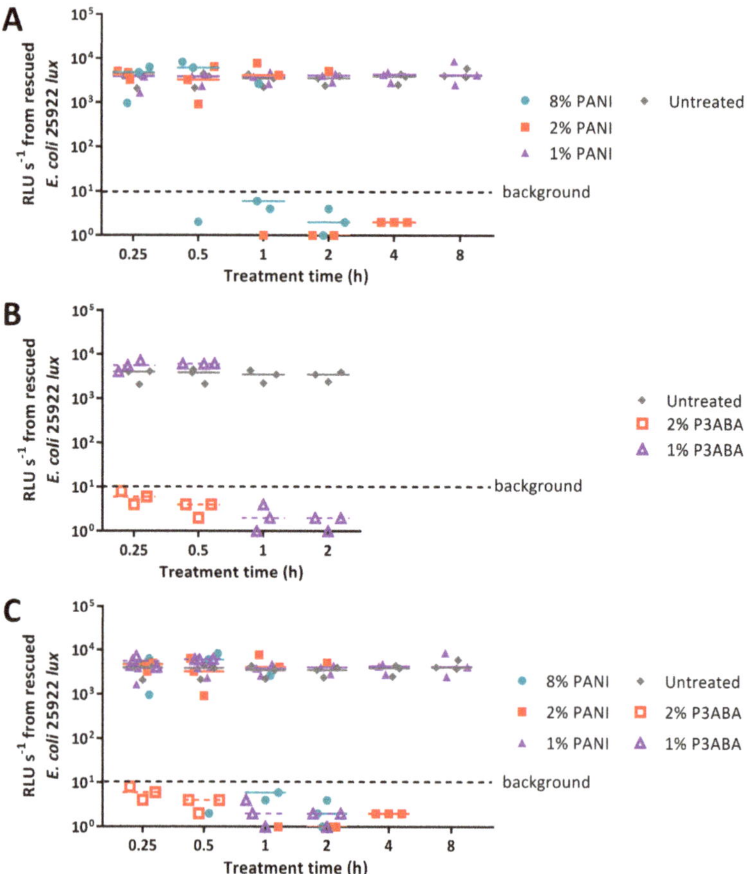

Figure 4. Sensitivity of *E. coli* 25922 *lux* to PANI and P3ABA in agar. (**A**) ~10^4 CFU of *E. coli* was exposed to 8% PANI, 2% PANI, and 1% PANI incorporated into LB agar for 0.25 h, 0.5 h, 1 h, 2 h, 4 h, and 8 h. (**B**) ~10^4 CFU of *E. coli* was exposed to 2% P3ABA and 1% P3ABA incorporated into LB agar for 0.25 h, 0.5 h, 1 h, and 2 h. Following treatment, the cells were rescued by washing the agar surface with LB broth and transferred to a 96 well plate. Each point represents the median of three technical replicates and each bar represents the median of each biological replicate. The data from A and B is combined for comparison in (**C**). The rescued cells were incubated at 37 °C for 16 h and light release was measured. The vertical axis shows the bioluminescence measurements (relative light units per second, RLU s^{-1}) from the recovered cells with each data point representing an independent experiment and the line representing the median. Background luminescence readings are ~10 RLU s^{-1}.

PANI in agar was less active against *S. aureus* 6538 than *E. coli* 25922 *lux*, reflecting the trend observed with suspension testing (Figures 2A, 4A, and 5A). Agar containing 2% and 8% PANI reduced *S. aureus* surface load to background levels within 8 h and 4 h, respectively (Figure 5A). The limit of detection for this assay is ~20 CFU (Figure A4). As observed for *E. coli*, 1% PANI in agar was inactive against *S. aureus* within the time constraints (8 h) of the experiment (Figures 4A and 5A). The viability of *S. aureus* cells following treatment with 8% PANI in agar for 2 h, 4 h, and 8 h, and following treatment with 2% PANI in agar for 8 h was significantly different from that of untreated cells (Friedman test, *p* value: less than 0.05, Dunn's multiple comparison test).

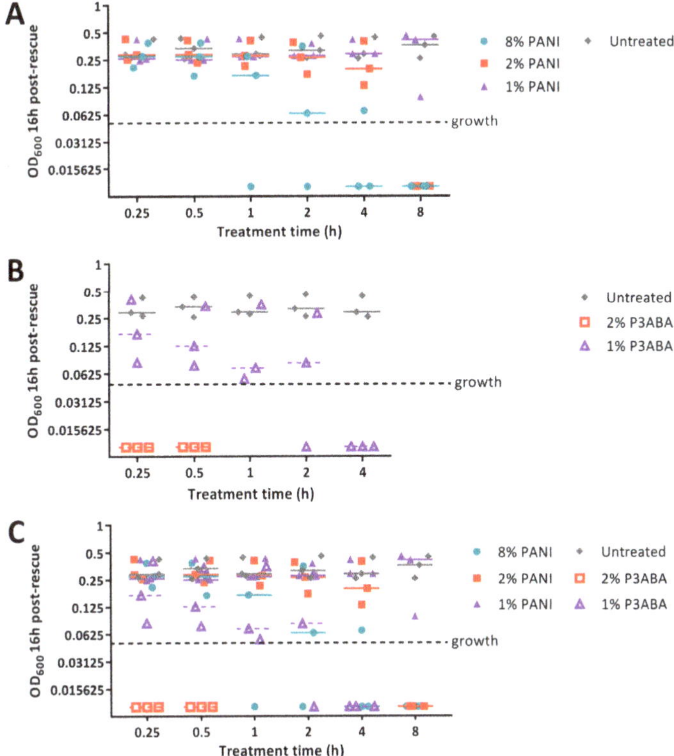

Figure 5. Sensitivity of *S. aureus* 6538 to PANI in agar. (**A**) ~10^4 CFU of *S. aureus* was exposed to 8% PANI, 2% PANI, and 1% PANI incorporated into LB agar for 0.25 h, 0.5 h, 1 h, 2 h, 4 h, and 8 h. (**B**) ~10^4 CFU of *S. aureus* was exposed to 2% P3ABA and 1% P3ABA incorporated into LB agar for 0.25 h, 0.5 h, 1 h, 2 h, and 4 h. Following treatment, the cells were rescued by washing the agar surface with LB broth and transferred to a 96 well plate. Each point represents the median of three technical replicates and each bar represents the median of each biological replicate. The data from A and B is combined for comparison in (**C**). The rescued cells were incubated at 37 °C for 16 h and optical density at 600 nm (OD_{600}) was measured. The vertical axis shows the OD_{600} measurements from the recovered cells with each data point representing an independent experiment and the line representing the median. OD_{600} readings above 0.05 are considered as growth.

Agar containing 2% P3ABA effectively decontaminated *S. aureus* on a surface following a 15 min treatment (Figure 5B), which was consistent with the activity against *E. coli* (Figure 5A). 1% P3ABA in agar was less active than the higher concentration, with the former requiring a 4 h exposure to reduce bacterial load (Figure 5B). The activity of 2% P3ABA in agar over 15 min and 1% P3ABA in agar over 4 h was statistically significant (Friedman test, *p* value: less than 0.05, Dunn's multiple comparison test).

The activity of PANI and P3ABA presented in Figures 4 and 5 demonstrates their efficacy against *E. coli* and *S. aureus* when incorporated into an absorbent surface.

2.3. Activity of Non-Absorbent Surfaces Containing PANI and P3ABA against E. coli and S. aureus

The activity of PANI and P3ABA as surface antimicrobials at non-absorbent surfaces to simulate surfaces that do not absorb water, such as walls and door handles, was investigated using SEBS films containing 5% PANI or 3% P3ABA [69]. The concentrations of the additive in these films are within the range typically used for incorporation into surfaces (0.1–5%) [69] and reflect the greater activity of P3ABA against *E. coli* and *S. aureus* compared to PANI as demonstrated in suspension (Figure 2) and in agar (Figures 4 and 5). The activity of PANI and P3ABA films was examined in a 'micro-surface testing assay' (MSTA), in which 10 μL of inoculum in LB broth is sandwiched between two pieces of film and recovered at particular time points in fresh LB broth in a 96 well plate [28,70]. Cell viability was determined by measuring the bioluminescence for *E. coli* 25922 *lux* and OD_{600} for *S. aureus* 6538.

The MSTA for testing 5% PANI and 3% P3ABA films against bacteria was optimised using *E. coli* 25922 *lux*. Following either 2 h or 24 h challenges on the films, bacteria were rescued by washing with LB broth and incubated in a 96-well plate for 16 h, after which bioluminescence was measured. The bacteria present in the remaining recovery broth were enumerated using plate counts to verify the ability of bioluminescence levels to infer cell number.

PANI and P3ABA films gave no reduction in bacterial viability for *E. coli* 25922 *lux* for after a 2 h challenge, which was indicated by both plate counts and bioluminescence readings (Figure 6A,B). Films containing P3ABA were more active than their PANI counterparts after 24 h exposure, with the former reducing the plate counts and bioluminescence levels by ~2 log relative to the untreated cells (Figure 6A,B). The activity of 3% P3ABA films against *E. coli* after 24 h treatment was statistically significant (2-way RM ANOVA; CFU/mL *p* value: less than 0.05; RLU s^{-1} *p* value: less than 0.05). The similarity in trends seen between the plate counts and bioluminescence measurements from *E. coli* 25922 *lux* that was treated with PANI and P3ABA films confirmed that the bioluminescence-based experimental approach to determining the activity of a non-absorbent surface was appropriate to use for further testing. The results presented show that non-absorbent surfaces containing P3ABA can reduce bacterial load after a 24 h exposure.

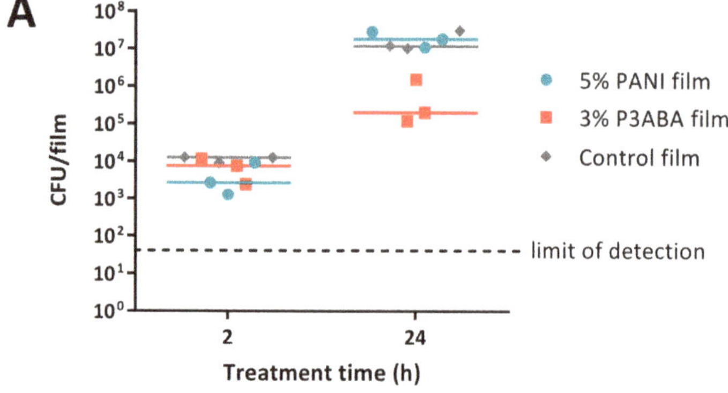

Figure 6. *Cont.*

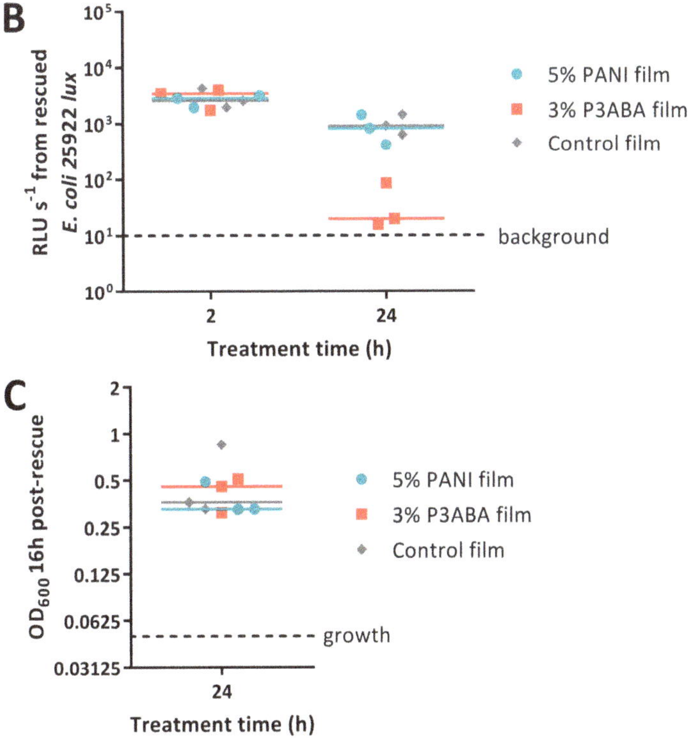

Figure 6. Sensitivity of *E. coli* 25922 *lux* and *S. aureus* 6538 to PANI and P3ABA films. ~10^4 CFU of *E. coli* (**A**,**B**) or *S. aureus* (**C**) in 10 μL LB broth was sandwiched between two pieces of PANI film, P3ABA film, or control film for 2 h (**A**,**B**) and 24 h (**A–C**). The cells were rescued by washing the film samples with LB broth and transferred to a 96 well plate (**B**,**C**). Each point represents the median of three technical replicates and each bar represents the median of each biological replicate. The rescued *E. coli* cells were also enumerated with plate counts (**A**). The cells in the 96 well plate were incubated at 37 °C for 16 h and light release (**B**) or OD_{600} (**C**) was measured. The vertical axes show the viable cell counts (**A**) and bioluminescence measurements (**B**) from the recovered *E. coli* cells, and OD_{600} measurements (**C**) from the recovered *S. aureus* cells, with each data point representing an independent experiment and the line representing the median. The limit of detection for the plate counts is 50 CFU/mL. Background luminescence readings are 10 RLU s^{-1}. OD_{600} readings above 0.05 are considered as growth.

Following from this, the optimised protocol for determining the activity of 5% PANI and 3% P3ABA films was used against *S. aureus* 6538 with an increase in OD_{600} above 0.05, indicating the presence of viable cells. *S. aureus* 6538 was treated for only 24 h as PANI and P3ABA films were not active against *E. coli* 25922 *lux* following 2 h treatments (Figure 6A,B) and *S. aureus* was less sensitive than *E. coli* to PANI and P3ABA in suspension (Figure 2). Both 5% PANI in films and 3% P3ABA in films displayed no activity against *S. aureus* 6538 inoculated in LB broth (Figure 6C).

2.4. Characterisation of the Action of PANI and P3ABA Films against E. coli and S. aureus

The activity of 5% PANI and 3% P3ABA films against *E. coli* and *S. aureus*, was poorer than expected. We hypothesised that this might be due to inoculating large numbers of bacteria in rich media. To test this hypothesis, films containing PANI and P3ABA were challenged with a range of concentrations of *E. coli* 25922 *lux* and *S. aureus* 6538. The test organisms were washed in saline to simulate a low nutrient environment. A 2 h exposure was used, as this contact time is more effective in disrupting transmission pathways. The influence of the presence of organic matter was examined by challenging PANI and P3ABA films with *E. coli* 25922 *lux* washed in LB broth or in 0.85% saline for 2 h.

Bioluminescence from all of the doses of *E. coli* exposed to films containing 5% PANI was reduced to below background levels (Figure 7A). Films containing 3% P3ABA exhibited similar levels of antimicrobial activity (Figure 7A). The antimicrobial activity of PANI and 3PABA films against each inoculum level tested was significantly different from the control film (2-way RM ANOVA, interaction of film type and CFU dose *p* value: less than 0.05). These results indicate that PANI and P3ABA films are active against *E. coli* in low nutrient conditions.

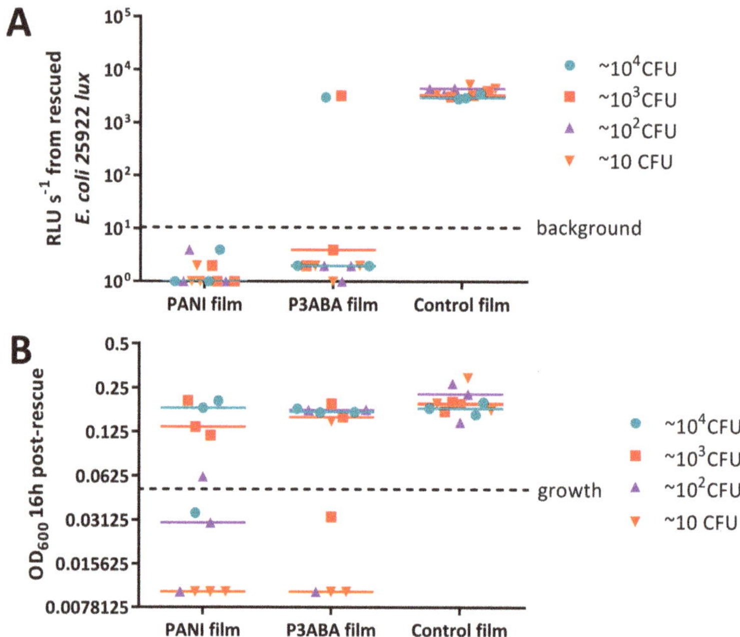

Figure 7. Activity of PANI and P3ABA films against a range of CFU doses of *E. coli* 25922 *lux* and *S. aureus* 6538. ~10 CFU–~10^4 CFU of *E. coli* (**A**) or *S. aureus* (**B**) in 10 µL 0.85% saline was sandwiched between two pieces of PANI film, P3ABA film, or control film for 2 h. The cells were rescued by washing the film samples with LB broth and transferred to a 96 well plate. The rescued cells were incubated at 37 °C for 16 h and light release (**A**) or OD$_{600}$ (**B**) was measured. The vertical axes show the bioluminescence measurements (**A**) or OD$_{600}$ (**B**) from the recovered cells with each data point representing an independent experiment and the line representing the median. Background luminescence readings are 10 RLU s^{-1}. OD$_{600}$ readings above 0.05 are considered as growth.

OD$_{600}$ values from ~10^4 CFU and ~10^3 CFU doses of *S. aureus* 6538 treated with 5% PANI and 3% P3ABA films were above OD$_{600}$ of 0.05, the threshold for growth, indicating that the films were not active against these higher CFU doses (Figure 7B). The OD$_{600}$ from lower doses of *S. aureus* that was exposed to PANI and P3ABA films did not increase above the threshold for growth implying killing of the inoculated cells occurred (Figure 7B). The activity of PANI films against ~10^2 and ~10 CFU, and P3ABA films against ~10 CFU was statistically significant (2-way RM ANOVA, film type *p* value: less than 0.05, CFU dose *p* value: less than 0.05). It can be concluded that PANI and P3ABA films are active against low inocula of *S. aureus* in saline.

The effect of the presence of organic matter on the surface activity of films containing PANI or P3ABA was determined by challenging *E. coli* 25922 *lux* in LB broth and 0.85% saline. Bioluminescence levels from *E. coli* 25922 *lux* recovered from 5% PANI films and 3% P3ABA films when inoculated in 0.85% saline were below background levels, whereas *E. coli* 25922 *lux* inoculated in LB broth released the same amount of light as bacteria that were recovered from control films (Figure 8). This indicates that *E. coli* in saline was much more sensitive to PANI and P3ABA films than *E. coli* in LB broth. It is possible that the constituents of LB broth interfere with the contact killing of *E. coli* on films containing PANI and P3ABA.

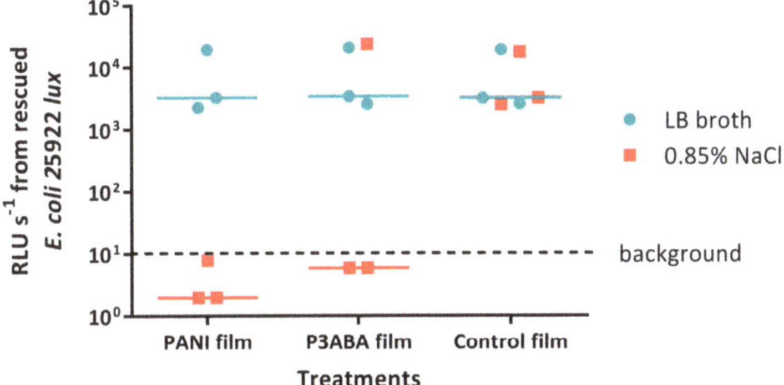

Figure 8. Activity of PANI and P3ABA films against *E. coli* 25922 *lux* in the presence and absence of organic matter. ~10^4 CFU *of E. coli* in 10 μL LB broth or 10 μL 0.85% saline was sandwiched between two pieces of PANI film, P3ABA film, or control film for 2 h. The cells were rescued by washing the film samples with LB broth and transferred to a 96 well plate. The rescued cells were incubated at 37 °C for 16 h and light release was measured. The vertical axis shows the bioluminescence measurements (RLU s^{-1}) from the recovered cells with each data point representing an independent experiment and the line representing the median. Background luminescence readings are ~10 RLU s^{-1}.

3. Discussion

PANI and P3ABA are promising additives to materials to create contamination resistance surfaces. Factors that may influence the antibacterial efficacy of the surface were explored over short treatment times (up to 4 h). Disrupting transmission pathways through surface decontamination can be best achieved with an antimicrobial agent that kills over a short period of time [71]. The longer a bacterium persists on a surface, the greater the opportunity to be spread [4]. Therefore, rapid decontamination times will decrease the chance that bacteria may be transferred to a new surface before sterilisation is achieved, and will decrease the likelihood of resistance developing [72].

The activity of PANI and P3ABA was determined against *E. coli* and *S. aureus*, representing important pathogens that are found in settings requiring antimicrobial surfaces, such as hospitals and food processing plants. Overall, *E. coli* had greater susceptibility to both PANI and P3ABA in suspension when compared to *S. aureus* (Figure 2). A similar trend was observed for PANI or P3ABA in agar (Figures 4 and 5) and in films (Figure 6). These results demonstrate that PANI and P3ABA are active against the model Gram-negative and Gram-positive bacteria, *E. coli* and *S. aureus*, respectively, in suspension and in different types of surfaces. The differing levels of activity tha were observed against *E. coli* and *S. aureus* highlight how a broad spectrum antimicrobial agent may be more or less effective against a range of bacteria and demonstrates why testing should be done against all the potential target organisms.

The effect of the presence of complex nutrients on the susceptibility of *E. coli* to PANI and P3ABA in suspension was examined. *E. coli* was more susceptible to the antimicrobial action of

PANI in suspension when incubated in minimal media when compared to LB broth (Figure 3A). The more efficacious activity of PANI in a low nutrient environment supports the incorporation of this antimicrobial agent in surfaces for applications that are associated with only minor contamination with organic matter. In contrast to this, P3ABA was more active against *E. coli* in rich media relative to minimal media (Figure 3B). The metabolic state of the cell may influence how it responds to bactericidal treatment [73]. The bactericidal action of antimicrobial agents is associated with increased respiration, while bacteriostatic action is characterised by suppressed cellular respiration [73]. The bacteriostatic effect reduces ATP demand and is often the dominant effect blocking bactericidal action [73]. Following from this, if cellular energy output is readily inhibited, such as in cells growing in energy poor conditions, antimicrobial action may result in the inhibition of growth rather than bactericidal killing [73]. Bacterial cells that are highly active may therefore be more susceptible to antimicrobial exposure because of accelerated respiration. The reduced sensitivity of *E. coli* cells to P3ABA in low nutrient conditions could be reflective of a predisposition to the bacteriostatic effect. The greater activity of P3ABA in the presence of nutrients that facilitate bacterial cell growth supports the use of P3ABA in surfaces in settings that are associated with contamination of organic matter, such as surfaces in the vicinity of patients with gastrointestinal infections, which are commonly contaminated with faecal matter containing the bacteria.

The feasibility of using PANI and P3ABA as additives to create antimicrobial surfaces was examined by determining the activity in suspension. Following confirmation of activity against *E. coli* and *S. aureus* (Figure 2), the activity of PANI and P3ABA as agents that are added to absorbent and non-absorbent surfaces was investigated. Overall, both PANI and P3ABA are most active in suspension, followed by in agar and then in films. *E. coli* treated with 0.5% PANI in suspension for 4 h was reduced in numbers by 2 log (Figure 3A), while 1% PANI in agar did not reduce the viable cell count, even after 8 h of treatment (Figure 4A). For surface incorporated PANI to achieve comparable activity to PANI in suspension, a higher concentration is required. This is demonstrated by total knockdown of *E. coli* after a 4 h exposure to 2% PANI in agar (Figure 4A); a result that was achieved by a concentration of 0.5% in suspension (Figure 3A). The reduction in activity of surface incorporated PANI and P3ABA is reflective of how immobilisation in a surface can affect bactericidal activity and how different surface matrixes may influence this in different ways [28].

In this study, the antibacterial activity of PANI and an fPANI were determined by the quantification of the viable cells remaining after a period of challenge, using either classical culture-based techniques, or measuring bioluminescence of genetically modified bacteria as a surrogate measure of viability. Future studies may be enhanced by coupling this type of analysis with scanning electron microscopy (SEM) of bacteria on surfaces and fluorescence microscopy after live/dead staining. SEM has previously allowed for visualisation of bacterial killing by fPANIs to the conclusion that the antimicrobial mode of action eventually leads to a loss of cell integrity [74,75]. Fluorescence microscopy of live/dead stained biofilms has allowed for the activity of another fPANI, polysulfanilic acid, to be followed, with the killing of bacteria being established in biofilms and the release of biomass from the surface, imaged [54]. In the study of bacterial attachment to surfaces real time imaging, e.g., using differential interference contrast microscopy [76] may allow for a better understanding of the interaction of bacteria with surfaces and the factors that influence resistance to colonisation.

It is believed that PANI and P3ABA exert antimicrobial action following contact with a bacterial cell [45,77]. Thus, the reduced contact that occurs between a bacterial cell and surface incorporated PANI and P3ABA (relative to in suspension) would mediate the decrease in antimicrobial efficacy. The least amount of contact between the antimicrobial agent and a bacterial cell would occur for non-absorbent surfaces, which mirrors the decreased activity that was observed for PANI and P3ABA in films. 2% PANI and 2% P3ABA in agar (Figure 5) were able to mediate knockdown of *S. aureus* in 8 h and 15 min, respectively, while 5% PANI and 3% P3ABA in films were unable to reduce bacterial cell numbers after a 24 h treatment (Figure 6C). In this example, higher concentration and treatment time did not ameliorate the reduction of activity for polymers that were incorporated into a non-absorbent surface. The results of this work demonstrate why it is important to test antimicrobial agents, first in suspension (associated with quick and reproducible results) before testing as part of a surface, which should reflect the final application [71].

In real world settings, antimicrobial surfaces may be challenged with a range of inocula. It is well known that the size of the inoculum that is used can influence the magnitude of antimicrobial activity in susceptibility testing [56]. In general, higher inocula need a higher concentration of antimicrobial agent and/or a longer treatment time to achieve knockdown [78]. The surface activity of PANI in film and P3ABA in film was affected by *S. aureus* inoculum size with activity demonstrated only for lower inocula (Figure 7B). The decreased surface activity in the presence of high numbers of bacteria may be mediated by the piling of bacterial cells on top of each other, thereby reducing direct contact with the antimicrobial agent for a portion of the population [79]. The results that are presented demonstrate the necessity to perform antimicrobial surface testing with appropriate inocula to simulate the potential challenges that would occur in the real world application. Surfaces in hospitals are considered to be contaminated when aerobic colony counts exceed 2.5 CFU/cm^2; however, sampling of objects in patient hospital rooms has demonstrated contamination with a range of bacterial loads (up to 10^4 CFU/m^2, equivalent to 10^2 CFU/cm^2), including 10^3 CFU/m^2 (equivalent to 10 CFU/cm^2) of MRSA on door handles [80–83]. Therefore, antimicrobial surfaces in hospitals would need to be active against up to 10^4 CFU/m^2 of contaminants to prevent bacterial spread.

Organic soiling of antimicrobial surfaces is a known cause of loss of activity and thus was investigated for surfaces containing PANI and P3ABA [25,71,84]. Surface activity of both PANI in film and P3ABA in film was decreased in the presence of organic matter (Figure 8). Organic matter can interfere with contact between the bacterial cell and the antimicrobial agent—particularly for charged proteins and polysaccharides that can disrupt charge based interactions—thus providing protection from antimicrobial action [28,85,86]. Additionally, contaminating organic matter may inactivate antimicrobial agents [86]. Typical organic contaminants on hospital surfaces include blood and faecal matter [25,71]. It is important that antimicrobial surfaces are tested in conditions, including contamination with organic matter, relevant to the application to verify that the surfaces will be sufficiently active in these settings [25].

While it is not ideal that a reduction in surface activity was observed, the loss of activity upon soiling is common and the effect of organic soiling can be reduced by regular cleaning. Therefore, antimicrobial surfaces need to be able to withstand any adverse environmental conditions that are associated with cleaning [27]. PANI and P3ABA have thermal stability up to 300 °C and environmental stability in the conducting form [45–48]. An fPANI containing surface was demonstrated to retain activity against *E. coli* and *S. aureus* after 10 repeated challenges if hydrogen peroxide, but not bleach, was the cleaning agent [87]. Future work will include examining the influence of current cleaning procedures on the activity of surface incorporated PANI and P3ABA.

P3ABA containing surfaces demonstrated potential as contamination resistant surfaces for applications. P3ABA as part of a non-absorbent surface reduced *E. coli* by 2 log after a 24 h incubation (Figure 6B), while an absorbent surface containing 2% P3ABA cleared the bacterial load after 15 min (Figure 4B). The P3ABA containing surfaces in this work indicate a superior performance than has been reported for triclosan, a popular additive claiming antimicrobial activity, which had no effect on the viable cell count of *E. coli* following a 24 h exposure [88]. Similarly, triclosan-incorporated plastic only inhibited *E. coli* O157:H7 after a 24 h incubation [89] and triclosan melt-mixed with 4.5% polystyrene inhibited *E. coli* Y 1090 for 5 h, after which the viable cell number increased [90]. Materials containing P3ABA may therefore have a future as a cost-effective antimicrobial surface to prevent or at least reduce the undesirable spread of micro-organisms.

4. Materials and Methods

4.1. Bacterial Strains and Growth Conditions

E. coli ATCC 25922 (referred to as *E. coli* 25922) and *S. aureus subsp. aureus* ATCC 6538 (referred to as *S. aureus* 6538) were used in this work because they are routinely used as control organisms to verify that antibiotic susceptibility results are accurate [56,91]. *E. coli* 25922 was tagged with an integrating plasmid (p16S*lux*) containing the bacterial luciferase (*lux*) operon (designated *E. coli* 25922 *lux*) [92,93]. *E. coli* 25922 *lux* was used for testing of surfaces containing PANI and P3ABA [28]. All strains were grown at 37 °C, with 200 rpm agitation where appropriate. The University of Auckland Institutional Biological Safety Committee approved the construction and use of genetically modified Enterobacteriaceae (GMO04-UA0027).

4.2. Media and Chemicals

PANI and P3ABA were synthesised via chemical oxidation of aniline and 3-aminobenzoic acid monomers, respectively [45]. Cell biology reagents were purchased from Sigma-Aldrich (New South Wales, Australia). Bacteria were cultured in LB broth (BD) or in minimal media. Minimal A medium was used to support growth in a minimal environment, providing only essential nutrients. A $5\times$ minimal A solution was made according to the following: 5 g $(NH_4)_2SO_4$, 22.5 g KH_2PO_4, 52.5 g K_2HPO_4, 2.5 g sodium citrate·$2H_2O$. After autoclaving, this solution was diluted to $1\times$ with sterile water and the following sterile solutions, per litre: 1 mL 1 M $MgSO_4$·$7H_2O$, 0.1 mL 0.5% thiamine plus the carbon source (10 mL of 40% succinate solution per litre).

4.3. Preparation of PANI and P3ABA Suspensions

PANI was finely ground using a mortar and pestle. This insoluble powder requires shaking at 200 rpm to stay in suspension. Reflecting the improved solubility of P3ABA, this polymer was suspended in broth by sonication (QSonica Q700 Sonicator, Newtown, CT, USA) at the following settings: amplitude 30, elapsed time 10 s, repeat $4\times$. Suspensions of PANI and P3ABA were prepared at 1% (w/v) for a final concentration of 0.5%.

4.4. Activity of PANI and P3ABA Suspensions against E. coli and S. aureus

Turbid overnight cultures of test bacteria were diluted to 10^6 CFU/mL in LB broth (*E. coli* 25922 *lux* and *S. aureus* 6538) or minimal A salts with 0.4% succinate (*E. coli* 25922 *lux*) [94]. The inocula were retrospectively enumerated on LB agar plates [64]. 500 μL of PANI suspension, P3ABA suspension, and growth media (untreated cells) were inoculated with 500 μL of diluted culture. At 0.5 h, 1 h, 2 h, and 4 h time points, each experimental sample was enumerated on LB agar plates. Following incubation, colonies were counted and CFU/mL was calculated. At least three biological replicates were obtained.

Linear regression analysis was used to compare the sensitivity of test strains to PANI or P3ABA suspensions. Specifically, the sensitivity of *E. coli* and *S. aureus* in LB broth to each suspension was compared and the sensitivity of *E. coli* in LB broth and in minimal media to each suspension was compared. Statistical analysis by linear regression was performed using GraphPad Prism software version 6 (GraphPad Software, Inc., La Jolla, CA, USA). Data was graphed in a scatter plot that was generated with viable cell counts post-treatment (CFU/mL) represented on the y-axis and time (h) represented on the x-axis. Linear regression was used to fit a straight line (regression line) through the data for the categorical factor (strain type or media type) generating the best-fit value of the slope and intercept. An analysis of covariance (ANCOVA) was used to compare the regression lines from the categorical factors to determine if there was a statistically significant difference in sensitivity.

4.5. Activity of Absorbent Surfaces Containing PANI and P3ABA against E. coli and S. aureus

Absorbent surfaces containing PANI or P3ABA can be modelled using agar, as drops of liquid containing bacteria will absorb into the agar surface [64]. Molten agar was mixed with varying amounts of PANI or P3ABA, which when left to set created absorbent surfaces containing the antimicrobial agents. PANI or P3ABA were established in agar at 1% and 2%; PANI was also established in agar at 8%. PANI and P3ABA containing absorbent surfaces were set up in triplicate in a 96 well plate by aliquoting 200 µL of each test agar and 200 µL of LB agar (for the untreated control) into individual wells. A turbid culture of test bacteria was diluted to 10^6 CFU/mL in broth and retrospectively enumerated. All of the test surfaces were inoculated with 10 µL diluted culture, resulting in 10^4 CFU in each well [55,56]. Agar samples for background readings received 10 µL LB broth.

At specified time points, bacterial cells were rescued in 200 µL fresh media in a 96 well plate [95]. Each type of absorbent surface was tested for the necessary time to achieve knockdown, therefore, highly active surfaces were tested only for the shorter treatment times. *E. coli* 25922 *lux* and *S. aureus* 6538 were challenged with PANI in agar for the following treatment times: 15 min, 30 min, 1 h, 2 h, 4 h, and 8 h. *E. coli* 25922 *lux* was exposed to P3ABA in agar for the following treatment times: 15 min, 30 min, 1 h, and 2 h. *S. aureus* 6538 was exposed to P3ABA in agar for the following treatment times: 15 min, 30 min, 1 h, 2 h, and 4 h. The viability of rescued *E. coli* 25922 *lux* was assessed after 16 h incubation by measuring bioluminescence using the VICTOR X Multilabel Plate Reader (Perkin Elmer, Foster City, CA, USA). The viability of rescued *S. aureus* 6538 was determined by measuring OD_{600} using the µQuant™ Microplate Spectrophotometer (BioTek Instruments, Winooski, VT, USA). Three biological replicates were obtained for each experiment.

The Friedman test was used to analyse the differences between untreated cells and those that were treated with PANI or P3ABA in agar. When a significant difference was identified (*p* value less than 0.05), specific groups were compared to each other using Dunn's multiple comparison test. Dunn's multiple comparison test was used to compare the treated and untreated cells at each time point, with a *p* value of less than 0.05 indicating a significant difference. Thus, comparisons were made between each treatment time for every concentration tested to identify a concentration-contact time combination that is associated with significant surface activity.

4.6. Activity of Non-Absorbent Surfaces Containing PANI and P3ABA against E. coli and S. aureus

Non-absorbent surface samples were prepared using SEBS films containing 5% PANI or 3% PANI or no additive (control film). The films were hole punched to generate ~5 mm diameter circles that fit into the wells of a 96 well plate. The film samples were disinfected by immersion in 70% ethanol for 10 min and dried in the Herasafe™ KS (NSF) Class II, Type A2 Biological Safety Cabinet (Thermo Scientific, Auckland, New Zealand) [25].

A turbid overnight culture of test bacteria was diluted to 10^6 CFU/mL in broth and enumerated. The activity of the PANI and P3ABA containing film samples was determined using the MSTA adapted from Japanese Industry Standard (JIS Z-2801) method [28,70]. A piece of film was placed in an empty well, inoculated with 10 µL of diluted culture, and a second piece of the same type of film was placed on top of the inoculum [28,70]. Film samples for background readings received 10 µL of LB broth. The film treatments were established in triplicate. At the specified time point(s) bacterial cells were rescued in 190 µL LB broth in a fresh 96 well plate [95]. The rescued cells were incubated at 37 °C in a sealed container with moist tissue for 16 h and the viability of rescued cells was determined [95]. Three biological replicates were obtained for each experiment.

For *E. coli* 25922 *lux*, cells were rescued after 2 h and 24 h treatments. The viability of cells post-treatment was assessed by using plate counts and measuring bioluminescence. To this end, a 100 µL aliquot of rescued cells was used to enumerate by drop counts and the remaining 100 µL of rescued cells was added to a dark OptiPlate-96 well microtitre plate containing 100 µL of LB broth for the measurement of bioluminescence using the VICTOR X Multilabel Plate Reader. For *S. aureus* 6538, cells were exposed to film treatments for only 24 h and the viability of cells post-treatment was assessed by incubating 200 µL of rescued cells in a 96 well plate for 16 h and measuring OD_{600} using the µQuant™ Microplate Spectrophotometer.

The activity of PANI and P3ABA in films against *E. coli* 25922 *lux* was analysed using a two-way repeated measures analysis of variation (2-way RM ANOVA). For both the plate counts and the bioluminescence data, the 2-way RM ANOVA determined how *E. coli* 25922 *lux* cell number was affected by two factors, treatment time (2 h and 24 h) and film type (PANI in film, P3ABA in film, no additive). A *p* value of less than 0.05 indicates that the cell number was significantly affected by at least one of the factors. When a significant difference was identified, treated cells were compared to the untreated control for each time point using Dunnett's multiple comparison test with a *p* value of less than 0.05, indicating a significant difference.

The Friedman test was used to analyse the differences between untreated *S. aureus* 6538 cells and those that were treated with PANI or P3ABA in film. The Friedman test is a nonparametric test that compares three or more matched groups—cells treated with 5% PANI in film, 3% P3ABA in film, and control film. A *p* value of less than 0.05 indicates that at least one of the groups differs from the rest. When a significant difference was identified, specific groups were compared to each other using Dunn's multiple comparison test. Dunn's multiple comparison test was used to compare the treated and untreated cells at each time point, with a *p* value of less than 0.05 indicating a significant difference.

4.7. Characterisation of the Action of PANI and P3ABA Films against E. coli and S. aureus

4.7.1. Challenge of PANI and P3ABA Films with a Range of CFU Doses of *E. coli* 25922 *lux* and *S. aureus* 6538 in Saline

Film punches were prepared and decontaminated, as described above. The MSTA was performed with bacterial challenges (10^4 CFU, 10^3 CFU, 10^2 CFU, and 10 CFU) prepared in 10 µL saline. The 10^6 CFU/mL culture was enumerated. Following a 2 h treatment, cells were rescued in 190 µL LB broth and incubated in a fresh 96 well plate for 16 h. Viability of bacteria was assessed by measuring bioluminescence for *E. coli* 25922 *lux* and by measuring OD_{600} for *S. aureus* 6538. The activity of PANI and P3ABA in films against a range of CFU doses of *E. coli* 25922 *lux* and *S. aureus* 6538 was analysed using a 2-way RM ANOVA.

4.7.2. Assay to Evaluate the Influence of the Presence of Organic Matter on the Activity of PANI and P3ABA Films against *E. coli* 25922 *lux*

Film punches were prepared and decontaminated, as described above. MSTA was performed with bacterial challenges (10^4 CFU) in 10 µL saline or 10 µL LB broth. The inocula were enumerated. Following a 2 h treatment, cells were rescued in 190 µL LB broth and incubated in a fresh 96 well plate for 16 h. The viability of rescued cells was determined by measuring the bioluminescence.

4.8. Appendix A Methods

4.8.1. Validation of Utilisation of *E. coli* 25922 *lux*

To examine if *E. coli* 25922 *lux* can be used for the testing of surfaces containing PANI and P3ABA, in place of the non-tagged version, the MIC and MBC of both strains were determined [55]. A range of concentrations of PANI and P3ABA in suspension were tested (0.03125–4%). The suspensions were established at $2\times$ the final desired concentration in 500 µL. The insolubility of PANI required each suspension to be set up separately by weighing the powder into 5 mL tubes and adding 500 µL LB broth. P3ABA suspensions were established from a stock solution using a doubling dilution series. 500 µL LB broth was aliquoted to set up an untreated control.

The PANI and P3ABA suspensions were inoculated with 500 µL of 10^6 CFU/mL of *E. coli* 25922 and *E. coli* 25922 *lux*. The MIC was defined as the lowest concentration of PANI or P3ABA that was able to inhibit the visible growth of test bacteria following a 24 h treatment [55,56]. Tubes that were observed by eye to have no visible growth were selected for MBC testing. For this, 20 µL of the experimental sample was spread onto six LB agar plates [55,56]. The spread plates were incubated at 37 °C for 16 h and the growth on these plates was determined. When countable colonies were present, the CFU/mL of the sample was calculated. The MBC was defined as the lowest concentration of PANI or P3ABA that either totally prevents growth or results in a ≥99.9% decrease in the initial inoculum following subculture on LB agar plates [55,96]. At least three biological replicates were obtained.

4.8.2. Determination of the Limit of Detection for *E. coli* 25922 *lux* and *S. aureus* 6538 Growing in a 96 Well Plate

The limit of detection of *E. coli* 25922 *lux* and *S. aureus* 6538 growing in a 96 well plate was examined by determining the lowest number of cells added to LB broth in a 96 well plate that can grow to detectable levels [95]. This was achieved by serially diluting an overnight culture in triplicate in a 96 well plate by transferring 20 µL of culture into wells containing 180 µL of LB broth. A range of inocula were established from ~10^9 CFU/mL to ~1 CFU/mL. The overnight culture was enumerated to confirm the cell numbers that were tested. The 96 well plate was incubated at 37 °C for 16 h in a sealed container with a moist tissue [95]. Growth of bacteria was assessed by measuring bioluminescence using the VICTOR X Multilabel Plate Reader for *E. coli* 25922 *lux* and by measuring OD_{600} using the µQuant™ Microplate Spectrophotometer for *S. aureus* 6538.

5. Conclusions

PANI and P3ABA both demonstrated bactericidal activity against *E. coli* and *S. aureus* in suspension and as part of an absorbent surface, with greater activity being observed with P3ABA. PANI in films was not active against *E. coli* or *S. aureus*, while P3ABA in films reduced the viability of *E. coli* after a 24 h treatment. The results that are presented in this work support the use of P3ABA to create contamination resistant surfaces.

Acknowledgments: The authors are grateful for research funding from both the New Zealand Ministry of Business, Innovation and Employment (MBIE) for research programmes UOAX0812 and UOAX1410, and the University of Auckland's Vice Chancellors Strategic Development Fund, grant number 23563. The authors are grateful for funding (University of Auckland, SMS Publication Bursary) covering the costs to publish in open access. The authors thank Sudip Ray, Adeline Le Cocq, Chris Wilcox and Walt Wheelwright for purified PANI and P3ABA.

Author Contributions: Julia Robertson and Simon Swift conceived and designed the experiments; Julia Robertson performed the experiments and analysed the data; Marija Gizdavic-Nikolaidis contributed materials and advised on the chemistry aspects; Julia Robertson wrote the paper.

Conflicts of Interest: The authors declare no conflict of interest. The founding sponsors had no role in the design of the study; in the collection, analyses, or interpretation of data; in the writing of the manuscript, and in the decision to publish the results.

Appendix A.

Appendix A.1. Validation of Utilisation of E. coli 25922 lux as a Proxy of E. coli 25922 for Investigation of PANI and P3ABA as Surface Antimicrobial Agents

Appendix A.1.1. PANI Has Similar Activity against *E. coli* 25922 and *E. coli* 25922 *lux* While P3ABA Is Less Active against the Latter

For *E. coli* 25922, a *lux*-tagged version was used as the released bioluminescence can be detected and serves as a marker of cell viability [28,92,93]. Utilisation of a bioluminescently-tagged strain is a practical alternative to enumeration by plate counts for future testing of these potential surface additives against slow growing bacteria, such as *Mycobacterium tuberculosis* [63]. *E. coli* 25922 *lux* has a chromosomal insertion of the bacterial luciferase (*lux*) operon (*luxCDABE*) into the 16S locus [93]. A cell expressing the *lux* operon (*luxCDABE*) will be in an altered state compared to the non-tagged version as cellular energy is diverted in order to generate the luminescence [28]. The bioluminescence reaction consumes reduced flavin mononucleotide ($FMNH_2$) and a long chain fatty aldehyde, and tetradecanoic acid is diverted from the fatty acid biosynthesis pathway to regenerate the aldehyde substrate [97]. Therefore, it is possible that the *lux*-tagged version of *E. coli* 25922 may have different sensitivities to PANI and P3ABA compared to the non-bioluminescent version.

Following from this, the activity of PANI and P3ABA in LB broth was determined against *E. coli* 25922 and *lux*-tagged *E. coli* 25922. The measure of activity used was the standard minimum inhibitory concentration (MIC) and minimum bactericidal concentration (MBC) [55]. For PANI in suspension, *E. coli* 25922 and the *lux*-tagged version had similar sensitivities (Figure A1). P3ABA had a similar MIC against *E. coli* 25922 and *E. coli* 25922 *lux* (Figure A2); however, the MBC of P3ABA against *E. coli* 25922 (0.125%) was lower than that for *E. coli* 25922 *lux* (1–4%), which was statistically significant (Mann-Whitney test, *p* value: less than 0.05). The difference in activity of P3ABA observed against *E. coli* 25922 and *E. coli* 25922 *lux* may be reflective of the metabolic burden of light production. The *lux*-tagged *E. coli* 25922 may be less susceptible to the bactericidal action of P3ABA over a 24 h treatment time. Overall, these results support the use of *lux*-tagged *E. coli* for testing of PANI and P3ABA.

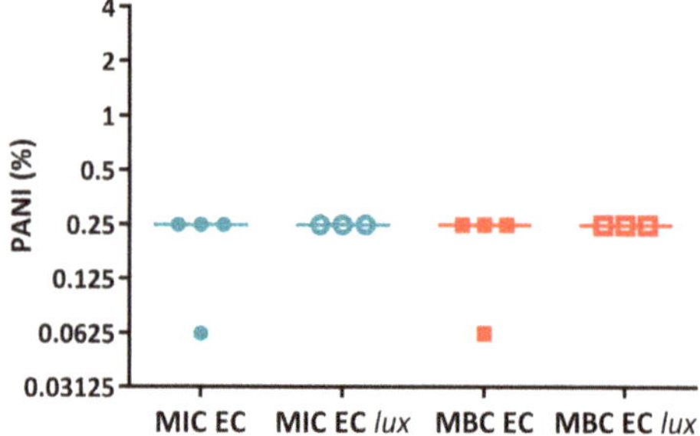

Figure A1. Activity of PANI against *E. coli* 25922 and *E. coli* 25922 *lux*. The MIC (circles) and MBC (squares) of PANI against *E. coli* 25922 (EC) and *E. coli* 25922 *lux* (EC *lux*) in LB broth. Data obtained from *E. coli* 25922 is represented by filled data points while data obtained from *E. coli* 25922 *lux* is represented by unfilled data points.

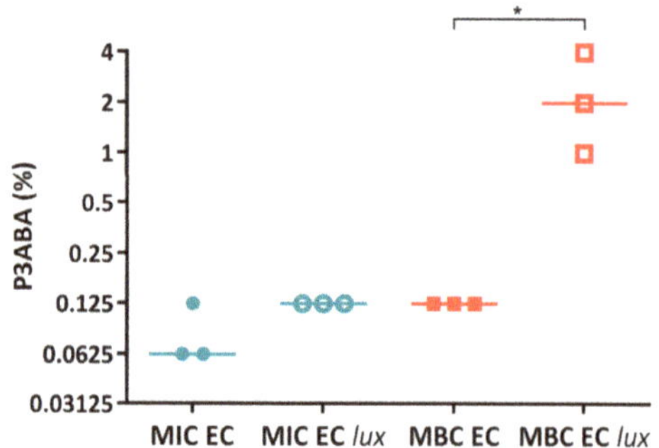

Figure A2. Activity of P3ABA against *E. coli* 25922 and *E. coli* 25922 *lux*. The MIC (circles) and MBC (squares) of PANI against *E. coli* 25922 (EC) and *E. coli* 25922 *lux* (EC *lux*) in LB broth. Data obtained from *E. coli* 25922 is represented by filled data points while data obtained from *E. coli* 25922 *lux* is represented by unfilled data points. Statistical significance is represented by * (Mann-Whitney test, *p* value: less than 0.05).

Appendix A.1.2. The Limit of Detection for *E. coli* 25922 *lux* and *S. aureus* 6538 Recovered in a 96 Well Plate Is 100 CFU/mL

Examination of the activity of PANI and P3ABA in surfaces involved recovery of challenged cells in a 96 well plate, which facilitated high-throughput testing of many concentrations and treatment times against one inoculum [95]. Therefore, it was necessary to determine the limit of detection of *E. coli* 25922 *lux* and *S. aureus* 6538 growing in this manner to enable interpretation of surface testing results. The limit of detection was examined by determining the lowest number of cells added to 180 µL of LB broth in a 96 well plate that can grow to detectable levels [95]. The limit of detection is presented as CFU/mL to relate back to the initial inoculum concentration. The absolute number of cells present in the 180 µL volume in the wells would be roughly 5-fold lower than the CFU/mL value.

After a 16 h incubation of *E. coli* 25922 *lux*, bioluminescence 2 log or more above background levels was detected in the wells inoculated with ~10^9 to ~10^3 CFU/mL indicating bacterial growth (Figure A3). For the well inoculated with ~10^2 CFU/mL, the median bioluminescence level was 2 log above background levels; although, two points were only slightly above background levels (Figure A3). Therefore, the limit of detection of *E. coli* 25922 *lux* in a 96 well plate was 100 CFU/mL, which would correspond to ~20 CFU in the well.

After a 16 h incubation of *S. aureus* 6538, growth was detected in wells inoculated with ~10^9 to ~10^2 CFU/mL (with growth defined as OD_{600} above 0.05) (Figure A4). The median OD_{600} value from starting inoculum of ~10 CFU/mL was above the threshold for growth; however, two points were below background levels (Figure A4). Therefore, the limit of detection of *S. aureus* 6538 in 96 well plate was 100 CFU/mL, which would correspond to ~20 CFU in the well.

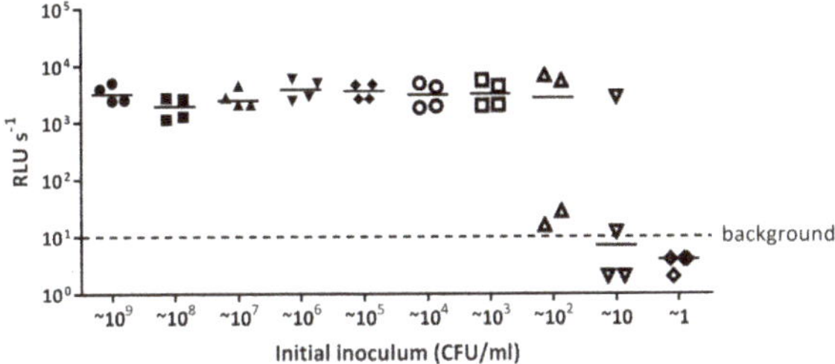

Figure A3. The limit of detection of *E. coli* 25922 *lux* when grown in a 96 well plate. *E. coli* 25922 *lux* was serially diluted in LB broth from ~10^9 CFU/mL to ~1 CFU/mL in a 96 well plate and incubated at 37 °C for 16 h. Light release (RLU s^{-1}) was measured from each dilution and detectable light levels above 10 RLU s^{-1} indicates growth. The data presented are from four independent experiments.

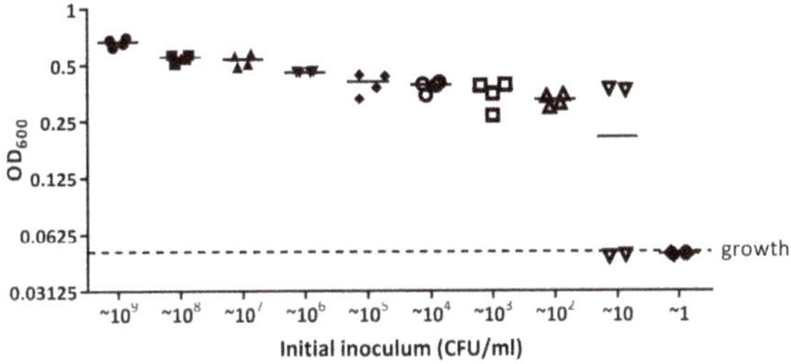

Figure A4. The limit of detection of *S. aureus* 6538 when grown in a 96 well plate. *S. aureus* 6538 was serially diluted in LB broth from ~10^9 CFU/mL to ~1 CFU/mL in a 96 well plate and incubated at 37 °C for 16 h. OD_{600} was measured from each dilution and OD_{600} readings above 0.05 are considered as growth. The data presented are from four independent experiments.

References

1. WHO. *Antimicrobial Resistance: Global Report on Surveillance*; WHO Press: Geneva, Switzerland, 2014.
2. Dancer, S.J. Controlling hospital-acquired infection: Focus on the role of the environment and new technologies for decontamination. *Clin. Microbiol. Rev.* **2014**, *27*, 665–690. [CrossRef] [PubMed]
3. Weber, D.J.; Anderson, D.; Rutala, W.A. The role of the surface environment in healthcare-associated infections. *Curr. Opin. Infect. Dis.* **2013**, *26*, 338–344. [CrossRef] [PubMed]
4. Kramer, A.; Schwebke, I.; Kampf, G. How long do nosocomial pathogens persist on inanimate surfaces? A systematic review. *BMC Infect. Dis.* **2006**, *6*, 130. [CrossRef] [PubMed]
5. Stiefel, U.; Cadnum, J.L.; Eckstein, B.C.; Guerrero, D.M.; Tima, M.A.; Donskey, C.J. Contamination of hands with methicillin-resistant Staphylococcus aureus after contact with environmental surfaces and after contact with the skin of colonized patients. *Infect. Control Hosp. Epidemiol.* **2011**, *32*, 185–187. [CrossRef] [PubMed]
6. Page, K.; Wilson, M.; Parkin, I. Antimicrobial surfaces and their potential in reducing the role of the inanimate environment in the incidence of hospital-acquired infections. *J. Mater. Chem.* **2009**, *19*, 3819–3831. [CrossRef]
7. Scott, E.; Bloomfield, S.F. The survival and transfer of microbial contamination via cloths, hands and utensils. *J. Appl. Bacteriol.* **1990**, *68*, 271–278. [CrossRef] [PubMed]
8. Rusin, P.; Maxwell, S.; Gerba, C. Comparative surface-to-hand and fingertip-to-mouth transfer efficiency of gram-positive bacteria, gram-negative bacteria, and phage. *J. Appl. Microbiol.* **2002**, *93*, 585–592. [CrossRef] [PubMed]
9. Gerhardts, A.; Hammer, T.R.; Balluff, C.; Mucha, H.; Hoefer, D. A model of the transmission of micro-organisms in a public setting and its correlation to pathogen infection risks. *J. Appl. Microbiol.* **2012**, *112*, 614–621. [CrossRef] [PubMed]
10. Otter, J.A.; Yezli, S.; French, G.L. The role played by contaminated surfaces in the transmission of nosocomial pathogens. *Infect. Control Hosp. Epidemiol.* **2011**, *32*, 687–699. [CrossRef] [PubMed]
11. French, G.L.; Otter, J.A.; Shannon, K.P.; Adams, N.M.T.; Watling, D.; Parks, M.J. Tackling contamination of the hospital environment by methicillin-resistant Staphylococcus aureus (MRSA): A comparison between conventional terminal cleaning and hydrogen peroxide vapour decontamination. *J. Hosp. Infect.* **2004**, *57*, 31–37. [CrossRef] [PubMed]
12. Bhalla, A.; Pultz, N.J.; Gries, D.M.; Ray, A.J.; Eckstein, E.C.; Aron, D.C.; Donskey, C.J. Acquisition of nosocomial pathogens on hands after contact with environmental surfaces near hospitalized patients. *Infect. Control Hosp. Epidemiol.* **2004**, *25*, 164–167. [CrossRef] [PubMed]
13. Williams, A.P.; Avery, L.M.; Killham, K.; Jones, D.L. Persistence of Escherichia coli O157 on farm surfaces under different environmental conditions. *J. Appl. Microbiol.* **2005**, *98*, 1075–1083. [CrossRef] [PubMed]
14. Carpentier, B.; Cerf, O. Review—Persistence of Listeria monocytogenes in food industry equipment and premises. *Int. J. Food Microbiol.* **2011**, *145*, 1–8. [CrossRef] [PubMed]
15. Otter, J.A.; Vickery, K.; Walker, J.T.; de Lancey Pulcini, E.; Stoodley, P.; Goldenberg, S.D.; Salkeld, J.A.G.; Chewins, J.; Yezli, S.; Edgeworth, J.D. Surface-attached cells, biofilms and biocide susceptibility: Implications for hospital cleaning and disinfection. *J. Hosp. Infect.* **2015**, *89*, 16–27. [CrossRef] [PubMed]
16. Yildiz, F.H. Processes controlling the transmission of bacterial pathogens in the environment. *Res. Microbiol.* **2007**, *158*, 195–202. [CrossRef] [PubMed]
17. Van Houdt, R.; Michiels, C.W. Biofilm formation and the food industry, a focus on the bacterial outer surface. *J. Appl. Microbiol.* **2010**, *109*, 1117–1131. [CrossRef] [PubMed]
18. Hota, B. Contamination, disinfection, and cross-colonization: Are hospital surfaces reservoirs for nosocomial infection? *Clin. Infect. Dis.* **2004**, *39*, 1182–1189. [CrossRef] [PubMed]
19. Guerrero, D.M.; Nerandzic, M.M.; Jury, L.A.; Jinno, S.; Chang, S.; Donskey, C.J. Acquisition of spores on gloved hands after contact with the skin of patients with Clostridium difficile infection and with environmental surfaces in their rooms. *Am. J. Infect. Control* **2012**, *40*, 556–558. [CrossRef] [PubMed]
20. Todd, E.C.D.; Greig, J.D.; Bartleson, C.A.; Michaels, B.S. Outbreaks where food workers have been implicated in the spread of foodborne disease. Part 5. Sources of contamination and pathogen excretion from infected persons. *J. Food Prot.* **2008**, *71*, 2582–2595. [CrossRef] [PubMed]
21. Malik, R.E.; Cooper, R.A.; Griffith, C.J. Use of audit tools to evaluate the efficacy of cleaning systems in hospitals. *Am. J. Infect. Control* **2003**, *31*, 181–187. [CrossRef] [PubMed]
22. French, G.L. The continuing crisis in antibiotic resistance. *Int. J. Antimicrob. Agents* **2010**, *36*, S3–S7. [CrossRef]

23. Robine, E.; Boulangé-Petermann, L.; Derangère, D. Assessing bactericidal properties of materials: The case of metallic surfaces in contact with air. *J. Microbiol. Methods* **2002**, *49*, 225–234. [CrossRef]

24. Escalada, M.G.; Russell, A.D.; Maillard, J.-Y.; Ochs, D. Triclosan-bacteria interactions: Single or multiple target sites? *Lett. Appl. Microbiol.* **2005**, *41*, 476–481. [CrossRef] [PubMed]

25. Ojeil, M.; Jermann, C.; Holah, J.; Denyer, S.P.; Maillard, J.-Y. Evaluation of new in vitro efficacy test for antimicrobial surface activity reflecting UK hospital conditions. *J. Hosp. Infect.* **2013**, *85*, 274–281. [CrossRef] [PubMed]

26. Suppakul, P.; Miltz, J.; Sonneveld, K.; Bigger, S.W. Active packaging technologies with an emphasis on antimicrobial packaging and its applications. *J. Food Sci.* **2003**, *68*, 408–420. [CrossRef]

27. Bastarrachea, L.J.; Denis-Rohr, A.; Goddard, J.M. Antimicrobial food equipment coatings: Applications and challenges. *Annu. Rev. Food Sci. Technol.* **2015**, *6*, 97–118. [CrossRef] [PubMed]

28. Green, J.-B.D.; Fulghum, T.; Nordhaus, M. Review of immobilized antimicrobial agents and methods for testing. *Biointerphases* **2011**, *6*, CL2–CL43. [CrossRef] [PubMed]

29. Siedenbiedel, F.; Tiller, J.C. Antimicrobial polymers in solution and on surfaces: Overview and functional principles. *Polymers* **2012**, *4*, 46–71. [CrossRef]

30. Cooksey, K. Effectiveness of antimicrobial food packaging materials. *Food Addit. Contam.* **2005**, *22*, 980–987. [CrossRef] [PubMed]

31. Hasan, J.; Crawford, R.J.; Ivanova, E.P. Antibacterial surfaces: The quest for a new generation of biomaterials. *Trends Biotechnol.* **2013**, *31*, 295–304. [CrossRef] [PubMed]

32. Bastarrachea, L.J.; Dhawan, S.; Sablani, S.S. Engineering Properties of Polymeric-Based Antimicrobial Films for Food Packaging: A Review. *Food Eng. Rev.* **2011**, *3*, 79–93. [CrossRef]

33. Vasilev, K.; Cook, J.; Griesser, H.J. Antibacterial surfaces for biomedical devices. *Expert Rev. Med. Devices* **2009**, *6*, 553–567. [CrossRef] [PubMed]

34. Fernández, A.; Soriano, E.; López-Carballo, G.; Picouet, P.; Lloret, E.; Gavara, R.; Hernández-Muñoz, P. Preservation of aseptic conditions in absorbent pads by using silver nanotechnology. *Food Res. Int.* **2009**, *42*, 1105–1112. [CrossRef]

35. Humphreys, H. Self-disinfecting and microbiocide-impregnated surfaces and fabrics: What potential in interrupting the spread of healthcare-associated infection? *Clin. Infect. Dis.* **2014**, *58*, 848–853. [CrossRef] [PubMed]

36. Mitchell, A.; Spencer, M.; Edmiston, C. Role of healthcare apparel and other healthcare textiles in the transmission of pathogens: A review of the literature. *J. Hosp. Infect.* **2015**, *90*, 285–292. [CrossRef] [PubMed]

37. Agarwal, A.; Weis, T.L.; Schurr, M.J.; Faith, N.G.; Czuprynski, C.J.; McAnulty, J.F.; Murphy, C.J.; Abbott, N.L. Surfaces modified with nanometer-thick silver-impregnated polymeric films that kill bacteria but support growth of mammalian cells. *Biomaterials* **2010**, *31*, 680–690. [CrossRef] [PubMed]

38. Zhou, B.; Hu, Y.; Li, J.; Li, B. Chitosan/phosvitin antibacterial films fabricated via layer-by-layer deposition. *Int. J. Biol. Macromol.* **2014**, *64*, 402–408. [CrossRef] [PubMed]

39. Cen, L.; Neoh, K.G.; Kang, E.T. Antibacterial activity of cloth functionalized with N-alkylated poly(4-vinylpyridine). *J. Biomed. Mater. Res.* **2004**, *71*, 70–80. [CrossRef] [PubMed]

40. Ohl, M.; Schweizer, M.; Graham, M.; Heilmann, K.; Boyken, L.; Diekema, D. Hospital privacy curtains are frequently and rapidly contaminated with potentially pathogenic bacteria. *Am. J. Infect. Control* **2012**, *40*, 904–906. [CrossRef] [PubMed]

41. Schweizer, M.; Graham, M.; Ohl, M.; Heilmann, K.; Boyken, L.; Diekema, D. Novel Hospital Curtains with Antimicrobial Properties: A Randomized, Controlled Trial. *Infect. Control Hosp. Epidemiol.* **2012**, *33*, 1081–1085. [CrossRef] [PubMed]

42. Lemmen, S.; Scheithauer, S.; Haefner, H.; Yezli, S.; Mohr, M.; Otter, J.A. Evaluation of hydrogen peroxide vapor for the inactivation of nosocomial pathogens on porous and nonporous surfaces. *Am. J. Infect. Control* **2015**, *43*, 82–85. [CrossRef] [PubMed]

43. Gizdavic-Nikolaidis, M.R.; Ray, S.; Bennett, J.R.; Swift, S.; Bowmaker, G.A.; Easteal, A.J. Electrospun poly(aniline-co-ethyl 3-aminobenzoate)/poly(lactic acid) nanofibers and their potential in biomedical applications. *J. Polym. Sci. Part A Polym. Chem.* **2011**, *49*, 4902–4910. [CrossRef]

44. Pandey, S.; Annapoorni, S.; Malhotra, B.D. Synthesis and Characterization of Poly(aniline-co-o-anisidine): A Processable Conducting Copolymer. *Macromolecules* **1993**, *26*, 3190–3193. [CrossRef]

45. Gizdavic-Nikolaidis, M.R.; Bennett, J.R.; Swift, S.; Easteal, A.J.; Ambrose, M. Broad spectrum antimicrobial activity of functionalized polyanilines. *Acta Biomater.* **2011**, *7*, 4204–4209. [CrossRef] [PubMed]

46. Gizdavic-Nikolaidis, M.R.; Bennett, J.R.; Zujovic, Z.; Swift, S.; Bowmaker, G.A. Characterization and antimicrobial efficacy of acetone extracted aniline oligomers. *Synth. Met.* **2012**, *162*, 1114–1119. [CrossRef]

47. Dhand, C.; Das, M.; Datta, M.; Malhotra, B.D. Recent advances in polyaniline based biosensors. *Biosens. Bioelectron.* **2011**, *26*, 2811–2821. [CrossRef] [PubMed]

48. Grennan, K.; Killard, A.J.; Hanson, C.J.; Cafolla, A.A.; Smyth, M.R. Optimisation and characterisation of biosensors based on polyaniline. *Talanta* **2006**, *68*, 1591–1600. [CrossRef] [PubMed]

49. Ni, H.; Jiang, T.; Hu, P.; Han, Z.; Lu, X.; Ye, P. Self-decontaminating properties of fluorinated copolymers integrated with ciprofloxacin for synergistically inhibiting the growth of Escherichia coli. *J. Biomater. Sci. Polym. Ed.* **2014**, *25*, 1920–1945. [CrossRef] [PubMed]

50. Mu, J.L.; Fan, W.J.; Shan, S.Y.; Hu, T.W.; Wang, Y.M.; Jia, Q.M. The Effects of Natural Dopant Acids on Morphologies and Antibacterial Activity of Polyaniline. *Adv. Mater. Res.* **2013**, *650*, 249–252. [CrossRef]

51. Humpolicek, P.; Kasparkova, V.; Saha, P.; Stejskal, J. Biocompatibility of polyaniline. *Synth. Met.* **2012**, *162*, 722–727. [CrossRef]

52. Qi, H.; Liu, M.; Xu, L.; Feng, L.; Tao, L.; Ji, Y.; Zhang, X.; Wei, Y. Biocompatibility evaluation of aniline oligomers with different end-functional groups. *Toxicol. Res.* **2013**, *2*, 427–433. [CrossRef]

53. Seshadri, D.T.; Bhat, N.V. Use of polyaniline as an antimicrobial agent in textiles. *Indian J. Fibre Text. Res.* **2005**, *30*, 204–206.

54. Gizdavic-Nikolaidis, M.R.; Pagnon, J.C.; Ali, N.; Sum, R.; Davies, N.; Roddam, L.F.; Ambrose, M. Functionalized polyanilines disrupt Pseudomonas aeruginosa and Staphylococcus aureus biofilms. *Colloids Surf. B. Biointerfaces* **2015**, *136*, 666–673. [CrossRef] [PubMed]

55. Andrews, J.M. Determination of minimum inhibitory concentrations. *J. Antimicrob. Chemother.* **2001**, *48* (Suppl. 1), 5–16. [CrossRef] [PubMed]

56. Wiegand, I.; Hilpert, K.; Hancock, R.E.W. Agar and broth dilution methods to determine the minimal inhibitory concentration (MIC) of antimicrobial substances. *Nat. Protoc.* **2008**, *3*, 163–175. [CrossRef] [PubMed]

57. Siani, H.; Maillard, J.-Y. Best practice in healthcare environment decontamination. *Eur. J. Clin. Microbiol. Infect. Dis.* **2015**, *34*, 1–11. [CrossRef] [PubMed]

58. Kram, K.E.; Finkel, S.E. Rich Medium Composition Affects Escherichia coli Survival, Glycation, and Mutation Frequency during Long-Term Batch Culture. *Appl. Environ. Microbiol.* **2015**, *81*, 4442–4450. [CrossRef] [PubMed]

59. Tao, H.; Bausch, C.; Richmond, C.; Blattner, F.R.; Conway, T. Functional genomics: Expression analysis of Escherichia coli growing on minimal and rich media. *J. Bacteriol.* **1999**, *181*, 6425–6440. [PubMed]

60. Zhang, J.; Greasham, R. Chemically defined media for commercial fermentations. *Appl. Microbiol. Biotechnol.* **1999**, *51*, 407–421. [CrossRef]

61. Sezonov, G.; Joseleau-Petit, D.; D'Ari, R. Escherichia coli physiology in Luria-Bertani broth. *J. Bacteriol.* **2007**, *189*, 8746–8749. [CrossRef] [PubMed]

62. Dong, T.; Schellhorn, H.E. Control of RpoS in global gene expression of Escherichia coli in minimal media. *Mol. Genet. Genom.* **2009**, *281*, 19–33. [CrossRef] [PubMed]

63. Robertson, J.; Dalton, J.; Wiles, S.; Gizdavic-Nikolaidis, M.; Swift, S. The tuberculocidal activity of polyaniline and functionalised polyanilines. *PeerJ* **2016**, *4*, e2795. [CrossRef] [PubMed]

64. Herigstad, B.; Hamilton, M.; Heersink, J. How to optimize the drop plate method for enumerating bacteria. *J. Microbiol. Methods* **2001**, *44*, 121–129. [CrossRef]

65. Liu, J.; Liu, C.; Liu, Y.; Chen, M.; Hu, Y.; Yang, Z. Study on the grafting of chitosan-gelatin microcapsules onto cotton fabrics and its antibacterial effect. *Colloids Surf. B Biointerfaces* **2013**, *109*, 103–108. [CrossRef] [PubMed]

66. Andreu, N.; Zelmer, A.; Fletcher, T.; Elkington, P.T.; Ward, T.H.; Ripoll, J.; Parish, T.; Bancroft, G.J.; Schaible, U.; Robertson, B.D.; et al. Optimisation of bioluminescent reporters for use with mycobacteria. *PLoS ONE* **2010**, *5*, e10777. [CrossRef] [PubMed]

67. Andreu, N.; Zelmer, A.; Sampson, S.L.; Ikeh, M.; Bancroft, G.J.; Schaible, U.E.; Wiles, S.; Robertson, B.D. Rapid in vivo assessment of drug efficacy against Mycobacterium tuberculosis using an improved firefly luciferase. *J. Antimicrob. Chemother.* **2013**, *68*, 2118–2127. [CrossRef] [PubMed]

68. Andreu, N.; Zelmer, A.; Wiles, S. Noninvasive biophotonic imaging for studies of infectious disease. *FEMS Microbiol. Rev.* **2011**, *35*, 360–394. [CrossRef] [PubMed]

69. Appendini, P.; Hotchkiss, J.H. Review of antimicrobial food packaging. *Innov. Food Sci. Emerg. Technol.* **2002**, *3*, 113–126. [CrossRef]

70. Japanese Standards Association. *JIS Z 2801:2000 Antimicrobial Products—Test for Antimicrobial Activity and Efficacy*; Japanese Standards Association: Tokyo, Japan, 2000.

71. Humphreys, P.N. Testing standards for sporicides. *J. Hosp. Infect.* **2011**, *77*, 193–198. [CrossRef] [PubMed]

72. Warnes, S.L.; Keevil, C.W. Lack of Involvement of Fenton Chemistry in Death of Methicillin-Resistant and Methicillin-Sensitive Strains of Staphylococcus aureus and Destruction of Their Genomes on Wet or Dry Copper Alloy Surfaces. *Appl. Environ. Microbiol.* **2016**, *82*, 2132–2136. [CrossRef] [PubMed]

73. Lobritz, M.A.; Belenky, P.; Porter, C.B.M.; Gutierrez, A.; Yang, J.H.; Schwarz, E.G.; Dwyer, D.J.; Khalil, A.S.; Collins, J.J. Antibiotic efficacy is linked to bacterial cellular respiration. *Proc. Natl. Acad. Sci. USA* **2015**, *112*, 8173–8180. [CrossRef] [PubMed]

74. Gizdavic-Nikolaidis, M.R.; Easteal, A.J.; Stepanovic, S. Bioactive Aniline Copolymers. International Patent Application No. PCT/NZ2008/000254, 26 September 2008.

75. Gizdavic-Nikolaidis, M.R.; Bowmaker, G.A.; Zujovic, Z. *The Synthesis, Physical Properties, Bioactivity and Potential Applications of Polyanilines*; Cambridge Scholars Publishing: Newcastle, UK, 2018; in press.

76. Fröls, S.; Dyall-Smith, M.; Pfeifer, F. Biofilm formation by haloarchaea. *Environ. Microbiol.* **2012**, *14*, 3159–3174. [CrossRef] [PubMed]

77. Shi, N.; Guo, X.; Jing, H.; Gong, J.; Sun, C.; Yang, K. Antibacterial effect of the conducting polyaniline. *J. Mater. Sci. Technol.* **2006**, *22*, 289–290.

78. Tan, C.; Smith, R.P.; Srimani, J.K.; Riccione, K.A.; Prasada, S.; Kuehn, M.; You, L. The inoculum effect and band-pass bacterial response to periodic antibiotic treatment. *Mol. Syst. Biol.* **2012**, *8*, 617. [CrossRef] [PubMed]

79. Warnes, S.L.; Green, S.M.; Michels, H.T.; Keevil, C.W. Biocidal efficacy of copper alloys against pathogenic enterococci involves degradation of genomic and plasmid DNAs. *Appl. Environ. Microbiol.* **2010**, *76*, 5390–5401. [CrossRef] [PubMed]

80. Oie, S.; Hosokawa, I.; Kamiya, A. Contamination of room door handles by methicillin-sensitive/methicillin-resistant Staphylococcus aureus. *J. Hosp. Infect.* **2002**, *51*, 140–143. [CrossRef] [PubMed]

81. Oie, S.; Suenaga, S.; Sawa, A.; Kamiya, A. Association between isolation sites of methicillin-resistant Staphylococcus aureus (MRSA) in patients with MRSA-positive body sites and MRSA contamination in their surrounding environmental surfaces. *Jpn. J. Infect. Dis.* **2007**, *60*, 367–369. [PubMed]

82. Schmidt, M.G.; Attaway, H.H.; Sharpe, P.A.; John, J.F.; Sepkowitz, K.A.; Morgan, A.; Fairey, S.E.; Singh, S.; Steed, L.L.; Cantey, J.R.; et al. Sustained reduction of microbial burden on common hospital surfaces through introduction of copper. *J. Clin. Microbiol.* **2012**, *50*, 2217–2223. [CrossRef] [PubMed]

83. O'Gorman, J.; Humphreys, H. Application of copper to prevent and control infection. Where are we now? *J. Hosp. Infect.* **2012**, *81*, 217–223. [CrossRef]

84. Noyce, J.O.; Michels, H.T.; Keevil, C.W. Use of copper cast alloys to control Escherichia coli O157 cross-contamination during food processing. *Appl. Environ. Microbiol.* **2006**, *72*, 4239–4244. [CrossRef] [PubMed]

85. Rabie, A.J.; McLaren, I.M.; Breslin, M.F.; Sayers, R.; Davies, R.H. Assessment of anti-Salmonella activity of boot dip samples. *Avian Pathol.* **2015**, *44*, 129–134. [CrossRef] [PubMed]

86. Fraise, A. Currently available sporicides for use in healthcare, and their limitations. *J. Hosp. Infect.* **2011**, *77*, 210–212. [CrossRef] [PubMed]

87. Honney, C. Colonisation Resistant Surfaces. MSc Thesis, University of Auckland, Auckland, New Zealand, 2014.

88. Møretrø, T.; Høiby-pettersen, G.S.; Habimana, O.; Heir, E.; Langsrud, S. Assessment of the antibacterial activity of a triclosan-containing cutting board. *Int. J. Food Microbiol.* **2011**, *146*, 157–162. [CrossRef] [PubMed]

89. Cutter, C.N. The effectiveness of triclosan-incorporated plastic against bacteria on beef surfaces. *J. Food Prot.* **1999**, *62*, 474–479. [CrossRef] [PubMed]

90. Kalyon, B.D.; Olgun, U. Antibacterial efficacy of triclosan-incorporated polymers. *Am. J. Infect. Control* **2001**, *29*, 124–125. [CrossRef] [PubMed]

91. Connolly, P.; Bloomfield, S.F.; Denyer, S.P. The use of impedance for preservative efficacy testing of pharmaceuticals and cosmetic products. *J. Appl. Bacteriol.* **1994**, *76*, 68–74. [CrossRef] [PubMed]

92. Ng, M.T. Development of a Series of Bioluminescent Sensor Strains for Rapid Screening of Antimicrobials. Ph.D. Thesis, University of Auckland, Auckland, New Zealand, 2013.
93. Riedel, C.U.; Casey, P.G.; Mulcahy, H.; O'Gara, F.; Gahan, C.G.M.; Hill, C. Construction of p16Slux, a novel vector for improved bioluminescent labeling of gram-negative bacteria. *Appl. Environ. Microbiol.* **2007**, *73*, 7092–7095. [CrossRef] [PubMed]
94. Seaver, L.C.; Imlay, J.A. Are respiratory enzymes the primary sources of intracellular hydrogen peroxide? *J. Biol. Chem.* **2004**, *279*, 48742–48750. [CrossRef] [PubMed]
95. Brewster, J.D. A simple micro-growth assay for enumerating bacteria. *J. Microbiol. Methods* **2003**, *53*, 77–86. [CrossRef]
96. Pankey, G.A.; Sabath, L.D. Clinical relevance of bacteriostatic versus bactericidal mechanisms of action in the treatment of Gram-positive bacterial infections. *Clin. Infect. Dis.* **2004**, *38*, 864–870. [CrossRef] [PubMed]
97. Meighen, E. Genetics of bacterial bioluminescence. *Annu. Rev. Genet.* **1994**, *28*, 117–139. [CrossRef] [PubMed]

Article

Electrochemically Enhanced Drug Delivery Using Polypyrrole Films

Sayed Ashfaq Ali Shah [1,2,†], Melike Firlak [1,†], Stuart Ryan Berrow [1], Nathan Ross Halcovitch [1], Sara Jane Baldock [1], Bakhtiar Muhammad Yousafzai [3], Rania M. Hathout [1,4,5,6] and John George Hardy [1,7,*]

[1] Department of Chemistry, Lancaster University, Lancaster, LA1 4YB, UK; ashfaqali66@yahoo.com (S.A.A.S.); m.firlak@gmail.com (M.F.); Stuart_Berrow@yahoo.co.uk (S.R.B.); n.r.halcovitch@lancaster.ac.uk (N.R.H.); s.baldock@lancaster.ac.uk (S.J.B.); rania.hathout@pharma.asu.edu.eg (R.M.H.)
[2] Department of Chemistry, Government Post Graduate College No. 1, Abbottabad 22010, Pakistan
[3] Department of Chemistry, Hazara University, Mansehra 21130, Pakistan; yousafzaibm@gmail.com
[4] Department of Pharmaceutics and Industrial Pharmacy, Faculty of Pharmacy, Ain Shams University, Cairo 11566, Egypt
[5] Bioinformatics Program, Faculty of Computer and Information Sciences, Ain Shams University, Cairo 11566, Egypt
[6] Department of Pharmaceutical Technology, Faculty of Pharmacy and Biotechnology, German University in Cairo, Cairo 11835, Egypt
[7] Materials Science Institute, Lancaster University, Lancaster, LA1 4YB, UK
[*] Correspondence: j.g.hardy@lancaster.ac.uk; Tel.: +44-1524-595-080
[†] These authors contributed equally to this work.

Received: 1 June 2018; Accepted: 26 June 2018; Published: 1 July 2018

Abstract: The delivery of drugs in a controllable fashion is a topic of intense research activity in both academia and industry because of its impact in healthcare. Implantable electronic interfaces for the body have great potential for positive economic, health, and societal impacts; however, the implantation of such interfaces results in inflammatory responses due to a mechanical mismatch between the inorganic substrate and soft tissue, and also results in the potential for microbial infection during complex surgical procedures. Here, we report the use of conducting polypyrrole (PPY)-based coatings loaded with clinically relevant drugs (either an anti-inflammatory, dexamethasone phosphate (DMP), or an antibiotic, meropenem (MER)). The films were characterized and were shown to enhance the delivery of the drugs upon the application of an electrochemical stimulus in vitro, by circa (ca.) 10–30% relative to the passive release from non-stimulated samples. Interestingly, the loading and release of the drugs was correlated with the physical descriptors of the drugs. In the long term, such materials have the potential for application to the surfaces of medical devices to diminish adverse reactions to their implantation in vivo.

Keywords: conducting polymers; electroactive polymers; medical devices; drug delivery; anti-inflammatory; antibiotic

---

1. Introduction

The market for global drug-delivery technologies is a multibillion-dollar industry, and there is a growing demand for drug-delivery devices in both developed and emerging economies (in part, driven by aging societies and rapid urbanization) [1,2]. The market for implantable medical devices is also a multibillion-dollar industry, with a similarly growing demand driven by the same factors (i.e., increasing geriatric populations and incidences of chronic diseases, coupled with the adoption of implantable medical devices) [3–5]. Medical devices are implanted in either hard tissues (e.g., orthopedic implants such as reconstructive joint replacements, dental implants, and spinal implants) or soft tissues (e.g., intraocular lenses, or the skin). The successful integration of such devices is dependent on the availability of sterile surgical conditions, patient health, etc., and their implantation is most commonly coupled with a course of condition- and patient-specific drugs [6,7].

Drug-delivery systems can be engineered to deliver drugs at rates controlled by specific features of the systems, particularly their chemical composition (e.g., inorganic/organic components, molecular weights of their constituents, crosslinking density of polymers, etc.) and the inclusion of components that respond to chemical stimuli (e.g., enzymes, ions, or pH) or physical stimuli (e.g., electromagnetic fields, or temperature) [8–15].

Responsiveness to electric fields is an inherent property of electrically conducting materials (e.g., metals), and certain molecules respond to the application of electric fields by orienting their dipoles with the applied field, whereas other molecules can undergo redox reactions in response to electricity. An exciting class of electrically conducting materials is that of organic electronic materials (OEMs). Various types of OEMs exist, including fullerenes (bucky balls or nanotubes), graphene/graphene oxide, or conjugated polymers (e.g., polyaniline, polypyrrole, or polythiophene). Some OEMs are commercially available, and their properties can be tailored (through chemical modification or the generation of composites) to suit the delivery of various drugs [16,17].

While OEM-based nanoparticles have promise for simultaneous imaging and drug delivery (i.e., theranostic applications) [18,19], nanoparticles are not the only morphology of materials that OEMs can be processed into, and it is also possible to manufacture OEM-based films, fibers, foams, and hydrogels [20–24]. The morphologies of these alternative materials are under investigation for their inclusion into new versions of a variety of clinically translated electronic interfaces for the body (e.g., cardiac pacemakers, cochlear implants, retinal prostheses, and electrodes for deep brain stimulation), or indeed, electronic interfaces for the peripheral nervous system (e.g., for the control of the bladder) [20–24]. The clinically translated examples of electronic interfaces for the body are all currently metal-based (typically connected to batteries implanted at the same time), and the mechanical properties of these metals are markedly different from the soft tissues in which they are implanted (known as a mechanical mismatch). Mechanical mismatches lead to inflammatory responses and the formation of scar tissue around the electronic interface [23,24]. Mismatches can potentially be diminished by coating the surface of the metals with relatively soft OEM-based materials [24], or indeed, the delivery of anti-inflammatories from OEM-based materials [25]. Moreover, it is noteworthy that the surgical procedures necessary to implant such devices are complex, and problems associated with microbial infections in the proximity of these devices can potentially be addressed through the delivery of antimicrobials [16,17,25,26].

Conjugated polymers have fascinating optoelectronic properties, and are consequently being developed for use in the electronic industry [27–30]. There is academic and industrial interest in their potential application in the biomedical industry for use as bioactuators, biosensors, drug-delivery devices, neural electrode coatings, or indeed, tissue scaffolds for tissue engineering [20–22,31]. Polyaniline, polypyrrole, and polythiophene derivatives are most commonly investigated for biomedical applications, and polyaniline- and polypyrrole-based systems were shown to be capable of delivery of a variety of drugs (including anions and, less frequently, cations) [16,17,25,26,32–36].

Prospects for the clinical translation of conjugated polymer-based drug-delivery systems are clearly dependent on their biocompatibility. Histological analyses of tissue in the vicinity of polypyrrole (PPY)-based materials implanted subcutaneously or intramuscularly in rats revealed immune cell infiltration comparable to Food and Drug Administration (FDA) approved poly(lactic-co-glycolic acid) [37], or FDA-approved poly(D,L-lactide-co-glycolide) [38]. Similarly low inflammatory responses were observed for PPY-based materials implanted at the interface of the coronary artery of rats after five weeks [39], or PPY-based sciatic-nerve guidance channels implanted in rats after eight weeks [40], and importantly, PPY-coated electrodes in rat brains after three or six weeks [41]. The implantation of poly(3,4-ethylenedioxythiophene) (PEDOT)-coated electrodes in rat brains resulted in a modest global tissue reaction of approximately the same magnitude as that for silicon probes [42], whereas there was no observable immune response after one week for PEDOT-based materials implanted subcutaneously [43]. The implantation of polyaniline (PANI)-based materials implanted subcutaneously in rats showed low levels of inflammation after four [44], or 50 weeks [45]. We recognize that differences in individual studies (e.g., composition/structure of the materials, animal/tissue models, and the methods used to evaluate immune responses) present challenges when attempting to directly compare the results of each study; however, conjugated polymer-based biomaterials have levels of immunogenicity that are comparable to other FDA-approved biomaterials, and have prospects for clinical translation in the long term.

With a view to the long-term development of surface coatings for medical devices (e.g., neural electrodes) to diminish adverse reactions to their implantation in vivo, we report the use of conducting polymer-based coatings that enhance the delivery of drugs upon the application of an electrical potential. As a simple model system, we used polypyrrole (PPY) loaded with clinically relevant drugs (either an anti-inflammatory, dexamethasone phosphate (DMP), or an antibiotic, meropenem (MER)), as depicted in Figure 1. The rationale behind the delivery of DMP was to address problems of local tissue inflammation in the proximity of the materials, whereas the rationale behind the delivery of MER was to help diminish the potential for microbial infections in the proximity of materials that might be associated with the complicated surgical procedures necessary to implant electronic interfaces for the body. The films were characterized using microscopic, spectroscopic, and electrochemical techniques; the delivery of the drugs into a biomedically relevant buffer (phosphate-buffered saline, PBS) was studied in vitro, and the correlation between drug loading/release was correlated with the physical descriptors of the drugs. Such materials have prospects for the preparation of conformal electroactive coatings for implantable biomaterials.

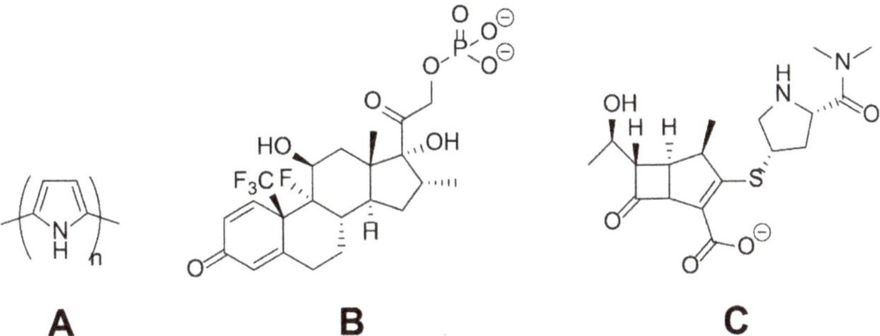

Figure 1. The chemical structures of the substances studied herein: (**A**) Polypyrrole (PPY); (**B**) dexamethasone phosphate (DMP); (**C**) meropenem (MER).

2. Results

Films of PPY loaded with an anionic drug (either DMP or MER, see Figure 1) were deposited onto the surface of indium tin oxide (ITO)-coated glass electrodes via electropolymerization. The electrochemical oxidation (at a potential of 1.0 V) and polymerization of pyrrole on the anode (the ITO-coated glass electrode) yielded a film where the positive charges on the backbone of the polypyrrole were counterbalanced by the presence of the anionic dopants from the electrolyte during electropolymerization (in this case one of the anionic drugs, DMP or MER).

The successful deposition of polypyrrole films onto the surface of the ITO electrodes was easily observable by eye (i.e., presence of a black film on the surface of a clear and colorless ITO electrode), and confirmed by scanning electron microscopy (SEM) and energy dispersive X-ray spectroscopy (EDX) data, as shown in Figure 2 (SEM and EDX imaging of DMP-doped films), Figure 3 (SEM and EDX imaging of MER-doped films), and Figure 4 (EDX data for DMP-doped and MER-doped films).

SEM revealed the surfaces of the films to have μm-scale roughness characteristic of electropolymerized PPY, with the features having a broad distribution of sizes, from very small particles of ca. 100–200 nm to much larger 1–20 μm "cauliflower-like" structures, akin to the features reported for films prepared via analogous electropolymerization methodologies in the literature. There were concomitant differences in electrical properties (i.e., impedance/resistance) relative to the surface-area-to-volume ratio of the materials [16,46]. EDX analysis [47] showed elemental signals characteristic of Au (instrumental background, characteristic Kα 2.123 keV; data not shown), Si (glass electrode, characteristic Kα 1.74 keV), and the elements associated with the drug-loaded polymer films: C and N (characteristic of polypyrrole) [48], F, O, and P (characteristic of DMP), and O and S (characteristic of MER).

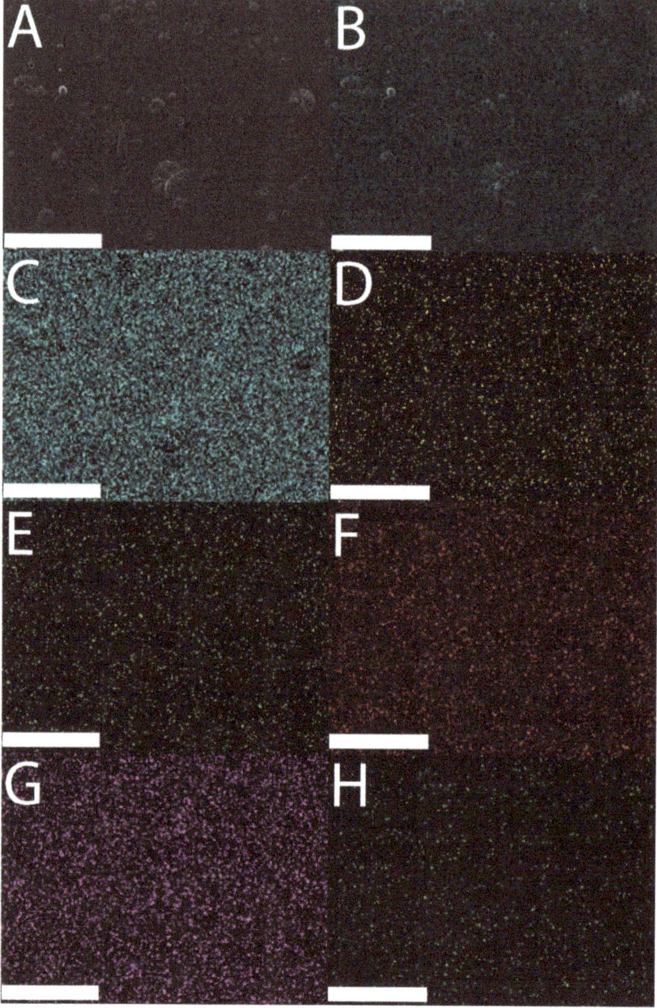

Figure 2. Scanning electron microscopy (SEM) and energy dispersive X-ray spectroscopy (EDX) images of a DMP-doped PPY film: (**A**) SEM image of a DMP-doped PPY film; (**B**) EDX layered image from a DMP-doped PPY film; (**C**) Kα emission of C; (**D**) Kα emission of F; (**E**) Kα emission of N; (**F**) Kα emission of O; (**G**) Kα emission of P; (**H**) Kα emission of Si. Scale bars represent 100 μm.

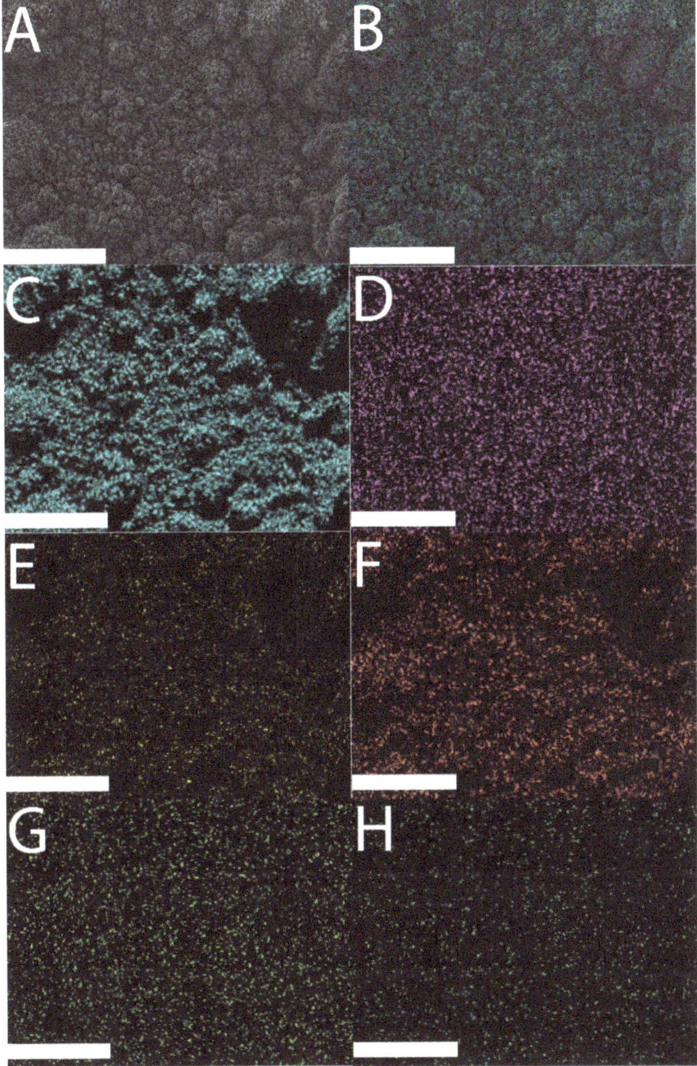

Figure 3. SEM and EDX images of an MER-doped PPY film: (**A**) SEM image of an MER-doped PPY film; (**B**) EDX layered image from an MER-doped PPY film; (**C**) Kα emission of C; (**D**) Kα emission of P; (**E**) Kα emission of F; (**F**) Kα emission of O; (**G**) Kα emission of N; (**H**) Kα emission of Si. Scale bars represent 100 μm.

The EDX maps in Figures 2 and 3 demonstrate the elemental composition of the films to be homogeneous over the surface of the films. Analysis of the EDX data for DMP-doped or MER-doped films (Figure 4A,B, respectively) showed elemental signals characteristic of carbon (Kα at 0.277 keV) and nitrogen (Kα at 0.392 keV) found on polypyrrole. For the DMP-doped films, there were peaks at 0.677 keV, 0.525 keV, and 2.014 keV, characteristic of the Kα of F, O, and P, respectively (Figure 4A). For the MER-doped films, there were peaks at 0.525 keV, 2.308 keV, and 2.622 keV, characteristic of the Kα of O, S, and Cl, respectively (indicative of the hydrochloride salt of MER, Figure 4B).

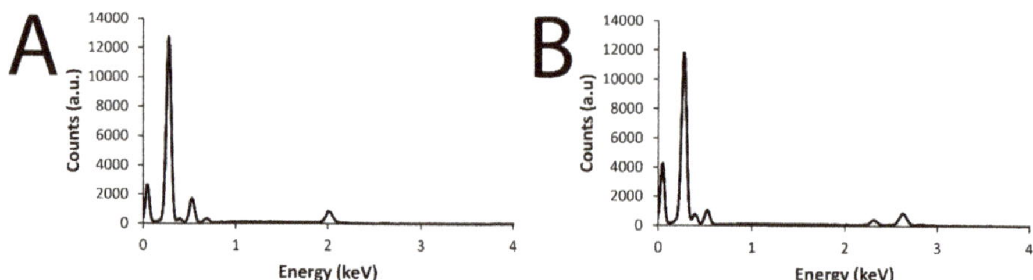

Figure 4. (**A**) EDX data from a DMP-doped PPY film; (**B**) EDX data from an MER-doped PPY film.

Examination of the films using Fourier-transform infrared (FTIR) spectroscopy in attenuated total reflection (ATR) mode also confirmed the presence of the drugs (DMP and MER) doping the PPY (Figure 5). The FTIR spectra of the PPY films showed absorptions at ca. 1480 cm^{-1} and ca. 1535 cm^{-1}, corresponding to the symmetric and asymmetric ring-stretching modes, respectively [49–53]. The FTIR spectra of DMP and DMP-doped PPY films showed absorptions at 989 cm^{-1} and 1197 cm^{-1}, corresponding to the characteristic symmetric and asymmetric stretching vibrations of the phosphate groups, while the absorption band at 1641 cm^{-1} corresponded to the C=O stretching vibration of DMP (Figure 5A,B). The FTIR spectra of MER and MER-doped PPY films showed absorptions characteristic of stretching vibrations of the C=O bond in the β-lactam ring of MER, located at 1863 cm^{-1} (Figure 5C,D).

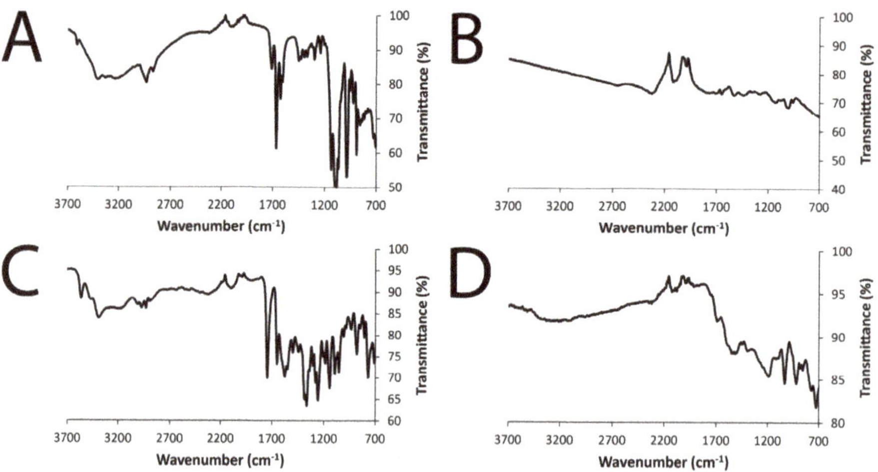

Figure 5. Fourier-transform infrared (FTIR) spectra collected in attenuated total reflection (ATR) mode: (**A**) DMP; (**B**) DMP-doped PPY film; (**C**) MER; (**D**) MER-doped PPY film.

X-ray diffraction (XRD) analysis of the DMP-doped and MER-doped films revealed some interesting structural information confirming the inclusion of the drugs in the films (Figure 6). PPY has a relatively amorphous structure with a broad peak in the region of 2θ = 20–30° in the XRD patterns [54] which is associated with the closest distance of approach of the planar aromatic rings of pyrrole (e.g., face-to-face pyrrole rings) [55]. Interestingly, doping PPY with DMP led to a peak shift to the region of 2θ = 15–28°, confirming that addition of DMP alters the packing of the PPY chains in the film [56]; likewise, doping PPY with MER also led to a peak shift to the region 2θ = 15–38°. Interestingly, some of the crystalline peaks of pure MER (which appear at 12.6°, 16.6°, 18.2°, 19°, 20°, 21.6°, 22.3°,

22.7°, 23.3°, 25.2°, 26.7°, 28.2°, 29°, 30°, 31.7°, 34.5°, 37.7°, 39°, and 44° [57]) can be identified in the XRD pattern (appearing, albeit weakly, at 2θ = 17.9°, 21.9°, 23.7°, 34.5°, and 38.0°).

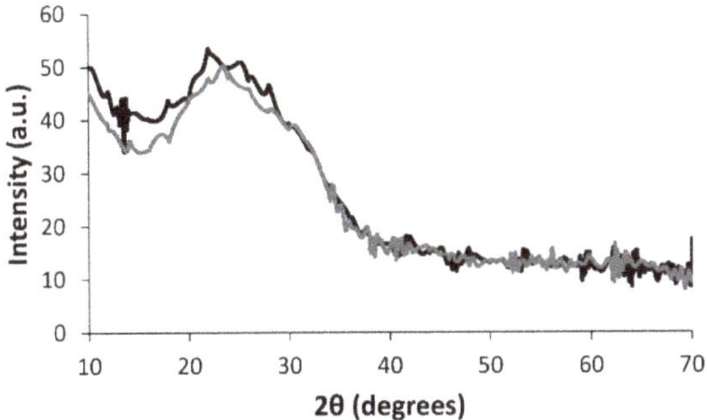

Figure 6. X-ray diffraction (XRD) data. (Black line) DMP-doped PPY film. (Gray line) MER-doped PPY film.

The electrochemical properties of the DMP-doped PPY and MER-doped PPY films were studied via cyclic voltammetry (CV, Figure 7), and electrochemical impedance spectroscopy (EIS, Figure 8). Cyclic voltammograms of DMP-doped PPY and MER-doped PPY films (1st, 5th, 10th, 15th, 20th, 25th, 30th, 35th, and 40th cycles) are displayed in Figure 7A,B, respectively. As evident from the CV curves, the charge-storage capacities of the films decreased steadily on repeated cycling, and the areas under the curves of the 35th and 40th cycles were almost the same. The currents evolved in the MER-doped PPY were higher than those in the DMP-doped PPY, and the oxidation and reduction peaks were more prominent. Occurrences of reduction and oxidation peaks correspond to the de-doping of the films (i.e., release of drug molecules), and the films were subsequently re-doped by other anions (either the anionic drug, or anions from the PBS buffer: $H_2PO_4^-$ and HPO_4^{2-}, and Cl^-).

EIS measurements were conducted to investigate the electron-transfer resistance (R_{et}) of the drug-doped PPY films, and the respective Nyquist plots are displayed in Figure 8. All plots have a semi-circular arc in the high-frequency range, followed by a vertical line along the imaginary axis corresponding to a diffusion process. The diameter of the suppressed semicircle gives the value of the electron-transfer resistance (R_{et}), which was the most directive and sensitive parameter reflecting the changes at the electrode–solution interfaces, and could be evaluated from the difference in the real part of the impedance between low frequency and high frequency [58]. The R_{et} values were 21.18 Ω and 76 Ω for DMP-doped PPY and MER-doped PPY films, respectively. The R_{et} of MER-doped PPY films was higher than that of DMP-doped PPY films, indicating that the electron transfer was more easily achieved at the DMP-doped PPY film interfaces. The diffusion resistance of DMP-doped PPY films was shorter than that of MER-doped PPY films, indicating a shorter ion-diffusion path length of the $[Fe(CN_6)]^{3-/4-}$, $H_2PO_4^-$, HPO_4^{2-}, and Cl^- ions into the interior of the film. Importantly, the EIS results are in good agreement with the CV results.

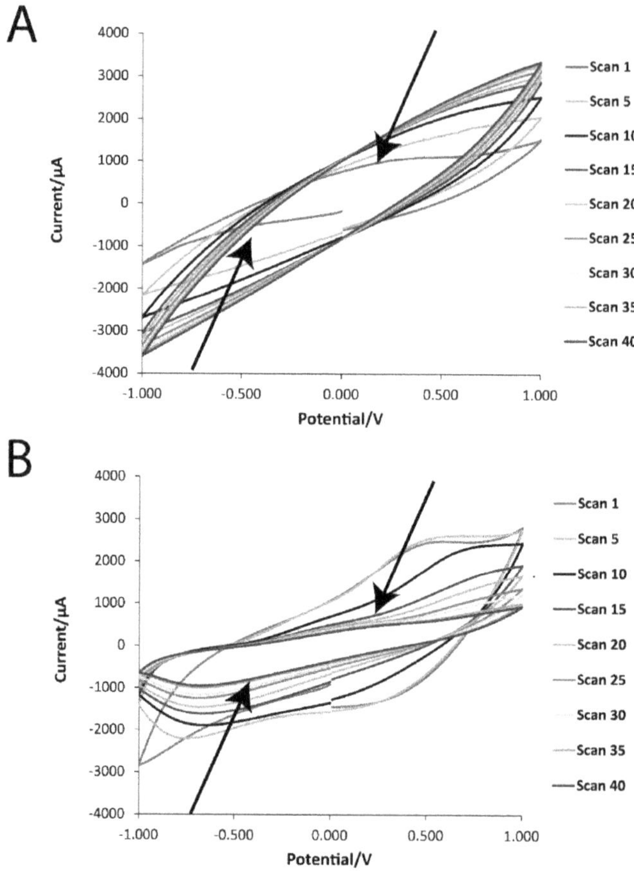

Figure 7. Cyclic voltammetry (CV) data of the films in phosphate-buffered saline (PBS; pH = 7.4) at a scan rate of 50 mV·s^{-1}: (**A**) DMP-doped PPY film; (**B**) MER-doped PPY film.

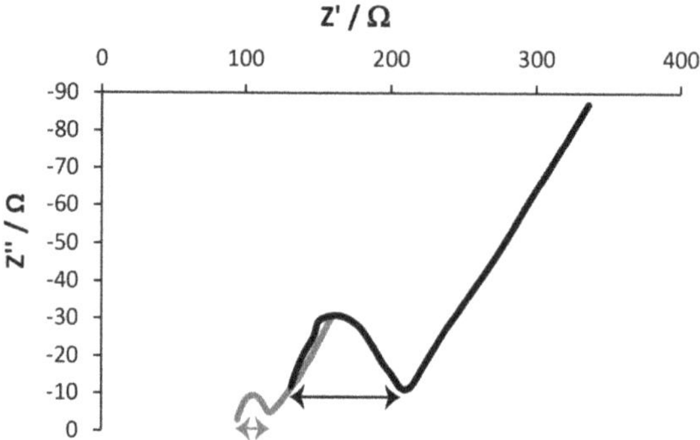

Figure 8. Nyquist plots derived from electrochemical impedance spectroscopy (EIS) data of the films in PBS (pH = 7.4). (Black line) DMP-doped PPY film. (Gray line) MER-doped PPY film.

The release of drugs doped into the PPY films was studied using UV spectroscopy (Figure 9), where the release was either passive (i.e., in the absence of an electrochemical stimulus) or electrochemically triggered (i.e., in the presence of one or more rounds of electrochemical stimulation—30 s of stimulation at a reducing potential of 0.6 V, followed by 10.5 min of rest (Figure 9A)). The quantity of the drug in solution was quantified at various time points, and the data are reported as cumulative release as a percentage of the total mass of the drug in the film (films were individually weighed; DMP-doped PPY films contained 12 wt % of DMP, and MER-doped PPY films contained 4 wt % of MER), and compared to passive drug release from unstimulated films every 11 min.

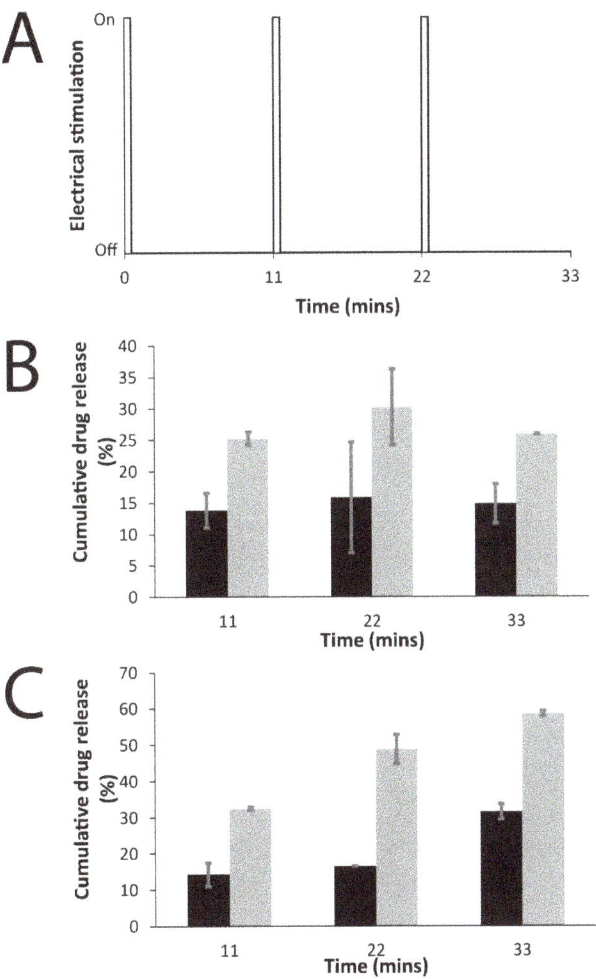

Figure 9. Electrochemically enhanced delivery of drugs from films in PBS (pH = 7.4) as determined by UV spectroscopy: (**A**) Electrical stimulation paradigm: three cycles of 30 s on, 10.5 min off; (**B**) cumulative release of DMP from DMP-doped PPY films, passive release (black bars), electrically stimulated release (gray bars); (**C**) cumulative release of MER from MER-doped PPY films, passive release (black bars), electrically stimulated release (gray bars).

For both DMP-doped and MER-doped films, the drugs were observed to passively diffuse from the films, as is the norm for drug-loaded polymers. The passive diffusion of DMP from the films was ca. 10–15% of the total DMP content of the films over the course of the experiment, whereas the passive diffusion of MER from the films was ca. 10–30% of the total MER content of the films over the course of the experiment. The amount of drug released from the electrically stimulated films was observed to be higher than that for the passive control samples at each time point measured. For the DMP-doped films (Figure 9B), there was an increase of ca. 10–15% in the amount of drug released at various time points; whereas for the MER-doped films (Figure 9C), there was a an increase of ca. 15–30% in the amount of drug released at various time points after each round of electrical stimulation.

Some physical descriptors (constitutional and electronic) for DMP and MER were calculated using the Molecular Operating Environment (MOE) software (version 2014.0901, Chemical Computing Group Inc., Montreal, QC, Canada). The selected descriptors were the dipole moment, LogP (octanol/water), molecular globularity, number of H-atom donors and acceptors, and molecular flexibility (Table 1). The dipole of DMP was lower than that of MER (1.7033 vs. 9.2305, respectively), as was the number of hydrogen-bond acceptors (8 vs. 9, respectively), hydrogen-bond donors (5 vs. 7, respectively), and flexibility (5.3661 vs. 8.8623, respectively). The globularity of DMP was higher than that of MER (0.1110 vs. 0.0265, respectively), as was the LogP (1.2640 vs. −0.5960, respectively).

Table 1. Physical descriptors of dexamethasone phosphate (DMP) and meropenem (MER).

Drug	Dipole	Number of Hydrogen Bond Acceptors	Number of Hydrogen Bond Donors	Globularity	Flexibility	LogP (octanol/water)	Molecular Weight (Da.)
DMP	1.7033	8	5	0.1110	5.3661	1.2640	472.4460
MER	9.2305	9	7	0.0265	8.8623	−0.5960	437.5140

3. Discussion

Films of PPY loaded with an anionic drug (either DMP or MER, see Figure 1) were deposited onto the surface of ITO-coated glass electrodes via electropolymerization, yielding films where the positive charges on the backbone of the polypyrrole were counterbalanced by the presence of the anionic dopants from the electrolyte during the electropolymerization reaction (in this case, one of the anionic drugs, DMP or MER). PPY films prepared via electropolymerization on flat electrodes typically display cauliflower-like morphologies [16,46].

Clearly, surface morphologies and surface-area-to-volume ratios of materials used for drug delivery play an important role in the rate of release of the payloads, and it was observed that mass transport from PPY films with low surface-area-to-volume ratios (e.g., the cauliflower-like morphology) released drugs more slowly than materials with high surface-area-to-volume ratios (e.g., nanowire-like PPY) [16,46,59].

The films were observed to be somewhat imperfect with cracks and inhomogeneities observable (visually or via SEM), and they occasionally delaminated from the underlying ITO electrode. Such problems (i.e., cracks/inhomogeneities and delamination) may be solved using alternative materials for the underlying electrode (e.g., gold, glassy carbon, etc.); through the use of composites, wherein the polymeric dopant forms an interpenetrating network with the conducting polymer binding them together [60–63], or through the development of heat/solution processable electroactive block copolymers, in which one block is electroactive and the other block is heat/solvent responsive [64,65]. Analysis via spectroscopy (EDX, FTIR, and UV-vis), XRD, and electrochemical techniques (CV and EIS) confirmed the presence of the drugs in the films, and subtle differences in the packing of the PPY chains in the films in the presence of each drug. Interestingly, measurements of the total mass of the drug in the films revealed differences in loading, with DMP-doped PPY films containing 12 wt % of DMP, and MER-doped PPY films containing 4 wt % of MER. The physical descriptors for DMP and MER (Table 1) offered an explanation as to why this was observed: the dipole of DMP was lower than that of MER (1.7033 vs. 9.2305, respectively), as was the number of hydrogen-bond acceptors (8 vs. 9, respectively) and hydrogen-bond donors (5 vs. 7, respectively). Furthermore, the LogP of DMP was higher than that of MER (1.2640 vs. -0.5960, respectively). Consequently, the more hydrophobic drug (DMP) was more readily loaded into the hydrophobic PPY matrix than the hydrophilic MER.

For both DMP-doped and MER-doped films, the drugs were observed to passively diffuse from the films, as is the norm for drug-loaded polymers. The passive diffusion of DMP from the films was ca. 10–15% of the total DMP content of the films over the course of the experiment, whereas the passive diffusion of MER from the films was ca. 10–30% of the total MER content of the films over the course of the experiment.

The application of a reducing potential of 0.6 V (for 30 s) to the drug-loaded PPY films resulted in the proactive release of the anionic drug from the films, and rest periods (10.5 min) between the applications of electrical potential offered opportunities for re-doping by other anions (either the anionic drug, or anions from the PBS buffer: $H_2PO_4^-$ and HPO_4^{2-}, and Cl^-). We observed the amount of drug released from the electrically stimulated films (including the rest period) to be higher than that for the passive control samples at each time point measured (every 11 min). For the DMP-doped films (Figure 9B), there was an increase of ca. 10–15% in the amount of drug released at various time points; whereas for the MER-doped films (Figure 9C), there was an increase of ca. 15–30% in the amount of drug released at various time points after electrical stimulation. The differences in the in vitro release behavior observed for DMP and MER, despite their close molecular weights, could be attributed to differences in their physical descriptors (such as the dipole moment, total number of hydrogen-bond donors and acceptors present, molecular flexibility, LogP, and molecular globularity), which confirmed MER to be more hydrophilic and polar than DMP, rendering MER easier to release via diffusion from the PPY matrix, and therefore, more responsive to electrical stimuli (Table 1) [66].

The clinically translated examples of electronic interfaces for the body are all currently metal-based. The mechanical properties of these metals are markedly different from the soft tissues in which they are implanted. This mechanical mismatch leads to inflammatory responses and the formation of scar tissue around the electronic interface. A topic of intense current research interest is the development of soft conductive OEM-based coatings for the surface of the metal electrodes, and it is attractive to be able to deliver bioactive substances from the electrode coating (e.g., anti-inflammatories such as DMP). It is also noteworthy that the surgical procedures necessary to implant such devices are complex, and problems associated with microbial infections in the proximity of these devices can potentially be addressed through the delivery of antimicrobials (e.g., MER) from the surface coatings. The release of the clinically relevant drugs (DMP or MER) loaded into the PPY films was observed to be enhanced by the application of an electrochemical stimulus, thereby demonstrating proof of concept that such materials may form a useful conformal coating on the surface of implantable medical devices, potentially diminishing adverse reactions to their implantation in vivo [16,24,60].

4. Materials and Methods

4.1. Materials

Unless otherwise stated, all chemicals and consumables were used as received without further purification. Pyrrole (Py, 98% reagent grade), meropenem trihydrate (MER) United States Pharmacopeia (USP) Reference Standard, dexamethasone 21-phosphate disodium salt (DMP), phosphate-buffered saline tablets (PBS, pH 7.4), and indium tin oxide (ITO)-coated glass electrodes were supplied by Sigma-Aldrich. Potassium ferrocyanide trihydrate (>99%) and potassium ferricyanide (>99%) were supplied by Acros Organics (Fisher Scientific, Hampton, NH, USA).

4.2. Preparation of Films via Electropolmerization

Indium tin oxide (ITO)-coated glass electrodes with dimensions of 2.5 cm in length and 1 cm in width were cut to size using a diamond pencil. The conductive sides of the glass electrodes were determined with a multimeter (Metrix MX 51, ITT Instruments, Paris, France), and a piece of wire (5 cm, tinned copper, RS Components, Northants, UK) was connected to the conductive side of the glass electrode with copper tape (Diamond Coating Ltd., Halesowen, West Midlands, UK), and wrapped with electrical insulating tape (Advance Tapes, Leicester, UK). Electropolymerizations were performed using a PalmSens EmStat 3+ potentiostat connected to a personal computer and PSTrace 7.4 software (PalmSens, Houten, The Netherlands). A three-electrode system was used with an Ag/AgCl reference electrode (CH Instruments, Inc. Austin, TX, USA), a platinum-mesh counter electrode (Sigma Aldrich, Gillingham, UK; used as a cathode during the electropolymerization), and an ITO-coated glass slide working electrode (used as an anode during the electropolymerization).

The PPY–drug films were deposited onto the ITO anode from solutions containing pyrrole (0.9 M), and either MER or DMP (1 M) in 4 mL of distilled water. Films were deposited onto the ITO working electrode by applying an oxidizing potential of 1.0 V versus the reference electrode for 30 min. After electropolymerization, the films were rinsed with distilled water (ca. 10 mL for ca. 15 s) to remove unreacted monomers and drugs, and were left to dry in air at laboratory temperature (21 °C) for 24 h.

4.3. Scanning Electron Microscopy (SEM) Studies

The surfaces of the films were analyzed with scanning electron microscopy (SEM) using a JEOL JSM-7800F SEM (JEOL UK, Welwyn Garden City, UK).

4.4. Fourier-Transform Infrared (FTIR) Spectroscopy Studies

Spectra were an average of 16 scans, and were obtained at a resolution of 1 cm^{-1} using an Agilent Technologies Cary 630 FTIR instrument (Agilent Technologies Ltd., Cheadle, UK).

4.5. X-ray Diffraction (XRD) Studies

X-ray diffractograms were collected using a Rigaku Smartlab powder diffractometer (Rigaku Ltd., Kent, UK) equipped with a DTex250 one-dimensional (1D) detector, irradiating the films at a wavelength of 0.15418 nm, from Cu Kα radiation. The Cu source was operated at 45 kV and 200 mA, and was fitted with parallel beam optics, with a scan range of $2\theta = 10–90°$.

4.6. Electrochemical Characterization of Films

Cyclic voltammetry (CV) measurements were performed using a PalmSens EmStat 3+ potentiostat connected to a personal computer using the PSTrace 7.4 software, whereas electrochemical impedance spectroscopy (EIS) measurements were performed using an Ivium-n-Stat Multichannel Electrochemical Analyzer. For CV and EIS measurements, a three-electrode system was used with an Ag/AgCl reference electrode (CH Instruments, Inc. Austin, TX, USA), a platinum-mesh counter electrode (Sigma Aldrich, Gillingham, UK), and an ITO-coated glass slide working electrode. The electrodes were in a biomedically relevant buffer (4 mL of phosphate-buffered saline [PBS] at pH 7.4).

For CV measurements, the potential was swept between −1.0 V and +1.0 V vs. the Ag/AgCl electrode at a scan rate of 0.05 Vs^{-1}.

For EIS measurements, the PBS also contained $[Fe(CN_6)]^{3-/4-}$ (5 mmol L^{-1}), and measurements were performed with an open-circuit potential of 230 mV, with an amplitude of applied potential perturbation of 10 mV in the frequency range of 0.1–105,000 Hz. The Nyquist plots were obtained to ascertain the electron-transfer resistance (R_{et}).

4.7. Drug-Delivery Studies

The electrochemically triggered release of MER or DMP from the films was achieved using a PalmSens EmStat 3+ potentiostat connected to a personal computer using the PSTrace 7.4 software (amperometric technique), and a three-electrode system (described above) in a biomedically relevant buffer (4 mL of PBS at pH 7.4). Prior to electrical stimulation, there was a quiet time of 20 s at a potential of 0.1 V, after which the PPY-coated ITO working electrodes were stimulated for 30 s with a reducing potential of 0.6 V. The films were allowed to rest for 10.5 min after each stimulation, during the last 30 s of which 10 µL of the solution was taken for quantification of drug release with UV spectroscopy (at either 242 nm for DMP, or 297 nm for MER) using a Nanodrop 2000c spectrophotometer (Thermo Fisher Scientific, Loughborough, UK). The PBS was not changed between rounds of stimulation, and the data are reported as cumulative release as a percentage of the total mass of the drug in the film (films were individually weighed; DMP-doped PPY films contained 12 wt % of DMP, and MER-doped PPY films contained 4 wt % of MER). These data were compared to the passive drug release from non-stimulated films every 11 min. To determine the total amount of drug in the films, the drug-doped films were stimulated at a reducing potential of 0.6 V for 60 min, and the medium was changed every 10 min. All reported data were normalized relative to the amount of released drug from a 1-mg-drug-doped PPY film with a surface area of 100 mm^2.

4.8. Calculating the Main Physical Descriptors of the Investigated Drugs

Physical descriptors (constitutional and electronic) were calculated. The selected descriptors were the dipole moment, LogP (octanol/water), molecular globularity, number of H-atom donors and acceptors, and the molecular flexibility. The descriptors were calculated using the MOE software version 2014.0901 (Chemical Computing Group Inc., Montreal, QC, Canada), and the builder tool in the same software was used to generate the three-dimensional (3D) structures of the investigated drugs from their isomeric simplified molecular-input line-entry system (SMILES) obtained from The PubChem Project® (National Institutes of Health, Bethesda, MD, USA).

5. Conclusions

Electroactive drug-loaded polymeric films represent an effective means of controlling the delivery of various types of drugs, and the development of new on–off therapies from surface coatings with drug-loaded conducting polymers applied to implantable devices. Such coatings have prospects for positive economic and health impacts in the short–medium term (e.g., as coatings for neural electrodes) [35,36], and societal impacts in the long term (e.g., enhanced quality of life). Likewise, the development of biodegradable alternatives to PPY has prospects for health impacts in the long term (e.g., as drug-delivery devices and scaffolds for tissue engineering) [37].

Author Contributions: Conceptualization, J.G.H. and M.F. Methodology, M.F. and J.G.H. Formal analysis, M.F. and J.G.H. Investigation, S.A.A.S., M.F., S.R.B., N.R.H., S.J.B., and R.M.H. Data curation, M.F., N.R.H., S.J.B., and J.G.H. Writing—original draft preparation, M.F. Writing—review and editing, all authors. Supervision, J.G.H. and B.M.Y. Project administration, J.G.H. Funding acquisition, J.G.H.

Funding: This research was funded by a variety of sources: a Higher Education Commission of Pakistan for an International Research Support Initiative Program (IRSIP) scholarship to support S.A.A.S. and B.M.Y., a Royal Society Newton International Fellowship to support M.F., a Lancaster University Faculty of Science and Technology Summer Internship to support S.R.B., a Lancaster University Faculty of Science and Technology Early Career Internal Grant to support R.M.H and J.G.H., and a Royal Society Research Grant (RG160449) and an EPSRC First Grant (EP/R003823/1) to support J.G.H.

Acknowledgments: We thank Naomi Stanhope for assistance with sample preparation.

Conflicts of Interest: The authors declare no conflict of interest. The sponsors had no role in the design of the study; in the collection, analyses, or interpretation of data; in the writing of the manuscript, and in the decision to publish the results.

References

1. Barbe, C.; Bartlett, J.; Kong, L.; Finnie, K.; Calleja, G.; Lin, H.Q.; Larkin, M.; Calleja, S.; Bush, A. Silica Particles: A novel drug-delivery System. *Adv. Mater.* **2004**, *16*, 1959–1966. [CrossRef]
2. Sahoo, S.K.; Labhsetwar, V. Nanotech approaches to drug delivery and imaging. *Drug Discov. Today* **2003**, *8*, 1112–1120. [CrossRef]

3.   Morgan, N.B. Medical shape memory alloy applications—The market and its products. *Mater. Sci. Eng. A* **2004**, *378*, 16–23. [CrossRef]

4.   Wei, X.; Liu, J. Power sources and electrical recharging strategies for implantable medical devices. *Front. Energy Power Eng.* **2008**, *2*, 1–13. [CrossRef]

5.   Khan, W.; Muntimadugu, E.; Jaffe, M.; Domb, A.J. Implantable Medical Devices. In *Focal Controlled Drug Delivery*; Domb, A., Khan, W., Eds.; Springer: Boston, MA, USA, 2014.

6.   Kurtz, S.M.; Devine, J.N. PEEK biomaterials in trauma, orthopedic, and spinal implants. *Biomaterials* **2007**, *28*, 4845–4869. [CrossRef] [PubMed]

7.   McAllister, B.S.; Haghighat, K. Bone augmentation techniques. *J. Periodontol.* **2007**, *78*, 377–396. [CrossRef] [PubMed]

8.   Qiu, Y.; Park, K. Environment-sensitive hydrogels for drug delivery. *Adv. Drug Deliv. Rev.* **2001**, *53*, 321–339. [CrossRef]

9.   Hoffman, A.S. Intelligent Polymers. In *Controlled Drug Delivery: Challenge and Strategies*; Park, K., Ed.; American Chemical Society: Washington, DC, USA, 1997; pp. 485–497.

10.   Bae, Y.H. Stimuli-Sensitive Drug Delivery. In *Controlled Drug Delivery: Challenge and Strategies*; Park, K., Ed.; American Chemical Society: Washington, DC, USA, 1997; pp. 147–160.

11.   Suzuki, M. Amphoteric polyvinyl alcohol hydrogel and electrohydrodynamic control method for artificial muscles. In *Polymer Gels*; DeRossi, D., Ed.; Plenum Press: New York, NY, USA, 1991; pp. 221–236.

12.   Kishi, R.; Ichijo, O.; Hirasa, O. Thermo-responsive devices using poly(vinylmethyl ether) hydrogels. *J. Intell. Mater. Syst. Struct.* **1993**, *4*, 533–537. [CrossRef]

13.   Kajiwara, K.; Ross-Murphy, S.B. Synthetic gels on the move. *Nature* **1992**, *355*, 208–209. [CrossRef]

14.   Osada, Y.; Okuzaki, H.; Hori, H. A polymer gel with electrically driven motility. *Nature* **1992**, *355*, 242–244. [CrossRef]

15.   Ueoka, Y.; Gong, J.; Osada, Y. Chemomechanical polymer gel with fish-like motion. *J. Intell. Mater. Syst. Struct.* **1997**, *8*, 465–471. [CrossRef]

16.   Zhao, Y.; Tavares, A.C.; Gauthier, M.A. Nano-engineered electro-responsive drug delivery systems. *J. Mater. Chem. B* **2016**, *4*, 3019–3030. [CrossRef]

17.   Clancy, K.F.A.; Hardy, J.G. Gene Delivery with Organic Electronic Biomaterials. *Curr. Pharm. Des.* **2017**, *23*, 3614–3625. [CrossRef] [PubMed]

18.   Qian, C.G.; Chen, Y.L.; Feng, P.J.; Xiao, X.Z.; Dong, M.; Yu, J.C.; Hu, Q.Y.; Shen, Q.D.; Gu, Z. Conjugated polymer nanomaterials for theranostics. *Acta Pharmacol. Sin.* **2017**, *38*, 764–781. [CrossRef] [PubMed]

19.   Repenko, T.; Rix, A.; Ludwanowski, S.; Go, D.; Kiessling, F.; Lederle, W.; Kuehne, A.J.C. Bio-degradable highly fluorescent conjugated polymer nanoparticles for bio-medical imaging applications. *Nat. Commun.* **2017**, *8*, 470. [CrossRef] [PubMed]

20.   Bendrea, A.D.; Cianga, L.; Cianga, I. Review paper: progress in the field of conducting polymers for tissue engineering applications. *J. Biomater. Appl.* **2011**, *26*, 3–84. [CrossRef] [PubMed]

21.   Guimard, N.K.; Gomez, N.; Schmidt, C.E. Conducting polymers in biomedical engineering. *Prog. Polym. Sci.* **2007**, *32*, 876–921. [CrossRef]

22.   Guiseppi-Elie, A. Electroconductive hydrogels: Synthesis, characterization and biomedical applications. *Biomaterials* **2010**, *31*, 2701–2716. [CrossRef] [PubMed]

23.   Hardy, J.G.; Lee, J.Y.; Schmidt, C.E. Biomimetic conducting polymer-based tissue scaffolds. *Curr. Opin. Biotechnol.* **2013**, *24*, 847–854. [CrossRef] [PubMed]

24.   Staples, N.A.; Goding, J.A.; Gilmour, A.D.; Aristovich, K.Y.; Byrnes-Preston, P.; Holder, D.S.; Morley, J.W.; Lovell, N.H.; Chew, D.J.; Green, R.A. Conductive Hydrogel Electrodes for Delivery of Long-Term High Frequency Pulses. *Front. Neurosci.* **2018**, *11*, 748. [CrossRef] [PubMed]

25.   Svirskis, D.; Travas-Sejdic, J.; Rodgers, A.; Garg, S. Electrochemically controlled drug delivery based on intrinsically conducting polymers. *J. Control. Release* **2010**, *146*, 6–15. [CrossRef] [PubMed]

26.   Pillay, V.; Tsai, T.-S.; Choonara, Y.E.; du Toit, L.C.; Kumar, P.; Modi, G.; Naidoo, D.; Tomar, L.K.; Tyagi, C.; Ndesendo, V.M.K. A review of integrating electroactive polymers as responsive systems for specialized drug delivery applications. *J. Biomed. Mater. Res. Part A* **2013**, *102*, 2039–2054. [CrossRef] [PubMed]

27.   Ma, H.; Liu, M.S.; Jen, A.K.-Y. Interface-tailored and nanoengineered polymeric materials for (opto)electronic devices. *Polym. Int.* **2009**, *58*, 594–619. [CrossRef]

28. Stenger-Smith, J.D. Intrinsically electrically conducting polymers, synthesis, characterization, and their applications. *Prog. Polym. Sci.* **1998**, *23*, 57–79. [CrossRef]

29. Gurunathan, K.; Murugan, A.V.; Marimuthu, R.; Mulik, U.P.; Amalnerkar, D.P. Electrochemically synthesised conducting polymeric materials for applications towards technology in electronics, optoelectronics and energy storage devices. *Mater. Chem. Phys.* **1999**, *61*, 173–191. [CrossRef]

30. Long, Y.Z.; Li, M.M.; Gu, C.Z.; Wan, M.X.; Duvail, J.L.; Liu, Z.W.; Fan, Z.Y. Recent advances in synthesis, physical properties and applications of conducting polymer nanotubes and nanofibers. *Prog. Polym. Sci.* **2011**, *36*, 1415–1442. [CrossRef]

31. Wallace, G.G.; Higgins, M.J.; Moulton, S.E.; Wang, C. Nanobionics: The impact of nanotechnology on implantable medical bionic devices. *Nanoscale* **2012**, *4*, 4327–4347. [CrossRef] [PubMed]

32. Wadhwa, R.; Lagenaur, C.F.; Cui, X.T. Electrochemically controlled release of dexamethasone from conducting polymer polypyrrole coated electrode. *J. Control. Release* **2006**, *110*, 531–541. [CrossRef] [PubMed]

33. Szunerits, S.; Teodorescu, F.; Boukherroub, R. Electrochemically triggered release of drugs. *Eur. Polym. J.* **2016**, *83*, 467–477. [CrossRef]

34. Uppalapati, D.; Sharma, M.; Aqrawe, Z.; Coutinho, F.; Rupenthal, I.D.; Boyd, B.J.; Travas-Sejdic, J.; Svirskis, D. Micelle directed chemical polymerization of polypyrrole particles for the electrically triggered release of dexamethasone base and dexamethasone phosphate. *Int. J. Pharm.* **2018**, *543*, 38–45. [CrossRef] [PubMed]

35. Ramtin, A.; Seyfoddin, A.; Coutinho, F.P.; Waterhouse, G.I.; Rupenthal, I.D.; Svirskis, D. Cytotoxicity considerations and electrically tunable release of dexamethasone from polypyrrole for the treatment of back-of-the-eye conditions. *Drug. Deliv. Transl. Res.* **2016**, *6*, 793–799. [CrossRef] [PubMed]

36. Seyfoddin, A.; Chan, A.; Chen, W.T.; Rupenthal, I.D.; Waterhouse, G.I.; Svirskis, D. Electro-responsive macroporous polypyrrole scaffolds for triggered dexamethasone delivery. *Eur. J. Pharm. Biopharm.* **2015**, *94*, 419–426. [CrossRef] [PubMed]

37. Schmidt, C.E.; Shastri, V.R.; Vacanti, J.P.; Langer, R. Stimulation of neurite outgrowth using an electrically conducting polymer. *Proc. Natl. Acad. Sci. USA* **1997**, *94*, 8948–8953. [CrossRef] [PubMed]

38. Wang, Z.; Roberge, C.; Dao, L.H.; Wan, Y.; Shi, G.; Rouabhia, M.; Guidoin, R.; Zhang, Z. In vivo evaluation of a novel electrically conductive polypyrrole/poly(D,L-lactide) composite and polypyrrole-coated poly(D,L-lactide-*co*-glycolide) membranes. *J. Biomed. Mater. Res. A* **2004**, *70*, 28–38. [CrossRef] [PubMed]

39. Mihardja, S.S.; Sievers, R.E.; Lee, R.L. The effect of polypyrrole on arteriogenesis in an acute rat infarct model. *Biomaterials* **2008**, *29*, 4205–4210. [CrossRef] [PubMed]

40. Durgam, H.; Sapp, S.; Deister, C.; Khaing, Z.; Chang, E.; Luebben, S.; Schmidt, C.E. Novel degradable co-polymers of polypyrrole support cell proliferation and enhance neurite out-growth with electrical stimulation. *J. Biomater. Sci. Polym. Ed.* **2010**, *21*, 1265–1282. [CrossRef] [PubMed]

41. George, P.M.; Lyckman, A.W.; LaVan, D.A.; Hegde, A.; Leung, Y.; Avasare, R.; Testa, C.; Alexander, P.M.; Langer, R.; Sur, M. Fabrication and biocompatibility of polypyrrole implants suitable for neural prosthetics. *Biomaterials* **2005**, *26*, 3511–3519. [CrossRef] [PubMed]

42. Ludwig, K.A.; Uram, J.D.; Yang, J.; Martin, D.C.; Kipke, D.R. Chronic neural recordings using silicon microelectrode arrays electrochemically deposited with a poly(3,4-ethylenedioxythiophene) (PEDOT) film. *J. Neural Eng.* **2006**, *3*, 59–70. [CrossRef] [PubMed]

43. Luo, S.C.; Mohamed Ali, E.; Tansil, N.C.; Yu, H.H.; Gao, S.; Kantchev, E.A.; Ying, J.Y. Poly(3,4-ethylenedioxythiophene) (PEDOT) nanobiointerfaces: Thin, ultrasmooth, and functionalized PEDOT films with in vitro and in vivo biocompatibility. *Langmuir* **2008**, *24*, 8071–8077. [CrossRef] [PubMed]

44. Mattioli-Belmonte, M.; Giavaresi, G.; Biagini, G.; Virgili, L.; Giacomini, M.; Fini, M.; Giantomassi, F.; Natali, D.; Torricelli, P.; Giardino, R. Tailoring biomaterial compatibility: In vivo tissue response versus in vitro cell behavior. *Int. J. Artif. Organs.* **2003**, *26*, 1077–1085. [CrossRef] [PubMed]

45. Wang, C.H.; Dong, Y.Q.; Sengothi, K.; Tan, K.L.; Kang, E.T. In-vivo response to polyaniline. *Synth. Met.* **1999**, *102*, 1313–1314. [CrossRef]

46. Ru, X.; Shi, W.; Huang, X.; Cui, X.; Ren, B.; Ge, D. Synthesis of polypyrrole nanowire network with high adenosine triphosphate release efficiency. *Electrochim. Acta* **2011**, *56*, 9887–9892. [CrossRef]

47. Available online: https://www.edax.com/resources/interactive-periodic-table (accessed on 16 June 2018).

48. Sirivisoot, S.; Pareta, R.; Webster, T.J. Electrically controlled drug release from nanostructured polypyrrole coated on titanium. *Nanotechnology* **2011**, *22*, 085101. [CrossRef] [PubMed]

49. Chougule, M.A.; Pawar, S.G.; Godse, P.R.; Mulik, R.N.; Sen, S.; Patil, V.B. Synthesis and Characterization of Polypyrrole (PPy) Thin Films. *Soft Nanosci. Lett.* **2011**, *1*, 3660. [CrossRef]

50. Hardy, J.G.; Cornelison, R.C.; Sukhavasi, R.C.; Saballos, R.J.; Vu, P.; Kaplan, D.L.; Schmidt, C.E. Electroactive Tissue Scaffolds with Aligned Pores as Instructive Platforms for Biomimetic Tissue Engineering. *Bioengineering* **2015**, *2*, 15–34. [CrossRef] [PubMed]

51. Hardy, J.G.; Hernandez, D.S.; Cummings, D.M.; Edwards, F.A.; Shear, J.B.; Schmidt, C.E. Multiphoton microfabrication of conducting polymer-based biomaterials. *J. Mater. Chem. B* **2015**, *3*, 5001–5004. [CrossRef]

52. Hardy, J.G.; Villancio-Wolter, M.K.; Sukhavasi, R.C.; Mouser, D.J.; Aguilar, D., Jr.; Geissler, S.A.; Kaplan, D.L.; Schmidt, C.E. Electrical stimulation of human mesenchymal stem cells on conductive nanofibers enhances their differentiation towards osteogenic outcomes. *Macromol. Rapid Commun.* **2015**, *36*, 1884–1890. [CrossRef] [PubMed]

53. Hardy, J.G.; Khaing, Z.Z.; Xin, S.; Tien, L.W.; Ghezzi, C.E.; Mouser, D.J.; Sukhavasi, R.C.; Preda, R.C.; Gil, E.S.; Kaplan, D.L.; et al. Into The Groove: Instructive Silk-Polypyrrole Films With Topological Guidance Cues Direct DRG Neurite Outgrowth. *J. Biomater. Sci. Polym. Ed.* **2015**, *26*, 1327–1342. [CrossRef] [PubMed]

54. Cai, L.; Jiang, H.; Wang, L. Enhanced photo-stability and photocatalytic activity of Ag_3PO_4 via modification with $BiPO_4$ and polypyrrole. *Appl. Surf. Sci.* **2017**, *420*, 43–52. [CrossRef]

55. Allen, N.S.; Murray, K.S.; Fleming, R.J.; Saunders, B.R. Physical properties of polypyrrole films containing trisoxalatometallate anions and prepared from aqueous solution. *Synth. Met.* **1997**, *87*, 237–247. [CrossRef]

56. Roozbahani, M.; Kharaziha, M.; Emadi, R. pH sensitive dexamethasone encapsulated laponite nanoplatelets: Release mechanism and cytotoxicity. *Int. J. Pharm.* **2017**, *518*, 312–319. [CrossRef] [PubMed]

57. Shaker, M.A.; Shaaban, M.I. Formulation of carbapenems loaded gold nanoparticles to combat multi-antibiotic bacterial resistance: In vitro antibacterial study. *Int. J. Pharm.* **2017**, *525*, 71–84. [CrossRef] [PubMed]

58. Zhao, D.M.; Zhang, X.H.; Feng, L.J.; Jia, L.; Wang, S.F. Simultaneous determination of hydroquinone and catechol at PASA/MWNTs composite film modified glassy carbon electrode. *Colloids Surf. B Biointerfaces* **2009**, *74*, 317–321. [CrossRef] [PubMed]

59. Jiang, S.; Sun, Y.; Cui, X.; Huang, X.; He, Y.; Ji, S.; Shi, W.; Ge, D. Enhanced drug loading capacity of polypyrrole nanowire network for controlled drug release. *Synth. Met.* **2013**, *163*, 19–23. [CrossRef]

60. Moulton, S.E.; Higgins, M.J.; Kapsa, R.M.I.; Wallace, G.G. Organic Bionics: A New Dimension in Neural Communications. *Adv. Funct. Mater.* **2012**, *22*, 2003–2014. [CrossRef]

61. Nguyen, D.N.; Yoon, H. Recent Advances in Nanostructured Conducting Polymers: From Synthesis to Practical Applications. *Polymers* **2016**, *8*, 118. [CrossRef]

62. Guarino, V.; Zuppolini, S.; Borriello, A.; Ambrosio, L. Electro-Active Polymers (EAPs): A Promising Route to Design Bio-Organic/Bioinspired Platforms with on Demand Functionalities. *Polymers* **2016**, *8*, 185. [CrossRef]

63. Kim, S.; Jeong, J.-O.; Lee, S.; Park, J.-S.; Gwon, H.-J.; Jeong, S.I.; Hardy, J.G.; Lim, Y.-M.; Lee, J.Y. Effective gamma-ray sterilization and characterization of conductive polypyrrole biomaterials. *Sci. Rep.* **2018**, *8*, 3721. [CrossRef] [PubMed]

64. Hardy, J.G.; Mouser, D.J.; Arroyo-Currás, N.; Geissler, S.; Chow, J.K.; Nguy, L.; Kim, J.M.; Schmidt, C.E. Biodegradable electroactive polymers for electrochemically-triggered drug delivery. *J. Mater. Chem. B.* **2014**, *2*, 6809–6822. [CrossRef]

65. Hardy, J.G.; Amend, M.N.; Geissler, S.; Lynch, V.M.; Schmidt, C.E. Peptide-directed assembly of functional supramolecular polymers for biomedical applications: electroactive molecular tongue-twisters (oligoalanine-oligoaniline-oligoalanine) for electrochemically enhanced drug delivery. *J. Mater. Chem. B.* **2015**, *3*, 5005–5009. [CrossRef]

66. Hathout, R.M.; El-Ahmady, S.H.; Metwally, A.A. Curcumin or bisdemethoxycurcumin for nose-to-brain treatment of Alzheimer disease? A bio/chemo-informatics case study. *Nat. Prod. Res.* **2017**, *12*, 1–10. [CrossRef] [PubMed]

 *materials*

Article

An Electrochemical Study on the Copolymer Formed from Piperazine and Aniline Monomers

Samiha Dkhili [1,2], **Sara López-Bernabeu** [2], **Chahineze Nawel Kedir** [2], **Francisco Huerta** [3], **Francisco Montilla** [2], **Salma Besbes-Hentati** [1] and **Emilia Morallon** [2,*]

[1] Laboratoire de Chimie des Matériaux, Faculté des Sciences de Bizerte, Zarzouna Université de Carthage, Jarzouna, Bizerte 7021, Tunisia; samihadkhili@yahoo.fr (S.D.); salma.hentati@fsb.rnu.tn (S.B.-H.)
[2] Departamento de Química Física e Instituto Universitario de Materiales, Universidad de Alicante, Ap. 99, E-03080 Alicante, Spain; sara.lopez@ua.es (S.L.-B.); kedir.nawel@hotmail.fr (C.N.K.); francisco.montilla@ua.es (F.M.)
[3] Departamento de Ingenieria Textil y Papelera, Universitat Politecnica de Valencia, Plaza Ferrandiz y Carbonell, 1, E-03801 Alcoy, Spain; frahuear@txp.upv.es
* Correspondence: morallon@ua.es; Tel.: +34-965-909590

Received: 23 May 2018; Accepted: 12 June 2018; Published: 14 June 2018

Abstract: A study on the electrochemical oxidation of piperazine and its electrochemical copolymerization with aniline in acidic medium is presented. It was found that the homopolymerization of piperazine cannot be achieved under electrochemical conditions. A combination of electrochemistry, in situ Fourier transform infrared (FTIR), and ex situ X-ray photoelectron spectroscopy (XPS) spectroscopies was used to characterize both the chemical structure and the redox behavior of an electrochemically synthesized piperazine–aniline copolymer. The electrochemical sensing properties of the deposited material were also tested against ascorbic acid and dopamine as redox probes.

Keywords: copolymer; polyaniline; piperazine; FTIR in situ

1. Introduction

Electrochemical devices based on conducting polymers, either working as sensors or as systems taking advantage of other electrocatalytic effects, have become a topic of growing interest for molecular electrochemistry during the last decades [1–6]. It is known that conducting polymers show the ability to incorporate catalytic molecules, and numerous works based on this particular property led to interesting applications in the field of bioelectrochemical sensing. The polymer constitutes an organic matrix where catalytic molecules, such as enzymes, may preserve its activity better, and where the conducting surroundings may electrically wire it to the metal electrode surface [7–10].

Besides the incorporation of catalytic species, the pristine conducting polymers (polyaniline, polypyrrole, etc.) can be also chemically modified to gain further catalytic capabilities. The most classical way to perform chemical modification is to copolymerize aniline or pyrrole, for example, with monomers that are able to provide the final material with the desired catalytic features. In this context, chemical derivatives of piperazine (diethylenediamine) constitute a promising group of catalytic molecules that have been successfully applied in chemical and electrochemical sensing [6,11–14]. In spite of this, few studies exist that are devoted to the exploration of the catalytic properties of polymer systems containing the parent piperazine molecule. Among them, it has been reported that a novel piperazine-functionalized mesoporous organic polymer exhibited highly catalytic activity and selectivity for some organic synthesis reactions in aqueous medium [15]. The electrochemical sensing ability of piperazine in combination with inorganic polymers has been explored recently in the selective detection of ascorbic acid [16]. The sensing system was a piperazine-functionalized mesoporous silica, and the results show that this type of hybrid material is a potential candidate for the construction of bioelectrochemical sensors.

The chemical copolymerization of aniline and piperazine was studied by Ramachandran et al. [17]. Although the catalytic properties of the obtained material were not analyzed, the copolymer showed electrochemical activity. The chemical structure proposed (see Scheme 1) seems constituted by alternated piperazine and aniline moieties, which are bound through aniline *ortho-* and *para-*positions. It was shown that charge delocalization in this polymer includes also oxidized piperazine centers, but extended conjugation was not observed. The electrical conductivity of the material was in the range of 10^{-7}–10^{-9} S cm^{-1}.

The goal of the present contribution is to study the electrochemical oxidation of piperazine in acidic medium and, additionally, its electrochemical copolymerization with aniline. A combination of in situ Fourier transform infrared (FTIR) spectroscopy and electrochemistry will be used to characterize the redox behavior of the copolymer, while X-ray photoelectron spectroscopy (XPS) will shed more light on the chemical structure of the electrochemically synthesized material in comparison with the chemically obtained one. Finally, the electrochemical sensing properties of the copolymer will be tested against ascorbic acid and dopamine.

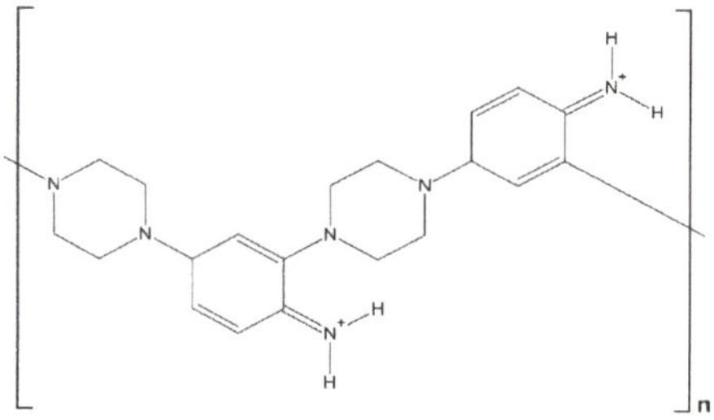

Scheme 1. Model structure proposed by Kabilan et al. [17] for the chemically synthesized piperazine–aniline copolymer in the doped state.

2. Experimental

The background electrolyte employed for the studies was perchloric acid (Merck Suprapur, Merck Group, Darmstadt, Germany), and the solutions were prepared with 18.2 MΩ cm water obtained from an Elga Labwater Purelab (Elga-Veolia, High Wycombe, UK) system. Piperazine (98.8%), aniline (99.5%), ascorbic acid (AA, 98.9%), and dopamine (DA, 97.9%) were purchased from Merck. Cyclic voltammetry experiments were carried out in a conventional three-electrode cell under N$_2$ atmosphere. The working electrode was a polycrystalline platinum sphere, and a platinum wire was used as the counter electrode. All of the potentials were measured against the reversible hydrogen electrode (RHE) immersed in the same electrolyte through a Luggin capillary. Cyclic voltammograms were recorded at a constant sweep rate of 0.05 V s^{-1} and at room temperature. The platinum electrodes were thermally cleaned and subsequently protected from the laboratory atmosphere by a droplet of ultrapure water.

A Nicolet 5700 spectrometer (Thermo Electron Scientific Instruments, Madison, WI, USA) equipped with an N$_2$-cooled mercury cadmium telluride detector was employed for the in situ FTIR experiments. The working Pt disc electrode was mirror-polished with alumina powder, and the spectroelectrochemical cell used a prismatic CaF$_2$ window beveled at 60° in order to increase the beam intensity reaching the infrared (IR) detector. All of the spectra were collected at the same 8 cm^{-1} resolution using deuterated water (99.9% D) as the solvent. The processed spectra have been presented in the standard mode $\Delta R/R$.

A VG-Microtech Multilab 3000 electron spectrometer (VG Microtech Ltd., Uckfield, UK) was employed to acquire the ex situ XPS spectra. The 300-W power radiation source was a non-monochromatized Mg-Kα, and the analysis was performed under 5×10^{-7} Pa pressure. The high-resolution spectra were acquired at 50-eV pass energy, and are presented as a combination of Lorentz (30%) and Gaussian (70%) curves. The C 1s line at 284.4 eV has been employed as the reference for the experimental binding energies, which were obtained with 0.2 eV accuracy.

The scanning electron micrographs were acquired by means of an ORIUS-SC600 Field Emission Scanning Electron Microscope (FE-SEM) (Gatan Inc., Pleasanton, CA, USA), which was equipped with a ZEISS microscope (Carl Zeiss Microscopy Ltd., Cambridge, UK).

3. Results and Discussion

3.1. Electrochemical Behavior of Piperazine on Pt

Cyclic voltammetry (CV) curves recorded for a polycrystalline platinum electrode immersed in 1 M of $HClO_4$ + 10 mM piperazine solution are illustrated in Figure 1. The electrode was immersed at a controlled potential of 0.1 V, and the response was firstly examined in the 0.05–0.5 V potential range (Figure 1a). Two oxidation peaks were observed at 0.18 and 0.30 V during the forward scan up to 0.5 V. Both features are related with the electrochemistry of partially blocked adsorption sites at the platinum surface in perchloric medium [18], which reveals that piperazine strongly adsorbs on this electrode. In a new experiment, the clean electrode was immersed at 0.1 V, and the potential was scanned up to 1.4 V to examine the anodic behavior of piperazine. The onset of oxidation occurs at about 0.5 V, but the electrochemical process appears more clear at potentials higher than, roughly, 1.0 V in the form of a broad current, with no well-defined peaks. The voltammetric profile recorded during the reverse scan does not reach the characteristic shape of a Pt electrode (Figure 1b, dashed line). This result indicates that some adsorbed species coming from piperazine oxidation still remain on the electrode surface and block part of Pt adsorption sites. However, as expected, no electropolymerization of piperazine is observed after continuous potential cycling, with the voltammetric profile being almost equivalent to that shown in the solid line of Figure 1b.

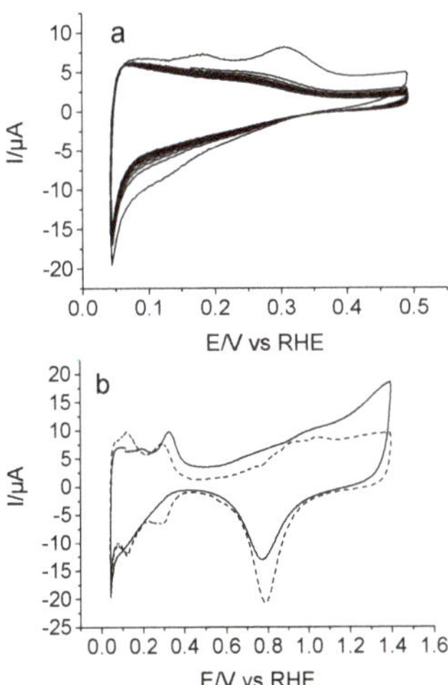

Figure 1. Cyclic voltammograms recorded for a Pt electrode in 1 M of $HClO_4$ solution containing 10 mM of piperazine. (**a**) Electrochemical behavior within the potential region 0.05–0.5 V (10 cycles); (**b**) Electrochemical response obtained during the first excursion of up to 1.4 V (solid line), and for clean Pt, in 1 M of $HClO_4$ free of piperazine (dashed line) in the same potential window. v = 50 mV s^{-1} in all cases.

In situ FTIR spectroscopy has been used to increase the understanding of the piperazine oxidation process. Figure 2 shows a set of spectra obtained for a Pt electrode immersed in 10 mM of piperazine

+ 0.1 M HClO$_4$, using D$_2$O as the solvent. The mirror-polished platinum electrode was transferred to the spectroelectrochemical cell, which was immersed at 0.1 V into the solution, and its surface was pressed against the CaF$_2$ window. In this case, the concentration of perchloric acid was 0.1 M in order to avoid the damage of the spectroscopic window. The reference spectrum was collected at 0.1 V, and then, the potential was stepped up to 1.4 V to collect several sample spectra. By referring each sample to the unique reference, the information on the redox transformations undergone by the piperazine as a function of the applied potential can be obtained. A positive-going absorption feature appears at 1502 cm^{-1} in the spectrum obtained at 0.4 V, whose intensity rises significantly at higher applied potentials. This means that the species giving rise to this vibrational mode disappears upon oxidation. The frequency of 1502 cm^{-1} is compatible with the –ND$_2^+$ stretching vibration of deuterated piperazine [19], which occurs because of the proton–deuterium exchange equilibrium in D$_2$O solvent. The electrochemical oxidation of piperazine at higher potential values results in the formation of different carbonyl groups within the piperazine ring, as deduced from the C=O stretching vibrations appearing at around 1570 and 1630 cm^{-1} [20,21].

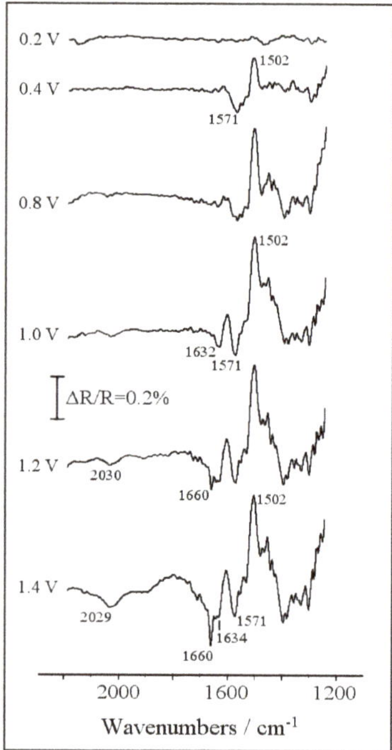

Figure 2. Set of in situ Fourier transform infrared (FTIR) spectra collected during the oxidation of 10 mM of piperazine in 0.1 M of HClO$_4$/D$_2$O solution. Reference potential: 0.1 V. Sample potential labeled for each spectrum: 100 interferograms at each potential.

Finally, the spectra collected at large anodic potentials display two additional negative bands at 1660 and 2030 cm^{-1}. The former seems related with the occurrence of C=O in amide species, while the frequency of the latter strongly suggests the formation of multiple C-N bonds, probably as isocyanates [22]. It is known that piperazine N-oxides obtained from the oxidation of piperazine show N-O stretching frequencies at around 1350 cm^{-1} [23]. The presence of this kind of structure cannot be ruled out during the electrochemical oxidation, because the frequency region between 1250 and 1450 cm^{-1} is altered in the spectra of Figure 2 due to the presence of diverse C-N, CH$_2$,

CND, and N-D absorptions. Anyway, from the results presented in this section, it is derived that the electrochemical oxidation of piperazine on Pt electrodes yields some kind of ketopiperazine species at moderate potentials and, at more positive potential values, the ring could open to produce both amide groups and isocyanates.

3.2. Electrochemical Copolymerization of Piperazine and Aniline

The results in the previous section strongly suggest that the copolymerization of piperazine with aniline should be carried out using as low a potential as possible, in order to minimize the irreversible oxidation of the former. However, it is known that anilinium cations (which originate at potentials beyond 1.2 V versus RHE) are needed to trigger the deposition of polyaniline-derived polymers. A compromise is then needed between the most favorable polymerization conditions to obtain a little degraded material, and the actual conditions to obtain a deposit. Figure 3 shows the experiment carried out to achieve electropolymerization under the established premises. Owing to the higher reactivity of aniline monomer, the copolymerization solution contained a piperazine:aniline relative concentration as large as 5:1 in 1 M of $HClO_4$. The first potential scan was carried out up to 1.3 V to generate an adequate amount of anilinium radicals, while the inversion potential was set at 0.9 V for the subsequent scans to ensure that the piperazine unbroken rings can be incorporated to the growing polymer. The development of new redox processes within the 0.05–0.9 V potential region evidences the growth of an electroactive polymeric species. After 10 potential cycles, the Pt electrode was removed from the solution, and its surface appeared covered by a dark blue film.

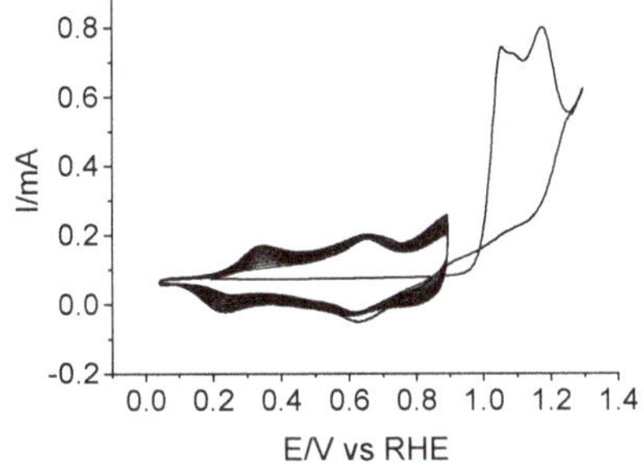

Figure 3. Cyclic voltammograms recorded for a Pt electrode during the electrochemical copolymerization of 0.5 M of piperazine and 0.1 M of aniline in 1 M of $HClO_4$ solution. The upper potential limit was set at 1.3 V for the first scan, and at 0.9 V for the subsequent ones. $v = 50$ mV s^{-1}.

The electrochemical behavior of the deposited copolymer was tested in an acidic background solution that was free of any monomer species, and the result is shown in Figure 4 (solid line). CV shows three redox transitions centered at around 0.33, 0.68, and 0.97 V. The first one can be assigned to a leucoemeraldine–emeraldine transformation similar to that of pristine polyaniline. The second one, which is broader and less intense, has been usually interpreted in terms of the presence of different quinoid structures [24,25]. For the copolymer studied here, the formation of ketopiperazines upon piperazine oxidation at very low anodic potentials (see Figure 2) demonstrates that the deposited material could incorporate a little amount of those previously formed quinoid structures. However, the high relative intensity of the voltammetric wave at 0.68 V strongly suggests a main contribution of active redox centers involving piperazine units which are oxidized *after* they are incorporated to the copolymer chain. Accordingly, the second redox peak may be related to the existence of redox transitions involving hydroxypiperazine $\rightleftharpoons$ ketopiperazine species within the copolymer structure [26]. With regard to the pair of redox peaks centered at 0.97 V in the CV of Figure 4, they can be clearly related to the emeraldine–pernigraniline transition of the copolymer, which is similar to that undergone by polyaniline under the same experimental conditions (dashed line).

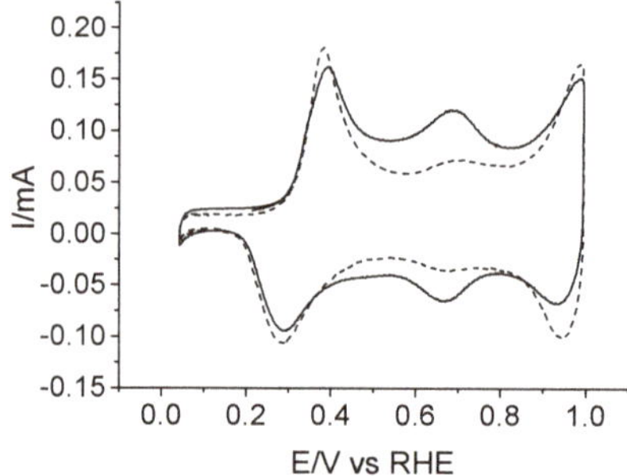

Figure 4. Electrochemical response in 1 M of HClO$_4$ medium of a Pt electrode covered with either polyaniline (dashed line) or an aniline–piperazine copolymer (solid line), which were deposited under the potential program used in Figure 3. v = 50 mV s^{-1}.

In order to monitor the redox behavior of the copolymer and analyze the chemical nature of the species involved in the redox transitions (particularly the existence of a keto-hydroxypiperazine transformation), in situ FTIR spectroscopy experiments were performed for the copolymer. The Pt-modified electrode was transferred to the IR spectroelectrochemical cell, which contained a perchloric acid solution that was free of monomers and prepared with D$_2$O to facilitate assignments in the 1500–1700 cm^{-1} spectral range. After some potential cycles within the stability window of the copolymer, the Pt surface was pressed against the prismatic CaF$_2$ window, and a reference spectrum was collected at 0.1 V. Finally, the potential was stepped to higher values to collect sample spectra, and the results are displayed in Figure 5. Three main positive bands at 1516, 1436, and 1212 cm^{-1}, and three clear negative bands at 1630, 1580, and 1170 cm^{-1} can be observed together with several features in the 1300–1400 cm^{-1} region. Some of the referred absorptions can be unambiguously assigned to the presence of a polyaniline skeleton. Particularly, the complete disappearance of the leucoemeraldine state at 0.6 V is evidenced by the vanishing of the aromatic C-C stretching mode at 1516 cm^{-1} and of the C-N-C stretching at 1212 cm^{-1} [27,28]. The formation of oxidized emeraldine (0.6 V) and pernigraniline (1.0 V) structures is also supported by the development of quinoid C=C stretching vibrations at 1580 cm^{-1} [28,29], by the -CH in-plane bending at oxidized aniline rings at 1170 cm^{-1}, and, finally, by the generation of several intermediate-order C-N vibrations in the 1300–1400 cm^{-1} frequency window [29,30]. On the other hand, the successful incorporation of piperazine structures to the polyaniline chain is evidenced by two representative bands, which cannot be observed for a pristine polyaniline. These absorptions correspond to the activation of the CH$_2$ bending upon oxidation (positive-going feature at 1435 cm^{-1} [31]), and to the carbonyl C=O stretching at 1630 cm^{-1}. This latter band supports the voltammetric result in Figure 3, and confirms that a significant fraction of piperazine rings are present in the form of electroactive ketopiperazines (C=O $\rightleftharpoons$ C-OH). The absence of additional vibrations at around 1660 and 2030 cm^{-1} shows that neither amide structures nor isocyanates are formed and, consequently, that piperazine was not overoxidized during the electropolymerization process under the experimental conditions employed. FTIR assignments are collected in the Supplementary Material (Table S1).

Additionally, the electrodeposited copolymer was examined by ex situ XPS in order to analyze its surface composition, and also to give support to the chemical structures suggested by in situ FTIR spectroscopy. A film grown after 10 voltammetric cycles as in Figure 3 was rinsed with ultrapure water, dried under nitrogen, stored in a dry place for 24 h, and then analyzed by XPS. Figure 6 shows the photoelectronic spectra of C 1s and N 1s core levels. The C 1s signal can be fitted with four peaks at 284.5, 285.4, 286.6, and 288.7 eV. Both the high energy level and the weak intensity of the latter contribution are compatible with the presence of a small amount of carbonyl carbon, which was probably associated to the ketopiperazine centers. On the other hand, there are two major signals that were undoubtedly associated to aromatic carbon, and hence to aniline rings. This is the main peak at 284.6 eV, which was attributed to plain aromatic carbon, and the signal at 285.4 eV, which was due to aromatic carbon bonded to neutral nitrogen. On the other hand, binding energies at around 286.6 eV are characteristic of carbon bonded to positive nitrogen [32], and consequently, this peak is compatible with the presence of piperazine rings within the polymer backbone. The best fit for the N 1s spectrum shows only two contributions at 399.6 and 401.6 eV, but unfortunately, it is not possible to distinguish signals coming from the piperazine and aniline environments. The peak at 399.6 eV is clearly attributed to neutral nitrogen, but it could be associated to any of the amine, imine or even amide groups, as these species do not show significantly different chemical shifts. In the same way, the higher binding energy signal at 401.6 eV is compatible with the presence of piperazine within the material. That peak is assigned to positively charged nitrogen atoms resulting from the protonation of imine centers (located exclusively at aniline rings) and secondary amine positions (at both piperazine and aniline rings) [32].

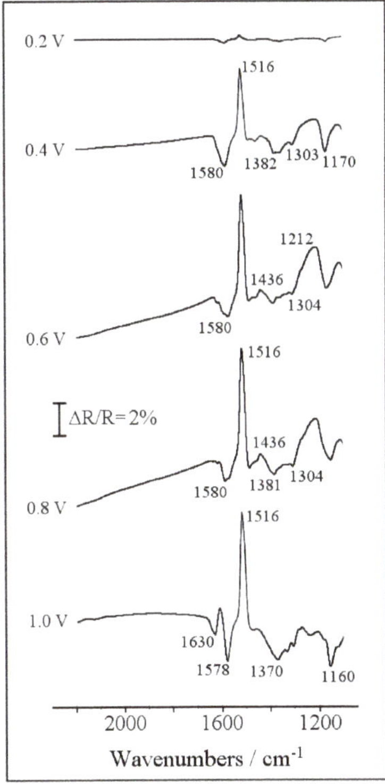

Figure 5. Set of in situ FTIR spectra collected during the oxidation of an electrochemically obtained poly(aniline-co-piperazine) film in 0.1 M of $HClO_4/D_2O$ test solution. Reference potential 0.1 V. Sample potential labeled for each spectrum: 100 interferograms at each potential.

Chemically obtained polyanilines are usually amorphous solids, but more or less ordered structures can also be obtained depending on the synthesis conditions [33]. In general, better ordering is observed for electrochemically-prepared thin films [34]. The surface morphology of the aniline–piperazine copolymer electrodeposited on platinum after 10 cycles has been examined by SEM. The top image in Figure 7 shows how this organic coating is completely distributed over the surface in the form of flat ribbon strings with a width of about 3–5 μm. These structures are quite different from those detected for unmodified polyaniline deposited on Pt under similar experimental conditions, for which a uniformly distributed, smooth film is obtained (Figure 7b). According to these observations, the presence of a significant amount of piperazine units is seen at the origin of the particular morphologic features shown by the copolymer material.

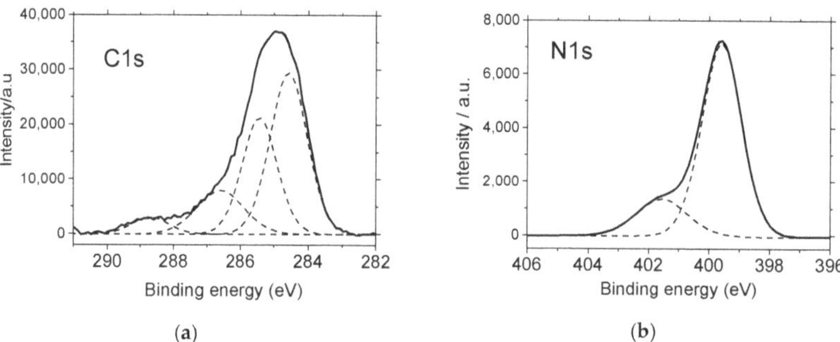

Figure 6. Deconvoluted (**a**) C 1s and (**b**) N 1s X-ray photoelectron spectroscopy (XPS) core-level spectra of an electrodeposited aniline–piperazine copolymer. The sample was obtained as in Figure 3.

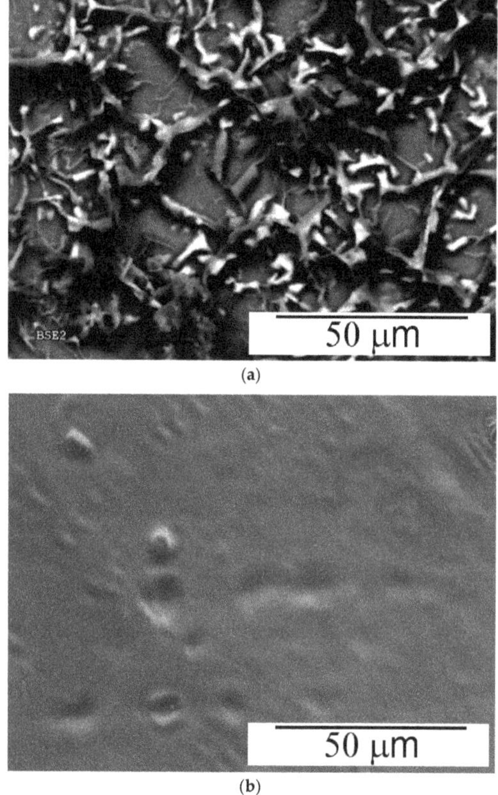

Figure 7. Scanning electron micrographs at 1000× magnification of copolymer (**a**) and polyaniline (**b**) deposited under the same experimental conditions.

3.3. Testing the Electrocatalytic Properties of the Copolymer towards Dopamine and Ascorbic Acid

Platinum electrodes coated with copolymer films have been employed to determine dopamine (DA) and ascorbic acid (AA) in synthetic samples. The sensitivity of the measurement and the catalytic performance of the copolymer have been evaluated in acidic medium. First, DA samples were prepared within a concentration range from 0.2 to 3.0 mM. Then, the oxidation current of this analyte was recorded for the Pt electrode covered with the copolymer at a potential of 0.85 V, which corresponds to the first anodic peak of the DA→DQ reaction (see Figure S1 in Supplementary Materials for a CV curve). This anodic peak potential is nearly the same as that reported in the literature for bare Pt electrodes in acidic medium [35,36]. Figure 8a shows how the oxidation current of dopamine increases almost linearly at increasing analyte concentrations. From that plot, it can be derived that the copolymer demonstrates quite a sensitive response, in the order of 29 μA mM^{-1}, within the range of concentrations studied. A similar electrocatalytic effect can be observed in Figure 8b for AA electrooxidation. For such a reaction, cyclic voltammograms recorded with a Pt electrode covered with the copolymer material (see Figure S2 in Supplementary Materials for a CV curve) shows the anodic peak centered at 0.89 V, which is a value slightly below that usually obtained for bare Pt surfaces under similar experimental conditions [37,38]. Also in this case, the determination of AA on the copolymer substrate shows linearly increasing faradaic responses with a sensitivity of about 22 μA mM^{-1}.

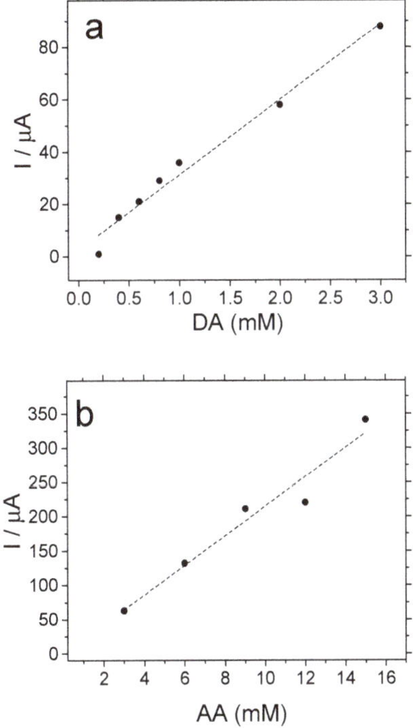

Figure 8. Peak oxidation currents recorded for a Pt electrode covered with a piperazine–aniline copolymer in 1 M of HClO$_4$ aqueous solutions containing an increasing concentration of analyte: (**a**) dopamine; (**b**) ascorbic acid.

4. Conclusions

The electrochemical oxidation of piperazine on platinum electrodes at moderate potentials (roughly below 1.0 V/RHE) preserves the ring structures and produces ketopiperazines as the main reaction product. In situ FTIR spectroscopy strongly suggested that ring opening and overoxidation occur at higher potentials to form both amides and isocyanates. As a result, it was observed that the homopolymerization of piperazine cannot be achieved in perchloric acid aqueous solution under electrochemical conditions.

On the contrary, piperazine can be successfully copolymerized with aniline in acidic medium. The deposited copolymer shows some electrochemical features similar to those of pristine polyaniline, particularly those related with leucoemeraldine-to-emeraldine and emeraldine-to-pernigraniline transitions. However, a key difference arises in the intermediate potential region between both transitions. As shown by in situ FTIR and XPS spectroscopies, the intermediate redox peak is a consequence of the incorporation of piperazine units to the copolymer structure. Most of these piperazine centers undergo electrochemical oxidation during the copolymerization potential scans and, as a result, a new reversible hydroxy $\rightleftharpoons$ ketopiperazine redox transformation seems to occur as the intermediate voltammetric feature centered at 0.68 V. It should be noted that, owing to the conservative potential program applied during the deposition process, any significant amount of overoxidation products was not incorporated to the copolymer structure. As a result, the deposited material is chemically stable, and presents a well-defined electrochemical behavior.

It was observed that the aniline–piperazine copolymer shows a linear response when applied to the electrochemical determination of dopamine or ascorbic acid in synthetic samples. The sensitivity of the measurement is in both cases high enough to assure the correct quantification of analytes. This electrode material has to be tested in real samples, but according to the results presented in this contribution, it shows potential application in DA and AA sensors, owing to its facile synthesis, high chemical stability, and reproducible linear response.

Supplementary Materials: The following are available online at http://www.mdpi.com/1996-1944/11/6/1012/s1, Figure S1: Linear Sweep Voltammogram showing the oxidation of 3 mM DA on a Pt electrode covered with the aniline-piperazine copolymer. DA oxidation peak is centered at 0.85 V, Figure S2: Linear Sweep Voltammogram showing the oxidation of 30 mM AA on a Pt electrode covered with the aniline-piperazine copolymer. AA oxidation peak is centered at 0.89 V, Table S1: Observed frequencies and proposed assignments for the vibrational bands derived from Figures 2 and 5.

Author Contributions: F.H., F.M. and E.M. conceived and designed the experiments; S.D., S.L.-B. and C.N.K. performed the experiments; S.D., C.N.K., F.H., F.M. and E.M. analyzed the data; S.D., F.H. and E.M. wrote the paper. All authors participated in the Investigation and in manuscript preparation. All authors contributed in Writing-Review & Editing of the manuscript and approved the final version.

Funding: This research was funded by the Spanish Ministerio de Economía y Competitividad and FEDER funds, grant MAT2016-76595-R. The stay of S. Dkhili at the University of Alicante was funded by the Ministry of Higher Education and Scientific Research of Tunisia.

Conflicts of Interest: The authors declare no conflict of interest.

References

1. Grieshaber, D.; MacKenzie, R.; Vörös, J.; Reimhult, E. Electrochemical Biosensors—Sensor Principles and Architectures. *Sensors* **2008**, *8*, 1400–1458. [CrossRef] [PubMed]
2. Janata, J.; Josowicz, M. Conducting polymers in electronic chemical sensors. *Nat. Mater.* **2003**, *2*, 19–24. [CrossRef] [PubMed]
3. Otero, T.F. Biomimetic Conducting Polymers: Synthesis, Materials, Properties, Functions, and Devices. *Polym. Rev.* **2013**, *53*, 311–351. [CrossRef]
4. Joulazadeh, M.; Navarchian, A.H. Ammonia detection of one-dimensional nano-structured polypyrrole/metal oxide nanocomposites sensors. *Synth. Met.* **2015**, *210*, 404–411. [CrossRef]
5. Le, T.-H.; Kim, Y.; Yoon, H. Electrical and Electrochemical Properties of Conducting Polymers. *Polymers* **2017**, *9*, 150. [CrossRef]
6. Yoon, H. Current Trends in Sensors Based on Conducting Polymer Nanomaterials. *Nanomaterials* **2013**, *3*, 524–549. [CrossRef] [PubMed]
7. Gerard, M.; Chaubey, A.; Malhotra, B.D. Application of conducting polymers to biosensors. *Biosens. Bioelectron.* **2002**, *17*, 345–359. [CrossRef]

8. Wu, Y.; Hu, S. Biosensors based on direct electron transfer in redox proteins. *Microchim. Acta* **2007**, *159*, 1–17. [CrossRef]

9. Nöll, T.; Nöll, G. Strategies for "wiring" redox-active proteins to electrodes and applications in biosensors, biofuel cells, and nanotechnology. *Chem. Soc. Rev.* **2011**, *40*, 3564–3576. [CrossRef] [PubMed]

10. López-Bernabeu, S.; Gamero-Quijano, A.; Huerta, F.; Morallón, E.; Montilla, F. Enhancement of the direct electron transfer to encapsulated cytochrome c by electrochemical functionalization with a conducting polymer. *J. Electroanal. Chem.* **2017**, *793*, 34–40. [CrossRef]

11. Gu, D.; Yang, G.; He, Y.; Qi, B.; Wang, G.; Su, Z. Triphenylamine-based pH chemosensor: Synthesis, crystal structure, photophysical properties and computational studies. *Synth. Met.* **2009**, *159*, 2497–2501. [CrossRef]

12. Li, S.; Ge, Y.; Piletsky, S.A.; Lunec, J. (Eds.) *Molecularly Imprinted Sensors: Overview and Applications*, 1st ed.; Elsevier: Amsterdam, The Netherlands, 2012.

13. Ghosh, K.; Tarafdar, D.; Samadder, A.; Khuda-Bukhsh, A.R. Piperazine-based simple structure for selective sensing of Hg^{2+} and glutathione and construction of a logic circuit mimicking an INHIBIT gate. *New J. Chem.* **2013**, *37*, 4206–4213. [CrossRef]

14. Sun, Z.; Li, H.; Guo, D.; Liu, Y.; Tian, Z.; Yan, S. A novel piperazine-bis(rhodamine-B)-based chemosensor for highly sensitive and selective naked-eye detection of Cu^{2+} and its application as an INHIBIT logic device. *J. Lumin.* **2015**, *167*, 156–162. [CrossRef]

15. Zhang, F.; Yang, X.; Jiang, L.; Liang, C.; Zhu, R.; Li, H. Piperazine-functionalized ordered mesoporous polymer as highly active and reusable organocatalyst for water-medium organic synthesis. *Green Chem.* **2013**, *15*, 1665–1672. [CrossRef]

16. Sachdev, D.; Maheshwari, P.H.; Dubey, A. Piperazine functionalized mesoporous silica for selective and sensitive detection of ascorbic acid. *J. Porous Mater.* **2016**, *23*, 123–129. [CrossRef]

17. Ramachandran, R.; Balasubramanian, S.; Aridoss, G.; Parthiban, P.; Yamuna, G.; Kabilan, S. Synthesis and studies of semiconducting piperazine–aniline copolymer. *Eur. Polym. J.* **2006**, *42*, 1885–1892. [CrossRef]

18. Horányi, G.; Bakos, I. Experimental evidence demonstrating the occurrence of reduction processes of ClO^{4-} ions in an acid medium at platinized platinum electrodes. *J. Electroanal. Chem.* **1992**, *331*, 727–737. [CrossRef]

19. Heacock, R.A.; Marion, L. The infrared spectra of secondary amines and their salts. *Can. J. Chem.* **1956**, *34*, 1782–1795. [CrossRef]

20. Wang, S.L.; Lin, S.Y.; Chen, T.F. Thermal-Dependent dehydration process and intramolecular cyclization of lisinopril dihydrate in the solid state. *Chem. Pharm. Bull.* **2000**, *48*, 1890–1893. [CrossRef] [PubMed]

21. Cheam, T.C.; Krimm, S. Vibrational analysis of crystalline diketopiperazine—I. Raman and i.r. spectra, Spectrochim. *Acta Part A Mol. Spectrosc.* **1984**, *40*, 481–501. [CrossRef]

22. Socrates, G. *Infrared and Raman Characteristic Group Frequencies: Tables and Charts*, 3rd ed.; John Wiley & Sons: Chichester, UK, 2004.

23. Pattar, V.P.; Magdum, P.A.; Patill, D.G.; Nandibewoor, S.T. Thermodynamic, kinetic and mechanistic investigations of Piperazine oxidation by Diperiodatocuprate(III) complex in aqueous alkaline médium. *J. Chem. Sci.* **2016**, *128*, 477–485. [CrossRef]

24. Shim, Y.-B.; Won, M.; Park, S. Electrochemistry of Conductive Polymers VIII. *J. Electrochem. Soc.* **1990**, *137*, 538. [CrossRef]

25. Cotarelo, M.; Huerta, F.; Quijada, C.; Cases, F.; Vázquez, J. The electrocatalytic behaviour of poly(aniline-co-4adpa) thin films in weakly acidic médium. *Synth. Met.* **2004**, *144*, 207–211. [CrossRef]

26. Owens, J.L.; Dryhurst, G. Electrochemical reduction of tetraketopiperazine. *Anal. Chim. Acta* **1976**, *87*, 37–50. [CrossRef]

27. Ping, Z.; Nauer, G.E.; Neugebauer, H.; Theiner, J.; Neckel, A. In situ Fourier transform infrared attenuated total reflection (FTIR-ATR) spectroscopic investigations on the base-acid transitions of leucoemeraldine. *Electrochim. Acta* **1997**, *42*, 1693–1700. [CrossRef]

28. Louarn, G.; Lapkowski, M.; Quillard, S.; Pron, A.; Buisson, J.P.; Lefrant, S. Vibrational properties of polyaniline—Isotope effects. *J. Phys. Chem.* **1996**, *100*, 6998–7006. [CrossRef]

29. Abidi, M.; López-Bernabeu, S.; Huerta, F.; Montilla, F.; Besbes-Hentati, S.; Morallón, E. The chemical and electrochemical oxidative polymerization of 2-amino-4-tert-butylphenol. *Electrochim. Acta* **2016**, *212*, 958–965. [CrossRef]

30. Quillard, S.; Berrada, K.; Louarn, G.; Lefrant, S.; Lapkowski, M.; Pron, A. In situ Raman spectroscopic studies of the electrochemical behavior of polyaniline. *New J. Chem.* **1995**, *19*, 365–374.

31. Hendra, P.J.; Powell, D.B. The infra-red and Raman spectra of piperazine. *Spectrochim. Acta* **1962**, *18*, 299–306. [CrossRef]

32. NIST X-ray Photoelectron Spectroscopy Database, Version 4.1 (Web Version), 2012. Available online: http://Srdata.Nist.Gov/Xps/ (accessed on 15 January 2018).

33. Langer, J.J. Polyaniline micro- and nanostructure. *Adv. Mater. Opt. Electron.* **1999**, *9*, 1–7. [CrossRef]

34. Yonezawa, S. Effects of the Solvent for the Electropolymerization of Aniline on Discharge and Charge Characteristics of Polyaniline. *J. Electrochem. Soc.* **1995**, *142*, 3309. [CrossRef]

35. Stern, D.A.; Salaita, G.N.; Lu, F.; McCargar, J.W.; Batina, N.; Frank, D.G.; Laguren-Davidson, L.; Lin, C.H.; Walton, N. Studies of L-DOPA and related compounds adsorbed from aqueous solutions at platinum(100) and platinum(111): Electron energy-loss spectroscopy, Auger spectroscopy, and electrochemistry. *Langmuir* **1988**, *4*, 711–722. [CrossRef]

36. Kavanoz, M.; Ülker, E.; Bük, U. A Novel Polyaniline–Poly(3-Methylthiophene)–Poly(3,3′-Diaminobenzidine) Electrode for the Determination of Dopamine in Human Serum. *Anal. Lett.* **2015**, *48*, 75–88. [CrossRef]

37. Březina, M.; Koryta, J.; Loučka, T.; Maršíková, D.; Pradáč, J. Adsorption and kinetics of oxidation of ascorbic acid at platinum electrodes. *J. Electroanal. Chem. Interfacial Electrochem.* **1972**, *40*, 13–17. [CrossRef]

38. Xing, X.; Bae, I.T.; Shao, M.; Liu, C.-C. Electro-oxidation of l-ascorbic acid on platinum in acid solutions: An in-situ FTIRRAS study. *J. Electroanal. Chem.* **1993**, *346*, 309–321. [CrossRef]

Article

Reactive Insertion of PEDOT-PSS in SWCNT@Silica Composites and its Electrochemical Performance

Halima Djelad [1,2], Abdelghani Benyoucef [1], Emilia Morallón [2] and Francisco Montilla [2,*]

[1]   Laboratoire des Sciences et Techniques de l'Eau, University of Mascara, Bp 763 Mascara 29000, Algeria;
      halima@ua.es (H.D.); a.benyoucel@univ-mascara.dz (A.B.)
[2]   Departamento de Química Física e Instituto Universitario de Materiales, Universidad de Alicante, Ap. 99,
      E-03080 Alicante, Spain; morallon@ua.es
*   Correspondence: francisco.montilla@ua.es

Received: 30 January 2020; Accepted: 3 March 2020; Published: 6 March 2020

Abstract: Hybrid silica-modified materials were synthesized on glassy carbon (GC) electrodes by electroassisted deposition of sol-gel precursors. Single-wall carbon nanotubes (SWCNTs) were dispersed in a silica matrix (SWCNT@SiO$_2$) to enhance the electrochemical performance of an inorganic matrix. The electrochemical behavior of the composite electrodes was tested against the ferrocene redox probe. The SWCNT@SiO$_2$ presents an improvement in the electrochemical performance towards ferrocene. The heterogeneous rate constant of the SWCNT@SiO$_2$ can be enhanced by the insertion of poly(3,4-Ethylendioxythiophene)-poly(sodium 4-styrenesulfonate) PEDOT-PSS within the silica matrix, and this composite was synthesized successfully by reactive electrochemical polymerization of the precursor EDOT in aqueous solution. The SWCNT@SiO$_2$-PEDOT-PSS composite electrodes showed a heterogeneous rate constant more than three times higher than the electrode without conducting polymer. Similarly, the electroactive area was also enhanced to more than twice the area of SWCNT@SiO$_2$-modified electrodes. The morphology of the sample films was analyzed by scanning electron microscopy (SEM).

Keywords: PEDOT-PSS; SiO$_2$; sol-gel; hybrid materials; ferrocene

---

1. Introduction

The need for sensors in molecular analysis has stimulated the development of new electrocatalytic materials and electrochemical devices. Most of these materials can be categorized as nanomaterials (metal nanoparticles, nanotubes, graphene, etc.) exhibiting novel electronic, optical or mechanical properties [1–3]. Single-wall carbon nanotubes (SWCNTs) comprise an interesting group of materials with applications in electrocatalysis that can be employed as sensing elements [4–7]. Electrodes modified with carbon nanotubes have been applied to the electrochemical detection of a large number of species (dopamine, uric acid, nicotinamide adenine dinucleotide, ascorbic acid, tyrosine, insulin, etc.). These nanotubes have been also used as transducers for direct electron transfer to redox enzymes [8–10]. A major drawback of carbon nanotubes is their strong tendency to aggregate when they are deposited on a substrate, because of the strong π–π attractive interaction between the tube walls. The aggregation of nanotubes produces the loss of some of the physicochemical properties in the macroscopic measurement. Therefore, the dispersion of the active material onto the supporting electrode, keeping the nanoscopic character of the electrocatalytic material, is a key point for the development of sensors with superior properties. The immobilization of the nanomaterial can be performed inside a porous inorganic matrix, such as silica, in a simple way following sol-gel methodologies. This technique provides an easy method to encapsulate chemical species in a stable host. These modifiers provide enhanced electrocatalytic activity related to an improvement of the heterogeneous electron transfer rate [11]. The electrocatalytic activity of single-wall carbon nanotubes (SWCNTs) dispersed within a SiO$_2$ matrix

(SWCNT@SiO$_2$) was examined against redox probes presented in previous works. This material improves heterogeneous rate transfer of the electrodes for all the common redox probes.

Following this approach, several silica nanocomposites functionalized with carbon materials have been employed as electrode modifiers in several applications [12]. The development of highly sensitive electrochemical sensors, e.g., biosensors, was achieved by the combination of conducting polymers with graphene or carbon nanotubes, which provide to the composite high electrical conductivity, active surface area and fast electron transfer rate [13–15].

However, most of SWCNTs incorporated in silica remain isolated from the underlying electrode with no direct electrical connection [16,17]. In this work, the electrochemical performance of carbon nanotubes dispersed in silica was tested against the ferrocene redox probe. This species is an outer-sphere redox probe. The electrochemical reaction happened without any adsorption step and showed low reorganization energy upon redox transitions [18]. For those reasons, ferrocene has been routinely used to investigate electron-transfer kinetics in chemically modified electrodes [19,20] since this redox probe is usually incorporated as a mediator in electrochemical biosensors [21–23].

The objective of the present work was to make a better electrical contact between the dispersed carbon nanotubes in silica gels by growing conductive molecular wires between the SWCNT and the supporting electrode. Due to the low solubility of the 3,4-ethylendioxythiophene (EDOT) monomer in aqueous solutions, it is necessary to add a surfactant to the solution. Poly (sodium 4-styrenesulfonate) (PSS) behaves as a surfactant but also as an electrolyte that supplies enough conductivity to the solution, remaining inserted in the polymer film as a doping agent [24]. These poly(3,4-Ethylendioxythiophene)-poly(sodium 4-styrenesulfonate) PEDOT-PSS films find applications as the transducers of biosensors for peroxides, or as mediators for redox enzymes [25,26].

We chose PEDOT-PSS since this polymer presents a poor electrocatalytic performance for the electron transfer to the ferrocene redox probe [11]. In that manner, if any electrocatalytic effect is observed for the composite material, this effect could be related only to the electrical wiring of the SWCNT with the electrode and not to the mere presence of the polymer. The morphology of the nanocomposite electrodes was characterized by electron microscopy and the electrochemical performance of the new nanocomposite electrode, SWCNT@SiO$_2$-PEDOT-PSS, was tested against a model redox probe. The effect of the nanotubes within the silica layer was studied in separate electrodes to confirm the wiring effect provided by the conducting polymer.

2. Materials and Methods

SWCNTs were purchased from Cheap Tubes Inc. (Brattleboro, VT, USA, purity 95%, 1–2 nm diameter) and were used without further purification. 3,4-Ethylendioxythiophene (EDOT), poly (sodium 4-styrenesulfonate) (PSS), ferrocenium hexafluorophosphate (Fc), tetraethyl orthosilicate and ethanol (EtOH) were purchased from (Sigma-Aldrich, Madrid, Spain). Potassium chloride, hydrochloric acid and sulfuric acid were purchased from Merck Company and all the solutions were freshly prepared with deionized water obtained from an Elga Labwater Purelab Ultra system.

Cyclic voltammetry (CV) experiments were carried out in a conventional three-electrode cell under N$_2$ atmosphere. A platinum wire was used as the counter electrode. The working electrode used was a glassy carbon (GC, geometric area = 0.07 cm^2, Carbone Lorraine, model V-25) rod. The current density was calculated from this geometric area. The GC electrode was submitted to the following cleaning procedure before each experiment. The GC was polished with fine emery paper and subsequently rinsed with ultrapure water. Potentials were measured against the reversible hydrogen electrode (RHE) immersed in the same electrochemical cell. An EDAQ EA163 model potentiostat coupled to an EG&G Parc Model 175 was used for both the synthesis and electrochemical testing of the samples. The surface morphology of modified GC electrodes was studied by scanning electron microscopy (SEM) and images were obtained using the field emission scanning electron microscope (FESEM).

Before deposition, GC electrodes were cleaned by polishing with alumina slurries and were rinsed with water. The precursor of silica was synthesized by a mixture of 2.69 mmol of TEOS, 8.2 mL of

EtOH and 5.8 mL of a solution of 0.01M HCl with 0.46 M KCl. This was stirred for 1 h in a closed vial. After 2 h, the resulting sol was submitted to evaporation by vacuum heating until the complete removal of the released ethanol from alkoxide hydrolysis was achieved.

Stable SWCNT aqueous suspensions were obtained as follows: 20 mg of SWCNT were poured into a vial containing 20 mL of 1% poly (4-styrenesulfonic acid) aqueous solution. The carbon nanotubes were dispersed and suspended by the application of an ultrasonic field by a VIRTIS probe (Virsonic 475, 475W maximum output power) at 1 min intervals for 1 h. To avoid overheating, samples were ice-cooled between sonication intervals.

For the preparation of SWCNT@SiO_2 on GC electrodes, 2.52 mL of the SWCNT solution were poured into the silica precursor solution. This mixture containing SWCNT and the hydrolyzed silica precursor was placed in an electrochemical glass cell which contained a platinum wire counter electrode and a reversible hydrogen reference electrode to proceed with the electroassisted deposition. Further details of the deposition method are provided in other research [11,27].

EDOT electropolymerization was carried out in an aqueous medium prepared by dissolving 1.46 g PSS in 10.0 mL ultrapure water; 53 µL EDOT monomer were then added and the resulting solution was stirred in an ultrasonic bath for 30 min.

At least 3 replicas of the synthesized electrodes were obtained. The different electrodes were tested with distinct redox probes, obtaining peak separation variations of less than 6 mV between the different samples. The most representative electrodes of each species are shown in this work.

3. Results and Discussion

3.1. Electrochemical Behavior of Modified Electrodes

The redox chemistry of ferrocene was studied using cyclic voltammetric (CV) with the modified electrodes. Ferrocene/ferricenium (Fc/Fc$^+$) is one of the most common outer-sphere redox probes and it is very sensitive to the active sites for electron transfer in SWCNTs as it may react through both nanotube walls and tips [7]. The test solution was prepared with 1.0 mM ferrocenium hexafluorophosphate (FcPF$_6$) in a 0.5 M sulfuric acid solution. The resultant stabilized CV curves are shown in Figure 1.

Electrodes were modified by silica films obtained by electroassisted deposition at a current density of 2.5 mA cm^{-2}. For these experiments, the total charge applied to the deposition of silica Q$_{silica}$ was 150 mC cm^{-2}.

Figure 1 shows the stabilized cyclic voltammogram of a GC/SiO_2 (150 mC·cm^{-2}) electrode immersed in a test solution of 0.5 M sulfuric acid solution containing Fc$^+$ at scan rate of 100 mV·s^{-1}. The stabilized voltammograms were obtained after 10 cycles between the upper and lower potential limits of the cyclic voltammogram. In the scan for positive potentials, we observed an oxidation peak at 0.53 V that corresponded to the oxidation peak of Fc to Fc$^+$. In the reverse scan, we observed a reduction peak at 0.40 V that corresponded to the reduction of Fc$^+$ to Fc. The peak potential separation between anodic and cathodic features (ΔE_p) was 130 mV. A fast, reversible, one-electron transfer would ideally have a ΔE_p = 59 mV at 298 K. The discrepancy from this ideal value was mainly attributed to slow electron transfers. Figure 1 also shows the stabilized cyclic voltammogram of a GC electrode modified with SWCNT@SiO_2 prepared in equivalent conditions to the previous electrode. The shape of the voltammogram was similar to the previous one, but the peak potential separation between anodic and cathodic features was 120 mV. This indicated that the response of the redox probe was more reversible in the present case than in SiO_2-modified electrodes in the absence of carbon nanotubes. It also indicated that these species can improve the electron transfer after their incorporation into the silica matrix.

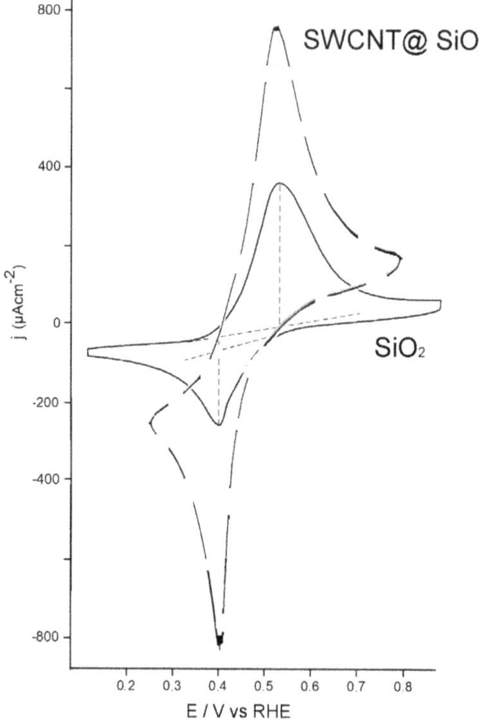

Figure 1. Stabilized cyclic voltammograms of a silica-modified electrode (solid line) and a Single-Wall Carbon Nanotubes in a silica matrix (SWCNT@SiO$_2$)-modified electrode (dashed line) in a solution of 1.0 mM FcPF$_6$ in 0.5M H$_2$SO$_4$. Scan rate of 100 mV s^{-1}.

Since SWCNT may remain electrically isolated within the dielectric silica matrix, a good strategy to improve the performance of this electrode is the growth of a conducting polymer through the silica functionalized electrodes. Figure 2 shows the electrochemical synthesis of PEDOT-PSS through a SWCNT@SiO$_2$-modified electrode.

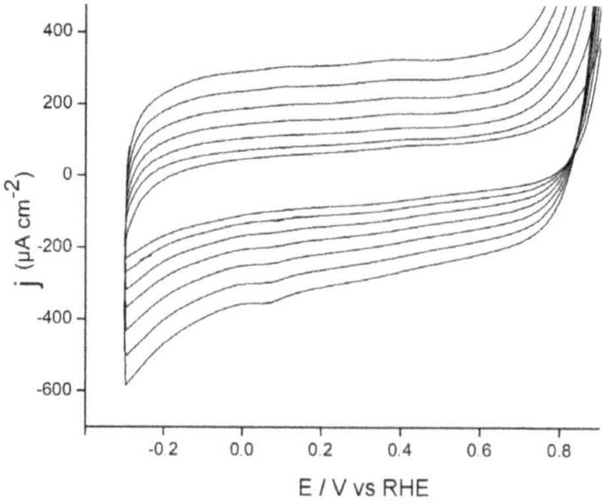

Figure 2. Cyclic voltammetric scans of an SWCNT@SiO$_2$ electrode in a solution of 3,4-ethylendioxythiophene (EDOT) in poly (sodium 4-styrenesulfonate) (PSS). Anodic limit of 1.0 V. Scan rate of 100 mV s^{-1}.

The first potential cycle was a featureless voltammetric profile and was recorded until a potential value above 0.80 V was reached. This point corresponded to the onset potential of EDOT monomer oxidation and, consequently, to the formation of PEDOT-PSS. The inversion potential was set at 1.0 V to obtain a suitable growth rate of the polymeric material. On subsequent potential scans, the presence of a current plateau in the potential region between −0.2 and 0.8 V, showing capacitive features and an increasing voltammetric charge, was observed. This feature was assigned to the growth of PEDOT-PSS across the silica matrix.

Following the synthesis process, the electrodes coated with the polymeric films were rinsed with water and then immersed in a solution containing Fc^+ with 0.5 M acid sulfuric solution. The voltammetric response of PEDOT-PSS electrosynthesized on the $SWCNT@SiO_2$ electrode is shown in Figure 3.

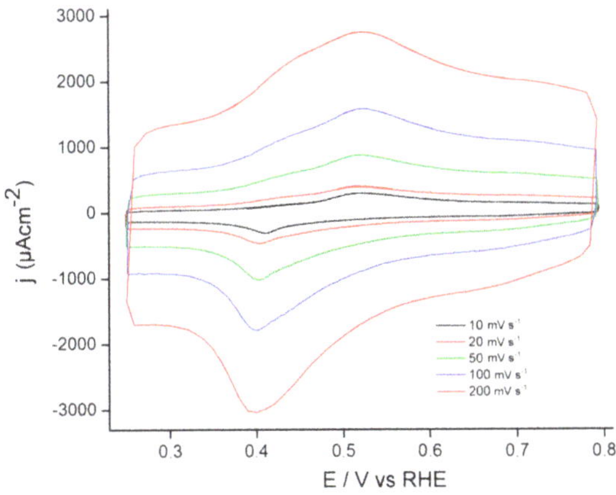

Figure 3. Stabilized cyclic voltammograms of an $SWCNT@SiO_2$-PEDOT-PSS electrode in a solution of 1.0 mM $FcFP_6$ in 0.5M H_2SO_4 at different scan rates.

The voltammetric responses of $SWCNT@SiO_2$-PEDOT-PSS presented clear current coming from capacitive processes of the conducting polymer at potentials lower than 0.4 V. The Fc/Fc^+ redox processes appeared at around 0.5 and 0.4 V for oxidation and reduction process, respectively, at the different scan rates.

To go into detail about the behavior of the modified electrodes, a kinetic analysis of their electrochemical performance was carried out. The kinetic reversibility of an electrochemical reaction can be evaluated from cyclic voltammetry experiments due to the Nicholson method, by making use of the values of peak potential separation at different scan rates. Figure 4A presents these values for the different electrodes immersed in the test solution of Fc^+.

As observed for SiO_2-modified electrodes at a low scan rate ($10\ mV \cdot s^{-1}$), the value of peak potential separation was 115 mV but when the scan rate was increased, this parameter sharply increased reaching values of around 140 mV at $200\ mV \cdot s^{-1}$. A similar trend was observed for the $SWCNT@SiO_2$ electrode, although the peak potential separation was lowered by the presence of the nanotubes inserted in the matrix. This indicated that a higher scan rate drove to a lower reversibility. The behavior of the $SWCNT@SiO_2$-PEDOT-PSS electrode is completely different, and we can observe a lower dependency of the reversibility with the scan rate.

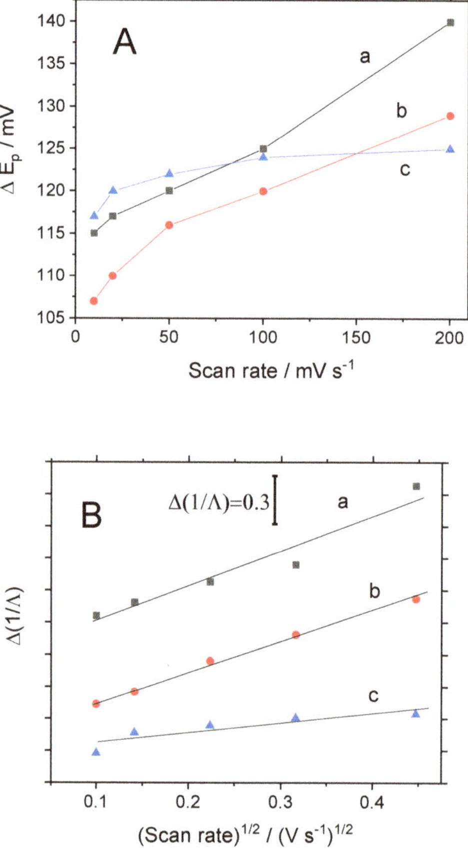

Figure 4. (**A**) Voltammetric peak potential separation between oxidation and reduction process of ferrocene as a function of the voltammetric scan rate for different electrodes. (**B**) Variation of the reciprocal of the Matsuda–Ayabe Λ parameter as a function of the square root of the voltammetric scan rate for different glassy carbon (GC)-modified electrodes. (**a**) SiO_2; (**b**) $SWCNT@SiO_2$; (**c**) $SWCNT@SiO_2$-PEDOT-PSS.

From the peak potential separation, we can obtain the Λ parameter defined by Matsuda and Ayabe [28]. It is usually assumed that an electrode process will be kinetically reversible for $\Lambda > 15$, quasireversible for $15 \geq \Lambda \geq 0.001$ and irreversible for Λ values lower than 0.001. In the present case, the silica-modified electrode presents values of Λ ranging from 0.69 (at 10 mV·s^{-1}) to 0.45 (at 200 mV·s^{-1}). The $SWCNT@SiO_2$ electrode presents Λ from 0.85 to 0.54 and the composite $SWCNT@SiO_2$-PEDOT-PSS have Λ values from 0.67 to 0.57. In all cases, this probe can be categorized as quasireversible for these electrodes. The relationship between the standard rate constant, k^0, for the electron transfer of the electrochemical reaction and the Λ parameter, was shown by Matsuda [28]:

$$\frac{1}{\Lambda} = \frac{1}{k^0}\left(\frac{nFD}{RT}\right)^{1/2} v^{1/2}$$

(1)

A representation of the reciprocal of Λ against the square root of the scan rate allows the determination of the heterogeneous rate constant from the slope of each curve for each material. Figure 4B shows this plot where a linear trend is observed for all the electrodes. From the slope, the values of k^0 were determined. The electrode modified with silica presented a value of 1.61×10^{-2} cm·s^{-1}, whereas the electrode with carbon nanotubes (SWCNT@SiO$_2$) presented a slightly higher electron transfer of 1.74×10^{-2} cm·s^{-1}. Finally, the greatest result was the composite electrode where the rate transfer was enhanced to a value of 5.47×10^{-2} cm·s^{-1}, indicating that the nanotubes were properly connected to the electrode support.

The electroactive area for electron transfer can be also determined from the voltammetric measurement. In this case, the classical Randles–Sevcik equation can be only applied to reversible systems:

$$j_p(rev) = 2.687 \times 10^5 ACn^{3/2}(Dv)^{1/2} \tag{2}$$

where $j_p(rev)$ is the current density for a reversible redox process (this current density is referred to the geometric area of the electrode), A is the real electroactive area for the electron transfer, (this is a unitless parameter also called the roughness factor), C is the concentration of the redox probe (in mol cm^{-3}), n is the number of electrons transferred, D is the diffusion coefficient (cm^2·s^{-1}) of the redox probes and v is the scan rate (V·s^{-1}). The application of this equation to both quasireversible and irreversible systems is only possible after the correction of the experimental peak current (I_p):

$$j(rev) = \frac{j_p}{k(\Lambda)} \tag{3}$$

where $k(\Lambda)$ is an adimensional parameter defined by Matsuda and Ayabe, which accounts for the kinetic factor governing the peak current.

Figure 5 presents the Randles–Sevcik plots of peak current vs. the square root of the scan rate. For the SiO$_2$-modified electrode, the Fc$^+$/Fc reaction occurs at an effective electrode area of A = 1.10, which is the real area that is very similar to the geometric area of the glassy carbon support. Upon the introduction of the electrocatalytic carbon nanotubes, the value of A reached 1.86, indicating that some nanotubes were directly connected to the electrode support. Finally, the SWCNT@SiO$_2$-PEDOT-PSS composite electrode presented a value of A = 2.39, which is indicative of a proper connection of some remaining nanotubes dispersed in the silica with the GC support.

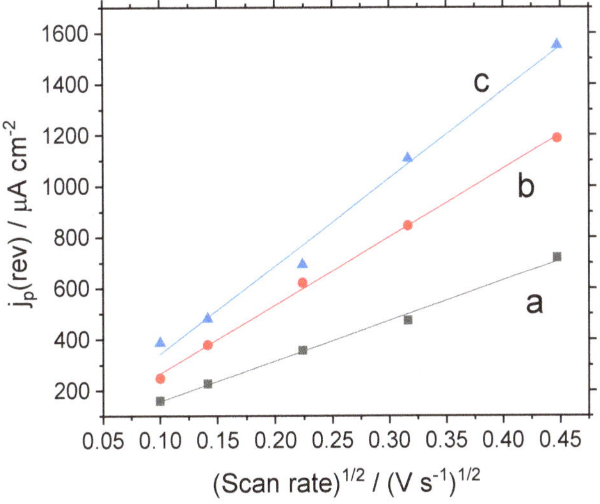

Figure 5. Randles–Sevcik plot for ferrocene oxidation with different GC-modified electrodes: (**a**) SiO$_2$; (**b**) SWCNT@SiO$_2$; (**c**) SWCNT@SiO$_2$-PEDOT-PSS.

3.2. Surface Characterizations by Scanning Electron Microscopy

The scanning electron micrographs of the SWCNT@SiO$_2$ electrode and modified electrode with PEDOT-PSS are shown in Figure 6.

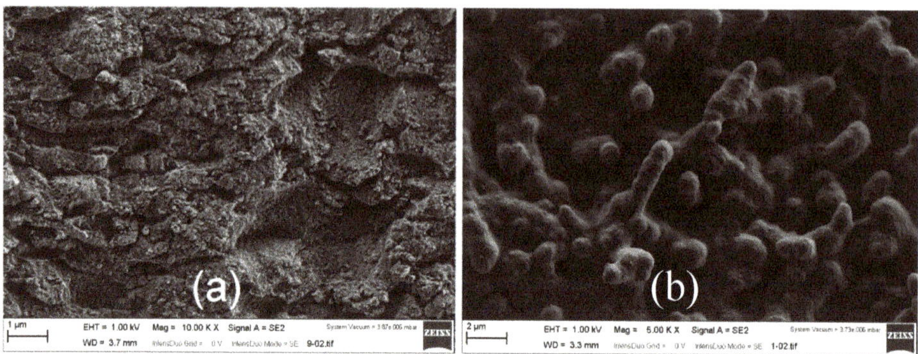

Figure 6. Scanning electron microscopy (SEM) images of GC-modified electrodes: (**a**) SWCNT@SiO$_2$; (**b**)SWCNT@SiO$_2$-PEDOT-PSS.

For the SWCNT@SiO$_2$ electrode, the electrochemically deposited layer looks homogeneous all over the surface, with randomly distributed pores showing an approximate diameter of around 2 μm. These results revealed the granular morphology and that the edges and angles of SWCNT became smooth and PEDOT-PSS was deposited onto the surfaces of the electrode. On the other hand, this resulting SWCNT@SiO$_2$-PEDOT-PSS electrode shows a rounded edge and broad (near one micron) dendritic structures, which provide the modified sample with an aspect quite different from the smoother, unmodified polymer films shown in Figure 6b. These striking architectures were formed by PEDOT-PSS emerging from the silica material and their shape is a consequence of the patterned growth of the PEDOT forced by the structure of silica.

4. Conclusions

The electrochemical behavior of the composite electrodes was tested against the ferrocene redox probe. The SWCNT@SiO$_2$ electrodes contain electrocatalytic nanotubes dispersed within its structure, as demonstrated by the improvement of the electrochemical performance in terms of heterogeneous rate constant and the electroactive area. However, the modest improvement of those parameters indicated that a major part of the SWCNT remains electrically isolated from the electrode support. PEDOT-PSS films were synthesized successfully by reactive electrochemical polymerization across SWCNT@SiO$_2$-modified electrodes. The SWCNT@SiO$_2$-PEDOT-PSS composite electrodes obtained a heterogeneous rate constant more than three times higher than the electrode without conducting polymer. Similarly, the electroactive area was also enhanced to almost double of the supporting GC electrode for the SWCNT@SiO$_2$-modified electrodes. A further increase of electroactive area was observed for the SWCNT@SiO$_2$-PEDOT-PSS composite electrodes.

Author Contributions: Conceptualization, E.M. and F.M.; methodology, H.D.; software, H.D.; validation, H.D., E.M. and F.M.; formal analysis, H.D.; investigation, H.D.; resources, H.D.; data curation, H.D.; writing-original draft preparation, A.B.; writing-review and editing, E.M. and F.M.; visualization, E.M. and F.M.; supervision, A.B. and F.M.; project administration, F.M.; funding acquisition, E.M. All authors have read and agreed to the published version of the manuscript.

Funding: This research was funded by the Directorate General of Scientific Research and Technological Development (DGRSDT) (Algeria) and by the Ministerio de Ciencia, Innovación y Universidades (MAT2016-76595-R) and by the Conselleria de Educación, Investigación, Cultura y Deporte, Generalitat Valenciana (PROMETEO/2018/087).

Conflicts of Interest: The authors declare no conflict of interest.

References

1. Sanghavi, B.J.; Wolfbeis, O.S.; Hirsch, T.; Swami, N.S. Nanomaterial-based electrochemical sensing of neurological drugs and neurotransmitters. *Microchim. Acta* **2015**, *182*, 1–41. [CrossRef]
2. Walcarius, A.; Minteer, S.D.; Wang, J.; Lin, Y.; Merkoçi, A. Nanomaterials for bio-functionalized electrodes: Recent trends. *J. Mater. Chem. B* **2013**, *1*, 4878. [CrossRef]
3. Wang, J. Nanomaterial-based electrochemical biosensors. *Analyst* **2005**, *130*, 421. [CrossRef] [PubMed]
4. Britto, P.J.; Santhanam, K.S.V.; Ajayan, P.M. Carbon nanotube electrode for oxidation of dopamine. *Bioelectrochem. Bioenerg.* **1996**, *41*, 121–125. [CrossRef]
5. Sieben, J.M.; Anson-Casaos, A.; Montilla, F.; Martinez, M.T.; Morallon, E. Electrochemical behaviour of different redox probes on single wall carbon nanotube buckypaper-modified electrodes. *Electrochim. Acta* **2014**, *135*, 404–411. [CrossRef]
6. Wang, J. Carbon-nanotube based electrochemical biosensors: A review. *Electroanalysis* **2005**, *17*, 7–14. [CrossRef]
7. Salinas-Torres, D.; Huerta, F.; Montilla, F.; Morallón, E. Study on electroactive and electrocatalytic surfaces of single walled carbon nanotube-modified electrodes. *Electrochim. Acta* **2011**, *56*, 2464–2470. [CrossRef]
8. Wang, Z.J.; Etienne, M.; Poller, S.; Schuhmann, W.; Kohring, G.W.; Mamane, V.; Walcarius, A. Dehydrogenase-Based Reagentless Biosensors: Electrochemically Assisted Deposition of Sol-Gel Thin Films on Functionalized Carbon Nanotubes. *Electroanalysis* **2012**, *24*, 376–385. [CrossRef]
9. Katz, E.; Willner, I. Biomolecule-functionalized carbon nanotubes: Applications in nanobioelectronics. *ChemPhysChem* **2004**, *5*, 1084–1104. [CrossRef]
10. Jacobs, C.B.; Peairs, M.J.; Venton, B.J. Carbon nanotube based electrochemical sensors for biomolecules. *Anal. Chim. Acta* **2010**, *662*, 105–127. [CrossRef]
11. Djelad, H.; Huerta, F.; Morallón, E.; Montilla, F. Modulation of the electrocatalytic performance of PEDOT-PSS by reactive insertion into a sol-gel silica matrix. *Eur. Polym. J.* **2018**, *105*, 323–330. [CrossRef]
12. Walcarius, A. Electrochemical applications of silica-based organic-inorganic hybrid materials. *Chem. Mater.* **2001**, *13*, 3351–3372. [CrossRef]
13. Shrivastava, S.; Jadon, N.; Jain, R. Next-generation polymer nanocomposite-based electrochemical sensors and biosensors: A review. *TRAC Trends Anal. Chem.* **2016**, *82*, 55–67. [CrossRef]
14. Kaur, G.; Adhikari, R.; Cass, P.; Bown, M.; Gunatillake, P. Electrically conductive polymers and composites for biomedical applications. *RSC Adv.* **2015**, *5*, 37553–37567. [CrossRef]
15. Wang, J.; Dai, J.; Yarlagadda, T. Carbon Nanotube–Conducting-Polymer Composite Nanowires. *Langmuir* **2005**, *21*, 9–12. [CrossRef] [PubMed]
16. Gamero-Quijano, A.; Huerta, F.; Salinas-Torres, D.; Morallón, E.; Montilla, F. Enhancement of the electrochemical performance of SWCNT dispersed in a silica sol-gel matrix by reactive insertion of a conducting polymer. *Electrochim. Acta* **2014**, *135*, 114–120. [CrossRef]
17. Gamero-Quijano, A.; Huerta, F.; Salinas-Torres, D.; Morallón, E.; Montilla, F. Electrocatalytic Performance of SiO2-SWCNT Nanocomposites Prepared by Electroassisted Deposition. *Electrocatalysis* **2013**, *4*, 259–266. [CrossRef]
18. McCreery, R.L. Advanced carbon electrode materials for molecular electrochemistry. *Chem. Rev.* **2008**, *108*, 2646–2687. [CrossRef]
19. Liu, G.; Liu, J.; Böcking, T.; Eggers, P.K.; Gooding, J.J. The modification of glassy carbon and gold electrodes with aryl diazonium salt: The impact of the electrode materials on the rate of heterogeneous electron transfer. *Chem. Phys.* **2005**, *319*, 136–146. [CrossRef]
20. Smalley, J.F.; Finklea, H.O.; Chidsey, C.E.D.; Linford, M.R.; Creager, S.E.; Ferraris, J.P.; Chalfant, K.; Zawodzinsk, T.; Feldberg, S.W.; Newton, M.D. Heterogeneous electron-transfer kinetics for ruthenium and ferrocene redox moieties through alkanethiol monolayers on gold. *J. Am. Chem. Soc.* **2003**, *125*, 2004–2013. [CrossRef]
21. Ronkainen, N.J.; Halsall, H.B.; Heineman, W.R. Electrochemical Biosensors. *Chemical Society Reviews.* **2010**, *39*, 1747–1763. [CrossRef] [PubMed]

22. Yoo, E.-H.; Lee, S.-Y. Glucose Biosensors: An Overview of Use in Clinical Practice. *Sensors* **2010**, *10*, 4558–4576. [CrossRef] [PubMed]

23. Setti, L.; Fraleoni-Morgera, A.; Ballarin, B.; Filippini, A.; Frascaro, D.; Piana, C. An amperometric glucose biosensor prototype fabricated by thermal inkjet printing. *Biosens. Bioelectron.* **2005**, *20*, 2019–2026. [CrossRef] [PubMed]

24. López-Bernabeu, S.; Huerta, F.; Morallón, E.; Montilla, F. Direct Electron Transfer to Cytochrome c Induced by a Conducting Polymer. *J. Phys. Chem. C* **2017**, *121*, 15870–15879. [CrossRef]

25. Park, J.; Kim, H.K.; Son, Y. Glucose biosensor constructed from capped conducting microtubules of PEDOT. *Sens. Actuators B Chem.* **2008**, *133*, 244–250. [CrossRef]

26. Santhosh, P.; Manesh, K.M.; Uthayakumar, S.; Komathi, S.; Gopalan, A.I.; Lee, K.-P. Fabrication of enzymatic glucose biosensor based on palladium nanoparticles dispersed onto poly(3,4-ethylenedioxythiophene) nanofibers. *Bioelectrochemistry* **2009**, *75*, 61–66. [CrossRef]

27. Porcel-Valenzuela, M.; Salinas-Castillo, A.; Morallón, E.; Montilla, F. Molecularly Imprinted Silica Films Prepared by Electroassisted Deposition for the Selective Detection of Dopamine. *Sens. Actuators B Chem.* **2015**, *222*, 63–70. [CrossRef]

28. Matsuda, H.; Ayabe, Y. Zur Theorie der Randles-Sevcikschen Kathodenstrahl-Polarographie. *Z. Elektrochem.* **1955**, *59*, 494–503.

 *materials* 

Article

Silica Modified with Polyaniline as a Potential Sorbent for Matrix Solid Phase Dispersion (MSPD) and Dispersive Solid Phase Extraction (d-SPE) of Plant Samples

Ireneusz Sowa [1,*], Magdalena Wójciak-Kosior [1,*], Maciej Strzemski [1], Jan Sawicki [1], Michał Staniak [1], Sławomir Dresler [2], Wojciech Szwerc [1], Jarosław Mołdoch [1,3] and Michał Latalski [4]

[1] Department of Analytical Chemistry, Medical University of Lublin, Chodźki 4a, 20-093 Lublin, Poland; maciej.strzemski@poczta.onet.pl (M.S.); 91chem91@gmail.com (J.S.); michal_staniak@wp.pl (M.S.); wojciech.szwerc@onet.eu (W.S.); jmoldoch@iung.pulawy.pl (J.M.)

[2] Department of Plant Physiology, Institute of Biology and Biochemistry, Maria Curie-Skłodowska University, Akademicka 19, 20-033 Lublin, Poland; dresler.slawomir@gmail.com

[3] Department of Biochemistry and Crop Quality, Institute of Soil Science and Plant Cultivation, State Research Institute, ul. Czartoryskich 8, 24-100 Puławy, Poland

[4] Children's Orthopaedics Department, Medical University of Lublin, Gębali 6, 20-093 Lublin, Poland; michall1@o2.pl

[*] Correspondence: i.sowa@umlub.pl (I.S.); kosiorma@wp.pl (M.W.-K.); Tel./Fax: +48-81-5357350 (I.S. & M.W.-K.)

Received: 1 March 2018; Accepted: 19 March 2018; Published: 22 March 2018

Abstract: Polyaniline (PANI) is one of the best known conductive polymers with multiple applications. Recently, it was also used in separation techniques, mostly as a component of composites for solid-phase microextraction (SPME). In the present paper, sorbent obtained by in situ polymerization of aniline directly on silica gel particles (Si-PANI) was used for dispersive solid phase extraction (d-SPE) and matrix solid–phase extraction (MSPD). The efficiency of both techniques was evaluated with the use of high performance liquid chromatography with diode array detection (HPLC-DAD) quantitative analysis. The quality of the sorbent was verified by Raman spectroscopy and microscopy combined with automated procedure using computer image analysis. For extraction experiments, triterpenes were chosen as model compounds. The optimal conditions were as follows: protonated Si-PANI impregnated with water, 160/1 sorbent/analyte ratio, 3 min of extraction time, 4 min of desorption time and methanolic solution of ammonia for elution of analytes. The proposed procedure was successfully used for pretreatment of plant samples.

Keywords: polyaniline; Si-PANI; d-SPE; MSPD; triterpenes; sample pretreatment

---

1. Introduction

Polyaniline (PANI) is one of the best known conductive polymers with broad application in many fields such as chemistry, physics, optics, materials and biomedical science. It was applied, e.g., as a component of sensors, diodes, solar batteries, electromagnetic shields, and materials for protection against corrosion [1–5]. Due to the unique properties of PANI such as simplicity of synthesis, mechanical and chemical flexibility, resistance on pH and temperature, hydrophobicity, π-conjugated structure, polar groups, and ion exchange ability [1,2,6,7], it also proved to be useful in extraction techniques, mainly as a component of various composites, e.g., with graphene [8], montmorillonite [9], cyclodextrin [10], polyester [11], etc. Metal fibers covered with these types of materials were used for solid-phase microextraction (SPME), magnetic solid phase extraction (MSPE), headspace solid phase microextraction (HS-SPME) and magnetic dispersive solid phase extraction (MDSPE) of different target compounds from the matrix. Polyaniline and sorbents modified with PANI were also successfully applied in solid phase extraction (SPE) [12–18]. However, reports on the applications of polyaniline based materials for plant samples are scarce. For instance, Arnnok et al. [19] used polyaniline-modified zeolite in DSPE of fruit and vegetables to isolate carbamate, organophosphate, sulfonylurea, pyrethroid and neonicotinoid; Alizadeh et al. [11] adapted polyester-polyaniline fiber for SPME of volatile organic compounds (VOCs) from lemon juice, and silica modified with polyaniline (Si-PANI) was applied as an SPE adsorbent for sample clean-up before HPLC analysis of triterpenes in plant extracts [18]. SPE is one of the most common techniques for pretreatment of biological materials; however, it has some drawbacks, e.g., the relatively time-consuming procedure, the high consumption of solvent, and the risk of losses of volatile analytes [20–22]. Therefore, recently, the other techniques have gained attention of researchers dispersive solid phase extraction (d-SPE) and matrix solid–phase extraction (MSPD). Both techniques are useful for pretreatment of samples with complex matrix, due to high efficiency combined with low cost, simplicity and high speed of process [23–27].

In the present study, Si-PANI was tested as a sorbent for d-SPE and MSPD of oleanolic, ursolic and betulinic acid from *Viscum album* L. and *Ocimum basilicum* L. Polyaniline was chosen for covering of silica because the mix mode retention mechanism allows to retain various group of analytes, both charged and uncharged, cationic and anionic forms. The extraction conditions such as: form of PANI, washing solution and eluent were optimized experimentally. The physico-chemical features of the sorbent such as polyaniline form and quality the deposition of PANI film on silica were verified by Raman spectroscopy and microscopy combined with automated procedure using computer image analysis.

2. Methods and Materials

2.1. Materials and Reagents

Silica gel Lichrospher 60 Si, aniline (for analysis EMSURE), ammonium peroxydisulphate (extra pure) used for synthesis of the adsorbent, solvents and reagents: ammonia solution, hydrochloric acid, ortho-phosphoric acid, ammonium acetate, methanol, acetonitrile (gradient grade for liquid chromatography) were from Merck (Darmstadt, Germany). Water was deionized using ULTRAPURE Milipore Direct-Q® 3UV–R (Merck). Standards of betulinic (BA) ($\geq$98%), oleanolic (OA) ($\geq$97%), and ursolic (UA) ($\geq$98.5%) acid were purchased from Sigma-Aldrich (St. Louis, MO, USA).

Extracts from *Viscum album* L. and *Ocimum basilicum* L were prepared by extraction of pulverized plant material (1.00 g) with methanol (2 × 20 mL) in ultrasonic bath (2 × 15 min). The obtained extracts were concentrated to 10 mL and filtered through Millex Samplicity Filters 0.20 μm (Merck).

2.2. Methodology

2.2.1. Synthesis and Characteristic of Si-PANI Sorbent

In situ polymerization of aniline was conducted directly on silica particles with the use of ammonium peroxydisulphate as an oxidation agent at a temperature of 0–2 °C. The procedure was described in detail in our previous publications [7,28]. The deposition of polyaniline on silica particles was verified by Raman analysis using a Thermo Scientific DXR confocal Raman microscope with the Omnic 8 software (Thermo Fisher Scientific Madison, Madison, WI, USA). The parameters for analysis of PANI distribution were as follows: excitation laser wavelength 780 nm, filters 780 nm, registered wavenumber range from 200 to 2000 cm^{-1}, laser power 10 mW and exposure time to 5 s per point.

Video images of silica and Si-PANI obtained with the use of confocal Raman microscope (50× magnification) were binarized and segmented using Fiji image processing and ImageJ software analysis [29,30]. The dimensions of analyzed beads were calculated on the basis of Feret diameter [31] using GrapPad Prism 5.0 (GrapPad Software, San Diego, CA, USA).

Si-PANI was deprotonated or protonated using 0.1 M methanolic solution of ammonia and HCl, respectively and the obtained sorbents were pre-washed with methanol (5 mL) and water (5 mL) to neutral pH of leakage.

2.2.2. Extraction Experiments

Optimisation of extraction procedure was conducted using a methanolic standard solution of oleanolic (OA), betulinic (BA) and ursolic acid (UA) at concentration of 0.1 mg/mL. All the experiments were performed in triplicate at ambient temperature.

Dispersive Solid Phase Extraction (d-SPE)

OA, UA and BA solution (1 mL) was mixed with various amounts of sorbent (100, 150, 200 and 250 mg), degassed, and dynamically shaken using vortex for 1–10 min. The suspension was subsequently centrifuged at 9000 rpm for 5 min, supernatant was removed and analyzed with the use of HPLC. The amount of retained compounds was calculated as a difference between applied amount and amount found in supernatant.

The analytes were eluted by shaking with 3 mL portions of various solvents, supernatants were filtered, analyzed with the use of HPLC and % of recovery was calculated.

Finally, the optimized d-SPE conditions were applied for sample pretreatment of plant extracts.

2.2.3. Matrix Solid Phase Dispersion (MSPD)

Pulverized plant material was mixed with Si-PANI, ground to powder (ca. 5 min), packed into a 3-mL polypropylene column, and retained by two polyethylene frits. Eluent was passed through the column using Millipore vacuum pump system (Merck) at the flow rate of 1 mL/min and the eluates were analyzed by HPLC.

2.2.4. HPLC Analysis

The HPLC analysis was conducted using a VWR Hitachi Chromaster 600 chromatograph with a spectrophotometric detector (DAD) and EZChrom Elite software (Merck) on a Discovery C18 reversed-phase column (25 cm × 4.0 mm i.d., 5 μm particle size) (Supelco, Sigma-Aldrich, St. Louis, MO, USA). Mobile phase consisted of acetonitrile-water—1% phosphoric acid (90:10:0.5 $v/v/v$). Flow rate of eluent was 1 mL/min and column temperature was 10 °C [18]. Chromatograms were recorded from 200 to 400 nm. The triterpenic acids were quantified at 205 nm.

3. Results and Discussion

3.1. Characteristics of Si-PANI

The morphology of Si-PANI particles was assessed using a confocal microscope and automated procedure of computer image analysis (Figure 1). Blue color of Si-PANI particles showed that polyaniline was successfully deposited on silica (Figure 1a). Moreover, no changes of particle shape were observed and this proved that the synthesis conditions did not cause the destruction of silica. The analysis of diameters (Figure 1b) showed the slight increase of Si-PANI particle diameter (average diameter was 10.7 μm) comparing to bare silica (average diameter was 10 μm) as a result of covering the surface with polyaniline film. The intensity of the PANI signal recorded during Raman analysis proved that polyaniline was deposited more intensively inside the adsorbent grain (Figure 1c). This may be explained by the fact that silica has a porous structure and the surface of grain inner pores is larger than the outside pores.

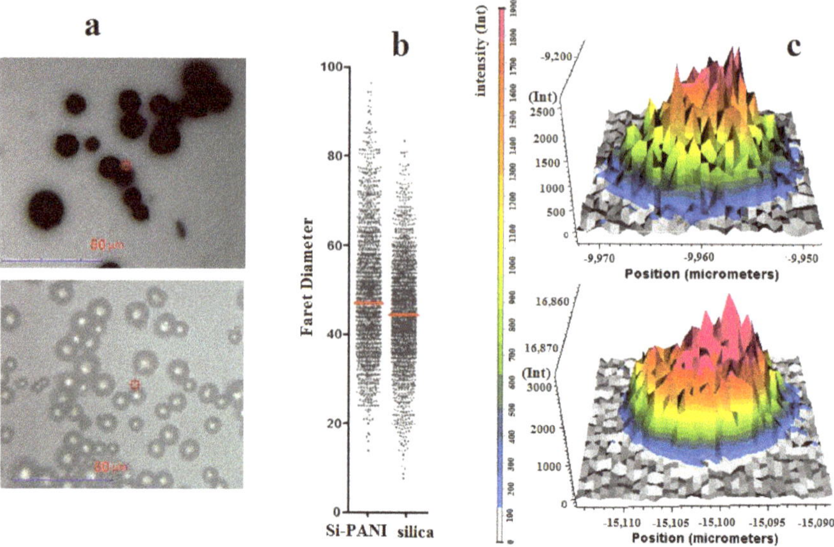

Figure 1. The morphology of Si-PANI particles: (**a**) microscope picture of silica and Si-PANI; (**b**) particle diameter distribution of silica and Si-PANI; (**c**) the examples of spatial distribution of polyaniline on the surface within the particle.

Polyaniline may occur in various forms [1]; therefore, in order to establish its form after protonation and deprotonation of Si-PANI bed, the Raman spectra were recorded (Figure 2). Both spectra matched the spectral pattern of the emeraldine [32]. Although minor shifts in some peak positions were observed, the differences between spectra were irrelevant.

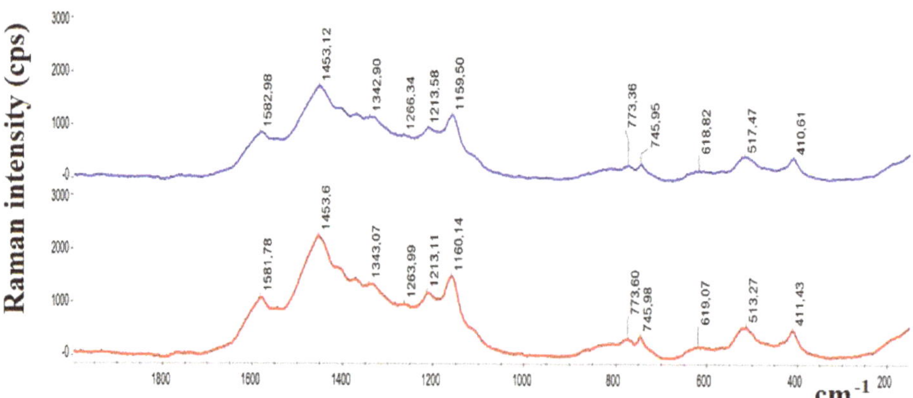

Figure 2. Smoothed Raman spectra of Si PANI sorbent: protonated (red line) and deprotonated (blue line).

3.2. Optimization of d-SPE Parameters

In order to establish the optimal conditions for d-SPE of triterpenic acids, the main parameters affecting the extraction efficiency and recovery of analytes were investigated using standard solution of UA, OA and BA.

3.2.1. Form of PANI and Impregnation Solution

The interaction of sorbent with analyte strongly affect the ability of trapping the analyte from solution. Four variants of experiment, using protonated (Si-PANI (+)) and deprotonated (Si-PANI) sorbent impregnated with methanol or water, were conducted to establish the optimal conditions to bond the highest amount of investigated compounds. The results are presented in Figure 3.

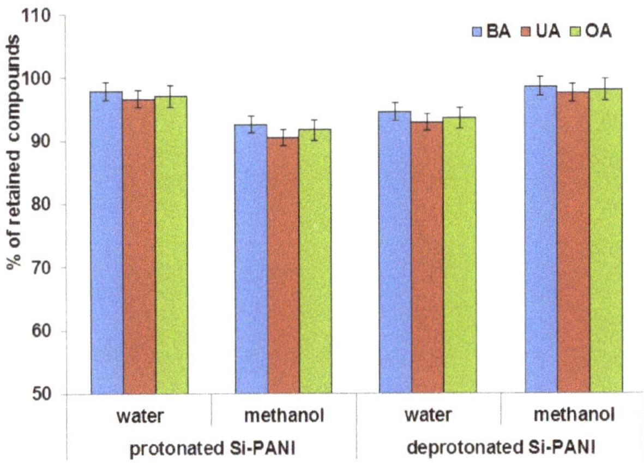

Figure 3. Percentage of retained analytes depending on Si-PANI form and impregnation solution.

As can be seen, the form of polyaniline and impregnation have an impact on efficiency of trapping the triterpenic acids from solution. Surprisingly, a high percentage of extraction efficacy was obtained both for protonated and deprotonated sorbent; however, the different impregnation solutions were required for particular form (water and methanol, respectively).

The structure of Si-PANI and Si-PANI (+) was modeled (Figure 4) to compare the charge density of the surface what could be helpful to explain the observed effect.

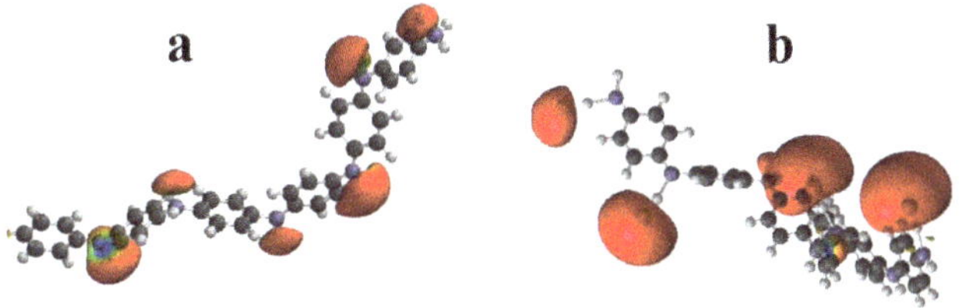

Figure 4. Modeled chain of polyaniline: (**a**) Deprotonated and (**b**) Protonated form.

Based on modeled structures, significant differences in distribution of charge on polyaniline were noted. As a result, the protonation of PANI with HCl, surface of sorbent was positively charged, the anions (Cl^-) were accumulate to compensate and the electrical double layer was formed (on Figure 4b the density of charge at Cl^- is visible). Presumably, Si-PANI (+) gained the ability to anion exchange. Impregnation with water occurred optimal for Si-PANI (+) because in water the ionization of analytes increased and then, the ability to exchange Cl^- on anionic analytes was possible (ion exchange mode of retention).

In turn, on deprotonated Si-PANI, the charges were focused on nitrogen (Figure 4a) and the retention was probably mostly caused by π-π interactions between aromatic rings of analytes and aromatic rings of sorbent which were enhanced by methanol [33].

3.2.2. Time of Extraction

The bonding of analyte in d-SPE strongly depends on time. The partition of analyte between solution and sorbent is dynamic process and appropriate time is necessary to obtain equilibrium [26,27]. Our investigation showed that the amount of bonded triterpenic acids increased up to 3 min and then remained constant (the plateau effect was observed) (Figure 5). No statistically significant differences between the investigated compounds or between both forms of sorbent were noted.

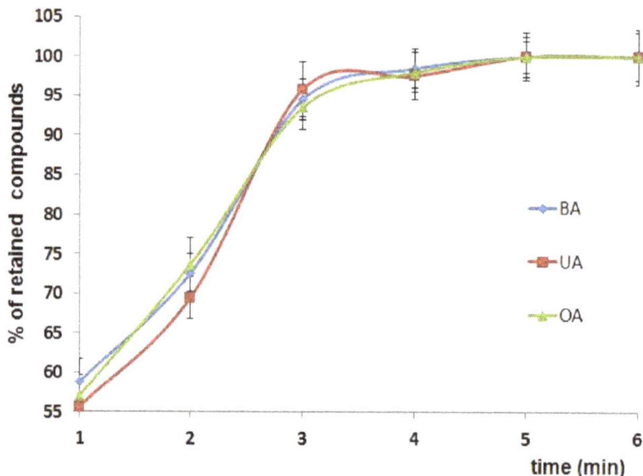

Figure 5. Effect of extraction time on percentage of retained analytes.

3.2.3. Ratio of Sorbent to Analyte

Since the number of active sites on adsorbent surface should be sufficient to trap the total amount of target compound, the sorbent/analyte ratio is a significant factor affecting the extraction efficiency. As can be seen on Figure 6 the ratio 160:1 (mg of sorbent/mg analytes) was found to be optimal for all investigated triterpenic acids. Moreover, we noticed that the curves of relationship between sorbent/analyte ratio and percentage of retained compound were similar for protonated and deprotonated Si-PANI.

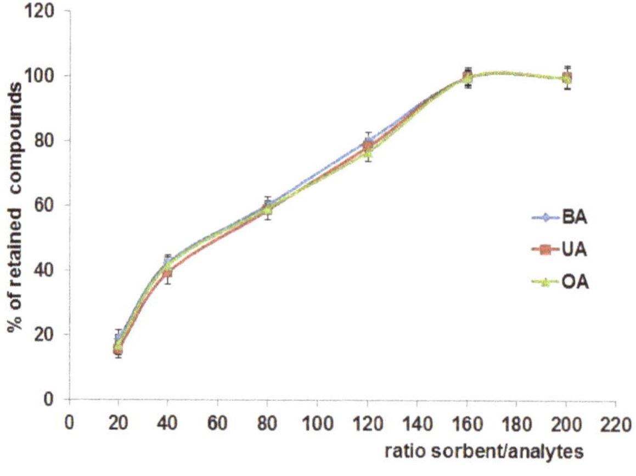

Figure 6. Effect of sorbent/analytes ratio on percentage of retained analytes.

3.2.4. Elution Solvent

The high recovery of analytes bonded with sorbent requires the selection of appropriate desorption solvent. Different factors should be taken into consideration, e.g., affinity of solvent to sorbent and analyte, its volatility and solubility of target compound.

Organic solvents with relatively low boiling point such as methanol, acetonitrile, ethanol and acetone are the most commonly used because they easily evaporate and therefore, the concentration of sample solution is possible without the risk of degradation the analyte. Sometimes, acids or buffers are added to change the ionization of investigated components or adsorbent and hereby, to decrease their interaction. As our study showed, the pure solvents had weak elution strength toward triterpenic acids bonded on Si-PANI. However, the additional low amount of ammonia, hydrochloric acid and ammonium acetate significantly increased the ability to elute of investigated triterpenes, probably as a result of ion exchange of bonded analyte on anion from mobile phase. The highest elution strength had 0.1 M methanolic solutions of investigated modifiers (Figure 7). The differences of elution from Si-PANI and Si-PANI (+) could be explain by stronger π-π interaction on deprotonated form. The higher % of recovery was obtained for protonated form of adsorbent (above 97%, 77%, and 80% for ammonia, HCl and ammonium acetate solution, respectively) comparing to deprotonated (ca. 89%, 55%, and 67% for ammonia, HCl and ammonium acetate solution, respectively).

Therefore, Si-PANI (+) was chosen for further experiments.

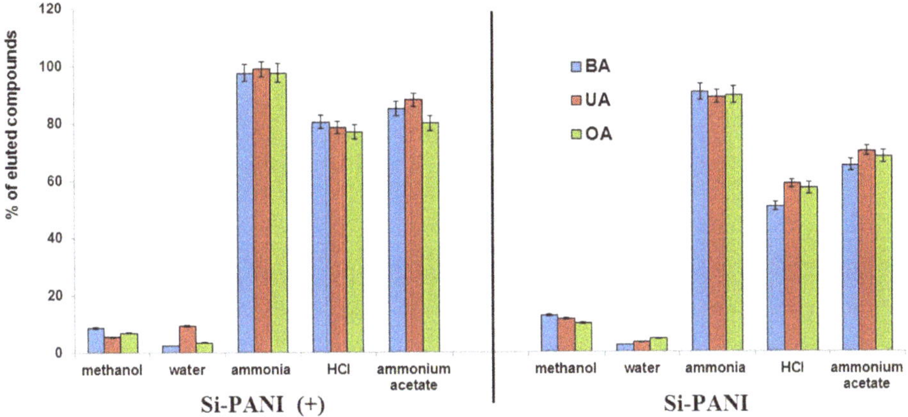

Figure 7. Effect of various solutions on the elution of retained analytes.

3.2.5. Desorption Time

Effect of time on desorption of analytes from sorbent is presented on Figure 8. Only slight differences of desorption kinetics between investigated compounds were noted. The amount of eluted analytes increased up to 4 min and then a plateau was observed.

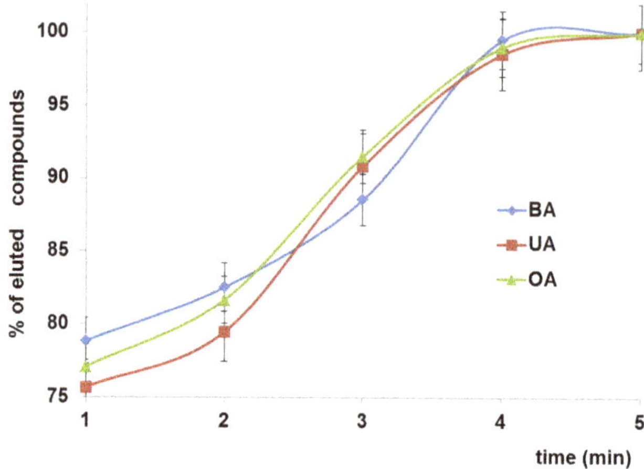

Figure 8. Effect of desorption time on percentage of eluted analytes.

3.3. Application of Si-PANI Sorbent for Pretreatment of Plant Material

Based on the conducted experiments, the optimal conditions for d-SPE of triterpenic acids were as follows: Si-PANI (+) impregnated with water, 3 min of extraction time, 2 mL of methanol water mixture (1:1, *v/v*) as washing solution, 4 min of desorption time and methanolic solution of ammonia as elution solvent. To verify the utility of the procedure, methanolic extracts from *Viscum album* L. and *Ocimum basilicum* L. were purified with the use of the above conditions. Moreover, MSPD was also conducted to assess the application potential of Si-PANI for isolation of triterpenic acids directly from raw plant material. The conditions for MSPD were established based on optimized d-SPE procedure and were as follows: Si-PANI (+) impregnated with water, 3 min of grinding time, 2 mL of methanol water mixture (1:1, *v/v*) as washing solution, and methanolic solution of ammonia as elution solvent.

The amount of triterpenic acids isolated using both techniques was determined by HPLC method. The validation parameters are summarized in Table 1 and the results of quantification are presented in Table 2.

Table 1. Validation parameters for determination of triterpenic acids ($n = 5$).

Parameters	Oleanolic Acid	Ursolic Acid	Betulinic Acid
Concentration range	0.05–1.00 mg/mL	0.005–1.00 mg/mL	0.002–0.10 mg/mL
Correlation coefficient (r)	0.9994	0.9998	0.9999
Linear regression equation	$y = 80,370,931x - 55,822$	$y = 13,2468,713x - 13,814$	$y = 101,240,680x - 28,874$
RSD values of peak area	0.64–1.32%	0.83–1.22%	0.41–0.78%
LOD (μg/mL)	0.13	0.14	0.12
LOQ (μg/mL)	0.43	0.46	0.40

Table 2. The content of investigated analytes obtained with the use of various extraction methods (mg analytes/g of dried material ± SD).

Compound	*Viscum Album* L.			*Ocimum Basilicum* L.		
	Without Purification	d-SPE	MSPD	Without Purification	d-SPE	MSPD
BA	0.82 ± 0.10	0.78 ± 0.04	0.57 ± 0.04	-	-	-
OA	6.95 ± 0.41	6.55 ± 0.30	4.81 ± 0.21	0.69 ± 0.09	0.64 ± 0.07	0.45 ± 0.03
UA	-	-	-	1.14 ± 0.11	1.04 ± 0.10	0.77 ± 0.06

As can be seen, the differences between a determined amount of investigated compounds in raw and d-SPE purified extracts were relatively low (5.1 and 5.9% for BA and OA, respectively in *Viscum album* and 7.5 and 9.2% for OA and UA, respectively in *Ocimum basilicum*). Moreover, the significant reduction of the accompanying matrix comparing to raw extracts was observed in three-dimensional (3D) HPLC chromatograms (Figure 9) and this proved the utility of the proposed method for sample pre-treatment of plant extracts.

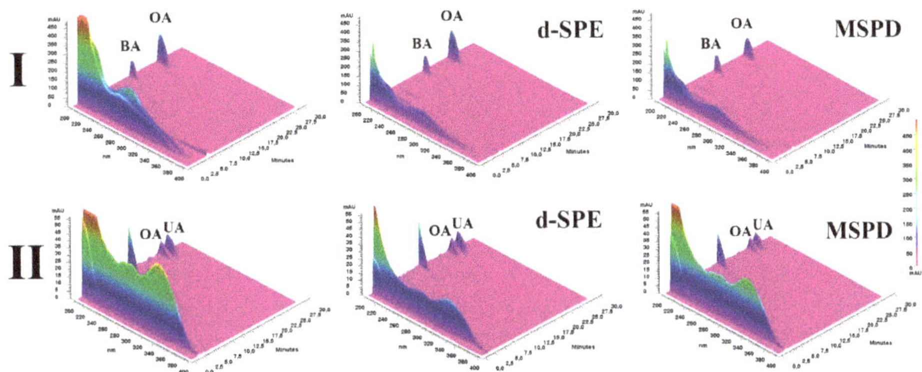

Figure 9. 3D HPLC chromatograms of investigated plant extracts obtained with the use of various extraction methods; I-*Viscum album* L. and II-*Ocimum basilicum* L.

Pre clean-up of the sample is an especially essential step before chromatographic analysis. It allows us to extend the column longevity because it prevents the clogging of inter-grain spaces and pores that may led to the reduction of an active surface of stationary phase and decreasing of chromatographic system efficacy (lower theoretical plate number, resolution and peak symmetry). It also prevents the excessive increase of pressure in the chromatographic system.

The reduction of the matrix was also observed for MSPD; however, the differences of quantified analytes in raw and MSPD extract were significant (in the range of 36–42%); thus, despite the simplicity, and relatively low time-consumption, this technique may be considered only for preliminary screening studies.

In our previous study [18], two commercially available sorbents were applied for SPE of triterpenic acids; however, they were less favorable due to weak sorption (octadecyl silica) or difficult elution (aminopropyl silica). The cost of Si-PANI is slightly higher than bare silica; however, it is lower than its other modifications and Si-PANI may be an alternative for these types of sorbents.

4. Conclusions

In the present study, Si-PANI sorbent obtained by in situ polymerization of aniline directly on silica gel particles (Si-PANI) was applied for dispersive solid phase extraction (d-SPE) and matrix solid–phase extraction (MSPD) of plant samples before HPLC analysis of triterpenic acids. The conditions of procedures were optimized experimentally. The sorbent was protonated (+) and deprotonated to change the ionization of surface and thus, to obtain the different sorption mechanism. Both on Si-PANI (+) and Si-PANI, analytes were strongly retained; however different impregnation solutions were required (water and methanol, respectively). Due to easier elution of investigated compounds, protonated sorbent was favorable. The other established optimal conditions were as follows: 3 min of extraction time, 2 mL of methanol water mixture (1:1, v/v) as washing solution, 4 min of desorption time and methanolic solution of ammonia as elution solvent. The optimized procedure was successfully used for pretreatment of *Viscum album* L. and *Ocimum basilicum* L samples using MSPD and d-SPE. The significant reduction of the matrix was observed for both techniques; however, after MSPD, the amount of determined analytes were lower than after d-SPE and this technique may be considered only for preliminary screening studies.

Author Contributions: Magdalena Wójciak-Kosior and Ireneusz Sowa conceived and designed the experiments; Maciej Strzemski, Jan Sawicki, Michał Staniak and Wojciech Szwerc performed the experiments; Magdalena Wójciak-Kosior, Ireneusz Sowa and Sławomir Dresler analyzed the data; Jarosław Mołdoch and Michał Latalski contributed reagents and materials; Magdalena Wójciak-Kosior and Ireneusz Sowa wrote the paper. All authors participated in manuscript preparation and approved the final version.

Conflicts of Interest: The authors declare no conflict of interest.

References

1. Bhadra, S.; Khastgir, D.; Singha, N.K.; Lee, J.H. Progress in preparation, processing and applications of polyaniline. *Progress Polym. Sci.* **2009**, *34*, 783–810. [CrossRef]
2. Wang, G.; Vivek, R.; Wang, J.-Y. Polyaniline nanoparticles: Synthesis, dispersion and biomedical applications. *Mini-Rev. Organ. Chem.* **2017**, *14*, 56–64. [CrossRef]
3. Liu, S.; Liu, L.; Meng, F.; Li, Y.; Wang, F. Protective performance of polyaniline-sulfosalicylic acid/epoxy coating for 5083 aluminum. *Materials* **2018**, *11*, 292. [CrossRef] [PubMed]
4. Silakhori, M.; Naghavi, M.S.; Metselaar, H.S.C.; Mahlia, T.M.I.; Fauzi, H.; Mehrali, M. Accelerated thermal cycling test of microencapsulated paraffin wax/polyaniline made by simple preparation method for solar thermal energy storage. *Materials* **2013**, *6*, 1608–1620. [CrossRef] [PubMed]
5. Zu, L.; Cui, X.; Jiang, Y.; Hu, Z.; Lian, H.; Liu, Y.; Jin, Y.; Li, Y.; Wang, X. Preparation and electrochemical characterization of mesoporous polyaniline-silica nanocomposites as an electrode material for pseudocapacitors. *Materials* **2015**, *8*, 1369–1383. [CrossRef] [PubMed]
6. Sowa, I.; Wójciak-Kosior, M.; Drączkowski, P.; Szwerc, W.; Tylus, J.; Pawlikowski, A.; Kocjan, R. Evaluation of pH and thermal stability of sorbent based on silica modified with polyaniline using high-resolution continuum source graphite furnace atomic absorption spectrometry and Raman spectroscopy. *Microchem. J.* **2015**, *118*, 88–94. [CrossRef]
7. Sowa, I.; Wójciak-Kosior, M.; Drączkowski, P.; Strzemski, M.; Kocjan, R. Synthesis and properties of a newly obtained sorbent based on silica gel coated with a polyaniline film as the stationary phase for non-suppressed ion chromatography. *Anal. Chim. Acta* **2013**, *787*, 260–266. [CrossRef] [PubMed]
8. Mehdinia, A.; Khani, H.; Mozaffari, S. Fibers coated with a graphene-polyaniline nanocomposite for the headspace solid-phase microextraction of organochlorine pesticides from seawater samples. *Microchim. Acta* **2014**, *181*, 89–95. [CrossRef]
9. Abolghasemi, M.M.; Parastari, S.; Yousefi, V. Microextraction of phenolic compounds using a fiber coated with a polyaniline-montmorillonite nanocomposite. *Microchim. Acta* **2015**, *182*, 273–280. [CrossRef]
10. Lei, Y.; He, M.; Chen, B.; Hu, B. Polyaniline/cyclodextrin composite coated stir bar sorptive extraction combined with high performance liquid chromatography-ultraviolet detection for the analysis of trace polychlorinated biphenyls in environmental waters. *Talanta* **2016**, *150*, 310–318. [CrossRef] [PubMed]
11. Alizadeh, M.; Pirsa, S.; Faraji, N. Determination of lemon juice adulteration by analysis of gas chromatography profile of volatile organic compounds extracted with nano-sized polyester-polyaniline fiber. *Food Anal. Methods* **2017**, *10*, 2092–2101. [CrossRef]
12. Bagheri, H.; Saraji, M. New polymeric sorbent for the solid-phase extraction of chlorophenols from water samples followed by gas chromatography-electron-capture detection. *J. Chromatogr. A* **2001**, *910*, 87–93. [CrossRef]
13. Bagheri, H.; Saraji, M. Conductive polymers as new media for solid-phase extraction: Isolation of chlorophenols from water sample. *J. Chromatogr. A* **2003**, *986*, 111–119. [CrossRef]
14. Bagheri, H.; Saraji, M.; Barceló, D. Evaluation of polyaniline as a sorbent for SPE of a variety of polar pesticides from water followed by CD-MEKC-DAD. *Chromatographia* **2004**, *59*, 283–289.
15. Sowa, I.; Pizoń, M.; Świeboda, R.; Kocjan, R.; Zajdel, D. Properties of chelating sorbent prepared by modification of silica gel with polyaniline and Acid Alizarin Violet N. *Sep. Sci. Technol.* **2012**, *47*, 1194–1198. [CrossRef]
16. Sowa, I.; Wójciak-Kosior, M.; Kocjan, R. The content of some trace elements in selected medicinal plants collected in the province of Lublin. *Acta Sci. Pol. Hortorum Cultus* **2012**, *11*, 15–22.

17. Sowa, I.; Wójciak-Kosior, M.; Kocjan, R. Application of SPE technique using a newly obtained sorbent based on silica gel covered with polyaniline to simultaneous determination of nitrate (III) and nitrate (V) anions in water samples. *Pol. J. Environ. Stud.* **2013**, *22*, 881–884.

18. Sowa, I.; Wójciak-Kosior, M.; Rokicka, K.; Kocjan, R.; Szymczak, G. Application of solid phase extraction with the use of silica modified with polyaniline film for pretreatment of samples from plant material before HPLC determination of triterpenic acids. *Talanta* **2014**, *122*, 51–57. [CrossRef] [PubMed]

19. Arnnok, P.; Patdhanagul, N.; Burakham, R. Dispersive solid-phase extraction using polyaniline-modified zeolite NaY as a new sorbent for multiresidue analysis of pesticides in food and environmental samples. *Talanta* **2017**, *164*, 651–661. [CrossRef] [PubMed]

20. Bladergroen, M.R.; Van Der Burgt, Y.E.M. Solid-phase extraction strategies to surmount body fluid sample complexity in high-throughput mass spectrometry-based proteomics. *J. Anal. Methods Chem.* **2015**, *2015*, 250131. [CrossRef] [PubMed]

21. Hu, K.; Qiao, Y.; Deng, Z.; Wu, M.; Liu, W. SPE-UHPLC-FLD method for the simultaneous determination of five anthraquinones in human urine using mixed-mode bis(tetraoxacalix[2]arene[2]triazine) modified silica as sorbent. *J. Anal. Methods Chem.* **2017**, *2017*, 1963908. [CrossRef] [PubMed]

22. Andrade-Eiroa, A.; Canle, M.; Leroy-Cancellieri, V.; Cerdà, V. Solid-phase extraction of organic compounds: A critical review (Part I). *Trends Anal. Chem.* **2016**, *80*, 641–654. [CrossRef]

23. Anastassiades, M.; Lehotay, S.J.; Stajnbaher, D.; Schenck, F.J. Fast and easymultiresidue method employing acetonitrile extraction/partitioning and dispersive solid-phase extraction for the determination of pesticide residues in produce. *J. AOAC Int.* **2003**, *86*, 412–431. [PubMed]

24. Fontana, A.R.; Camargo, A.; Martinez, L.D.; Altamirano, J.C. Dispersive solid-phase extraction as a simplified clean-up technique for biological sample extracts. Determination of polybrominated diphenyl ethers by gas chromatography-tandem mass spectrometry. *J. Chromatogr. A* **2011**, *1218*, 2490–2496. [CrossRef] [PubMed]

25. Wang, T.; Chen, Y.; Ma, J.; Jin, Z.; Chai, M.; Xiao, X.; Zhang, L.; Zhang, Y. A polyethyleneimine-modified attapulgite as a novel solid support in matrix solid-phase dispersion for the extraction of cadmium traces in seafood products. *Talanta* **2018**, *180*, 254–259. [CrossRef] [PubMed]

26. Gao, L.; Wei, Y. Fabrication of a novel hydrophobic/ion-exchange mixed-mode adsorbent for the dispersive solid-phase extraction of chlorophenols from environmental water samples. *J. Sep. Sci.* **2016**, *39*, 3186–3194. [CrossRef] [PubMed]

27. Pashaei, Y.; Ghorbani-Bidkorbeh, F.; Shekarchi, M. Superparamagnetic graphene oxide-based dispersive-solid phase extraction for preconcentration and determination of tamsulosin hydrochloride in human plasma by high performance liquid chromatography-ultraviolet detection. *J. Chromatogr. A* **2017**, *1499*, 21–29. [CrossRef] [PubMed]

28. Sowa, I.; Kocjan, R.; Wójciak-Kosior, M.; Świeboda, R.; Zajdel, D.; Hajnos, M. Physicochemical properties of silica gel coated with a thin layer of polyaniline (PANI) and its application in non-suppressed ion chromatography. *Talanta* **2013**, *115*, 451–456. [CrossRef] [PubMed]

29. Schindelin, J.; Arganda-Carreras, I.; Frise, E.; Kaynig, V.; Longair, M.; Pietzsch, T.; Preibisch, S.; Rueden, C.; Saalfeld, S.; Schmid, B.; et al. Fiji: An open-source platform for biological-image analysis. *Nat. Methods* **2012**, *9*, 676–682. [CrossRef] [PubMed]

30. Rasband, W.S. ImageJ. U.S. National Institutes of Health: Bethesda, Maryland, USA, 1997–2014. Available online: https://imagej.nih.gov/ij/ (accessed on 22 March 2018).

31. Merkus, H.G. *Particle Size Measurements: Fundamentals, Practice, Quality*; Springer: Dordrecht, The Netherlands, 2009; p. 15. ISBN 978-1-4020-9016-5.

32. Nascimento, G.M.; Temperini, M.L.A. Studies on the resonance Raman spectra of polyaniline obtained with near-IR excitation. *J. Raman Spectrosc.* **2008**, *39*, 772–778. [CrossRef]

33. Yang, M.; Fazio, S.; Munch, D.; Drumm, P. Impact of methanol and acetonitrile on separations based on $\pi-\pi$ interactions with a reversed-phase phenyl column. *J. Chromatogr. A* **2005**, *1097*, 124–129. [CrossRef] [PubMed]

 *materials*

Article

Basic Blue Dye Adsorption from Water Using Polyaniline/Magnetite (Fe₃O₄) Composites: Kinetic and Thermodynamic Aspects

Amir Muhammad [1], Anwar-ul-Haq Ali Shah [1,*], Salma Bilal [2,3,*] and Gul Rahman [1]

[1] Institute of Chemical Sciences, University of Peshawar, Peshawar 25120, Pakistan; amirics2015@gmail.com (A.M.); gul_rahman47@uop.edu.pk (G.R.)

[2] National Centre of Excellence in Physical Chemistry, University of Peshawar, Peshawar 25120, Pakistan

[3] TU Braunschweig Institute of Energy and Process Systems Engineering, Franz-Liszt-Straße 35, 38106 Braunschweig, Germany

[*] Correspondence: anwarulhaqalishah@uop.edu.pk (A.-u.-H.A.S.); s.bilal@tu-braunschweig.de or dresalmabilal@gmail.com (S.B.); Tel.: +92-919216652 (A.-u.-H.A.S.); +49-531-39163651 or +92-919216766 (S.B.)

Received: 4 May 2019; Accepted: 28 May 2019; Published: 30 May 2019

Abstract: Owing to its exciting physicochemical properties and doping–dedoping chemistry, polyaniline (PANI) has emerged as a potential adsorbent for removal of dyes and heavy metals from aqueous solution. Herein, we report on the synthesis of PANI composites with magnetic oxide (Fe_3O_4) for efficient removal of Basic Blue 3 (BB3) dye from aqueous solution. PANI, Fe_3O_4, and their composites were characterized with several techniques and subsequently applied for adsorption of BB3. Effect of contact time, initial concentration of dye, pH, and ionic strength on adsorption behavior were systematically investigated. The data obtained were fitted into Langmuir, Frundlich, Dubbanin-Rudiskavich (D-R), and Tempkin adsorption isotherm models for evaluation of adsorption parameters. Langmuir isotherm fits closely to the adsorption data with R^2 values of 0.9788, 0.9849, and 0.9985 for Fe_3O_4, PANI, and PANI/Fe_3O_4 composites, respectively. The maximum amount of dye adsorbed was 7.474, 47.977, and 78.13 mg/g for Fe_3O_4, PANI, and PANI/Fe_3O_4 composites, respectively. The enhanced adsorption capability of the composites is attributed to increase in surface area and pore volume of the hybrid materials. The adsorption followed pseudo second order kinetics with R^2 values of 0.873, 0.979, and 0.999 for Fe_3O_4, PANI, and PANI/Fe_3O_4 composites, respectively. The activation energy, enthalpy, Gibbs free energy changes, and entropy changes were found to be 11.14, −32.84, −04.05, and −0.095 kJ/mol for Fe_3O_4, 11.97, −62.93, −07.78, and −0.18 kJ/mol for PANI and 09.94, −74.26, −10.63, and −0.210 kJ/mol for PANI/Fe_3O_4 respectively, which indicate the spontaneous and exothermic nature of the adsorption process.

Keywords: Basic Blue 3 dye (BB3), polyaniline/Fe_3O_4 composite; Freundlich; Langmuir; Tempkin and Dubbanin-Radushkavitch adsorption isotherm

---

1. Introduction

The use of organic synthetic dyes has increased dramatically and uncontrollably in last few decades. Different types of dyes are frequently employed in plastics, paper, cosmetics, leather, and textile industries for coloring purposes [1–3]. These dyes are released in water as effluents, which are of low biological oxygen demand (BOD) and high chemical oxygen demand (COD) [4]. Some of these dyes, such as azo-dyes, are toxic and carcinogenic in nature. Their addition into nearby streams and rivers contaminates water and greatly upsets the biological activities of aquatic life [5,6]. It is highly desirable to explore efficient technologies for remediation and separation of these potential pollutants from effluents.

Various protocols and techniques, such as reverse osmosis, precipitation, coagulation, membrane filtration, chemical oxidation, electrochemical methods, ion exchange, and adsorption are used to remove these dyes and other hazardous materials from polluted water [7,8]. However, adsorption is the most frequently used technique to remove dyes from water, because this technique, in addition to easiness and low cost, causes low generation of residues and the adsorbent used may be regenerated and reused [9–11]. Several adsorbents, such as rice husk, sawdust, activated carbon, orange peel, and chitosan, have been used to remove dyes from aqueous environment [12–15]. However, the major drawback of the use of these materials is that they must be activated either physically or chemically before use. Physically these adsorbents are usually activated at very high temperature, which needs high energy. After removal of dyes, desorption must be carried out to regenerate the adsorbent, which is sometimes complicated and mostly generates secondary pollutants [16], while if thrown without treatment, they will cause water pollution. These complications make the use of these materials very expensive and time consuming, and threatening to the environment. Although activated carbon has been known as the most efficient adsorbent owing to its high specific surface area, its use is also restricted due to the non-selectivity and regeneration issues. Therefore, there is a need for the development of an environment-friendly material that is easy to regenerate [17].

In recent years, some conducting polymers, such as polyaniline, polythiophene, polypyrrole, and their composites with other materials have attracted much interest because of their conducting behavior and fascinating physicochemical properties. Such materials have been successfully applied in solar cells, fuel cells, sensors, super-capacitors, and for corrosion protection in organic coating [18–20]. Polypyrrole/TiO_2, polypyrrole/graphene oxide/Fe_3O_4, and polyaniline/magnetite have also been applied as adsorbents to remove dyes and heavy metals from aqueous environments [21–23]. Polyaniline, which exists in various oxidation states, is environmentally stable and a good conducting material with excellent electrochemical properties and can be easily prepared with less cost [24–26]. PANI and its composites with other materials, such as TiO_2, MnO_2, Fe_2O_3, SeO_2, SiO_2, Ag, Cd, and Zn, have been applied in sensors, biosensors, rechargeable batteries, fuel cells, and solar cells [27–31]. Some of these composites have also been used as adsorbents to remove heavy metals and dyes from aqueous environments [32,33]. Janaki et al. [34] removed Coomassie brilliant blue, congo red, and methylene blue from aqueous solution using polyanilline/chitosan composites. Sultana et al. [35] synthesized copper ferrite nanoparticles doped polyanilline for removal of direct yellow-27 from aqueous solution. Ayad and Al-Naser [1] applied polyanilline nanotube base as an adsorbent to remove methylene blue from an aqueous environment.

Magnetic materials such as Fe_3O_4 have attracted special attraction from scientists because of their numerous applications, such as in drug delivery systems [36], magnetic resonance imaging (MRI) [37,38], efficient hyperthermia for removal of cancer [39], clinical diagnosis [40], and removal of heavy metals form aqueous solution [41,42]. Fe_3O_4 can be prepared by a number of methods, including hydrothermal method [43], chemical co-precipitation method [44], sol-gel [45], gas phase [46], liquid phase [47], and micro emulsion methods [48]. Polyaniline/magnetite(Fe_3O_4) composites have the advantage of being stable at high temperatures and can be synthesized easily from low cost materials, which make them superior over the other existing natural/synthetic and biodegradable polymers for the adsorption of dyes. They can be regenerated easily after adsorption and due to their conductive nature, electrochemical study of these materials after adsorption can be carried out. Several reports are available on the use of PANI/iron-oxide-based materials as adsorbents for dyes; a comparison of adsorption properties of these materials with the present work is made in Table S1 of Supplementary Information.

The present study is aimed at investigating the adsorption capacity of Fe_3O_4 and PANI by synthesizing PANI/Fe_3O_4 composites for the removal of Basic blue 3 dye from aqueous solution. For comparison, PANI and Fe_3O_4 were also synthesized and tested for dye removal efficiency. Chemical co-precipitation protocol was adopted for the preparation of Fe_3O_4 in basic medium in the temperature range of 85–90 °C. PANI and PANI/Fe_3O_4 composites were synthesized by chemical oxidation method using $FeCl_3$ as an oxidant. The synthesized Fe_3O_4, PANI and composites were characterized with Fourier transforms infrared spectroscopy (FTIR), scanning electron microscopy (SEM), X-ray diffraction (XRD), energy Dispersive X-Ray spectroscopy (EDX), and surface area measurements. Batch adsorption experiments were carried out to study the effect of pH, initial concentration of dye, contact time, and temperature on the adsorption phenomenon by using UV-Visible spectrophotometer. The resulted data were fitted into Friundlich, Langmuir, Tempkin, and The Dubinin-Radushkevitch (D-R) adsorption models. Kinetics and thermodynamic aspects of the adsorption of Basic blue 3 dye on these materials were also investigated.

2. Experimental

2.1. Materials

Aniline (Across) was distilled before use under vacuum. Basic blue 3 dye, $FeCl_3 \cdot 6H_2O$ (Sigma-Aldrich, St. Louis, MO, USA), $FeSO_4 \cdot 7H_2O$ (Merck, Kenilworth, NJ, USA), Na_2SO_4 (Panreac Quimica SA, Barcelona, Spain), and Dodecyl benzene sulphonic acid, DBSA, (Across) were used as received. All chemicals used were of analytical grade.

2.2. Synthesis of PANI

PANI was synthesized via chemical oxidation method by adding 0.3 mol (0.82 mL) aniline in 30 mL double distilled water. Then, 0.02 mol (0.25 mL) Do-decylbenzene sulphonic acid (DBSA) prepared in 40 mL double distilled water was added as an emulsifying agent as well as a dopant. Afterwards, 0.01 M $FeCl_3 \cdot 6H_2O$ solution (30 mL) was added dropwise to this mixture as an oxidant. The solution was stirred on a magnetic stirrer for about 12 h. Initially, the solution was a milky white color, but after an hour the solution turned light green and then dark green after 3 hours. Finally, the product was extensively washed with acetone and double distilled water till the filtrate became clear and dried in an oven at 60 °C for 24 h.

2.3. Synthesis of Fe_3O_4

Chemical co-precipitation method was used to synthesize Fe_3O_4 by mixing $FeCl_3 \cdot 6H_2O$ and $FeSO_4 \cdot 7H_2O$ in a molar ratio of 2:0.5. DBSA was used as the emulsifying agent. The reaction was performed in basic medium (pH 10) in the temperature range of 85–90 °C. Then, 5 M ammonia solution (60 mL) was added as precipitating agent, which turned the reaction mixture black. The mixture was continuously stirred for about 2 h. Then, it was washed with plenty of distilled water and ethanol until the filtrate became clear. The resulting black precipitate was dried in an oven at 80 °C for 10 h and finally annealed in a furnace at 600 °C for 5 h [49]. The schematic representation of the process is presented in Scheme 1.

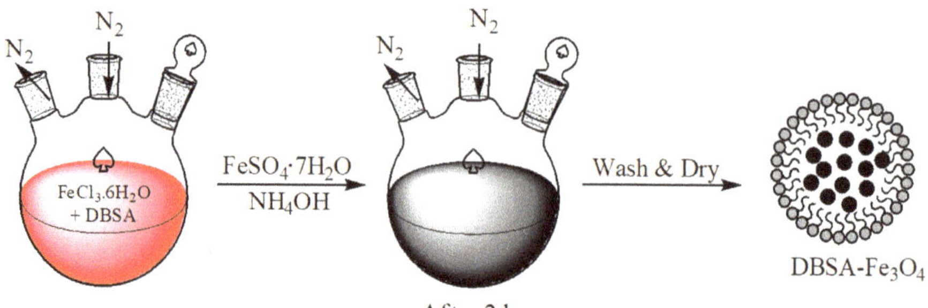

Scheme 1. Synthesis of Fe_3O_4.

2.4. Synthesis of PANI/Fe$_3$O$_4$ Composites

Chemical oxidation method was used to synthesize PANI/Fe$_3$O$_4$ composites. First, 0.2 g Fe$_3$O$_4$ was mixed with 1.818 mL of aniline suspended in double distilled water (50 mL) and DBSA (0.5 mL). The mixture was stirred for about 30 min and followed by addition of 0.15 M FeCl$_3$·6H$_2$O prepared in 40 mL double distilled water as oxidizing agent. Initially the reaction mixture was milky white due to DBSA but turned reddish brown after addition of Fe$_3$O$_4$ particles. When oxidant was added a light green color appeared within 20 min, which changed into dark black after about 2 h. After 8 h continuous stirring, the synthesized product was washed with acetone and plenty of double distilled water. Finally, the clean precipitate was dried in an oven at 60 °C for 24 h. The schematic representation of the process in provided in Scheme 2.

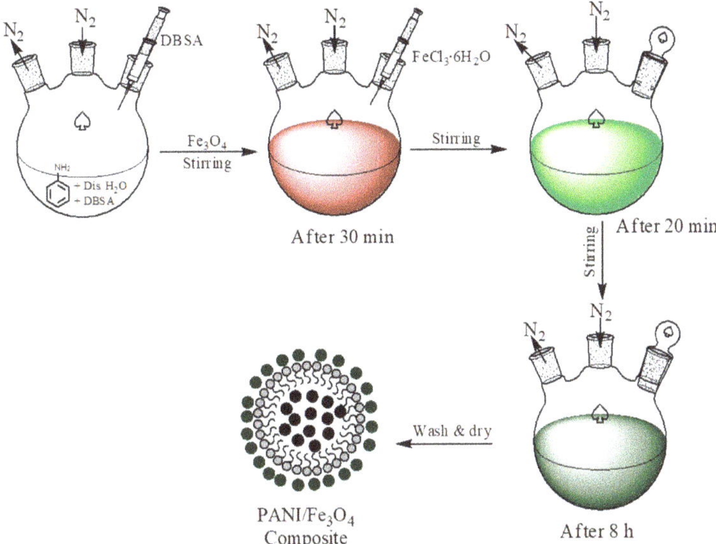

Scheme 2. Synthesis of PANI/Fe$_3$O$_4$ Composite.

2.4.1. Batch Adsorption Study for Removal of BB3 Dye

Basic blue 3 dye solution of desired concentrations ranging 0.01–110 (mg/L) were prepared in 20 mL volume by dilution method from the respective stock solution. To these solutions, Fe$_3$O$_4$ was added and shacked in a shaker at a speed of 150 rpm for 90 min. These solutions were then filtered and the concentration of dye was determined using a carry-50 UV-Visible spectrophotometer. The amount of dye adsorbed was determined by using the following equation [50].

$$q_e = \frac{(C_i - C_e)}{m} \times V \tag{1}$$

where q_e (mg/g) is the amount of dye adsorbed at equilibrium, Ci and Ce are the initial concentration and the concentration of dye present at equilibrium, respectively, m (g) is the amount of adsorbent added, and V (L) is the volume of solution. The effects of contact time, pH, initial concentration of dye, temperature, and ionic strength on the adsorption process were studied. The data obtained were used to calculate the kinetics and thermodynamic parameters. The same procedure was adopted for studying adsorption of Basic blue 3 dye on PANI and PANI/Fe$_3$O$_4$ composites.

After adsorption of BB3 dye on PANI/Fe$_3$O$_4$ composite, it was collected in filter paper with plenty of double distilled water to run out the adsorbed dye. After removal of BB3, the PANI/Fe$_3$O$_4$ composite was washed with 0.1 M HCl, to remove the remaining dye from the surface. In this way composites were regenerated and could be reused

2.4.2. Characterization

The surface morphologies of Fe_3O_4, PANI, and PANI/Fe_3O_4 composites were studied with scanning electron microscopy (SEM) using a JSM-6490 (JEOL, Tokyo, Japan) electron microscope. FTIR spectra of the Fe_3O_4, PANI, and Fe_3O_4/PANI composites were recorded with IRAffinity-1S Shimadzu Fourier Transform Infrared Spectrophotometer (Shimadzu, Tokyo, Japan) in the spectral range of 400 to 4000 cm^{-1}. X-ray diffraction (XRD) were recorded with by using Cu Kα radiations (λ = 1.5405 Å) through JEOL JDX-3532 (JEOL, Tokyo, Japan). UV-Visible spectrophotometer (Perkin Elmer, Buckinghamshire, UK) was used to find out the concentration of dye in the solution and to check the amount of dye adsorbed on the composite. Energy-dispersive X-ray (EDX) spectrophotometer model (Oxford, UK) Inca 200 was used for determination of elemental composition. BET surface areas of PANI, Fe_3O_4, and composite before and after adsorption were determined in N_2 atmosphere by adsorption–desorption method with surface area analyzer model 2200 e Quanta Chrome (Boynton Beach, FL, USA).

3. Results

3.1. Characterization

After synthesis, different techniques were used in order to know about the structural and morphological features and to get insights into the formation of composites and their adsorption properties. For comparison purposes, the same studies were carried out in parallel for Fe_3O_4 and PANI alone.

3.1.1. SEM Study

The surface morphology of Fe_3O_4, PANI, and PANI/Fe_3O_4 composites were studied with scanning electron microscopy. The SEM image (Figure 1a) shows that Fe_3O_4 consists of finite spherical shape with average particle size of 0.25 μm, which tends to form aggregates. It is somewhat porous in texture and becomes rough after adsorption of BB3 (Figure 1b). The adsorption of dye on the surface of Fe_3O_4 decreases its porosity, as reported elsewhere [51]. Shreepathi and Holze reported fibrous morphology of PANI prepared in different concentrations of DBSA [52]. The SEM image of PANI synthesized in this work shows cauliflower-like surface morphology, which after adsorption of dye changes into clusters of small ball-like structures, shown in Figure 1c,d. The SEM image of PANI/Fe_3O_4 depicts surface characteristics of both PANI and Fe_3O_4. Close observation of the composite morphology indicates adherence of Fe_3O_4 particles on the surface of PANI. The average size of composite particles was 0.28 μm. The development of magnetic micro and nanoparticle composites with PANI has been reported to greatly enhance adsorption characteristics of the hybrid materials [53–56].

3.1.2. UV-Vis Spectroscopic Study

UV-Vis spectroscopy is widely used for studying optical properties of materials. UV-Visible spectra of Fe_3O_4, PANI, and PANI/Fe_3O_4 composites were recorded in ethanol and chloroform. Figure 2A shows the UV-Vis spectra of Fe_3O_4, PANI, and PANI/Fe_3O_4 composites before adsorption of BB3. In Fe_3O_4 spectrum, the band at 441.9 is due to the surface plasmon resonance effect (SPR). The surface plasmon resonance phenomenon occurs due to interactions between incident radiations and valence electrons of the metal atom in Fe_3O_4 and causes the valence electron of the metal to oscillate with the frequency of the electromagnetic source [57]. The other band at 570.7 nm arises due to the presence of DBSA moieties in the synthesized magnetic oxide particles, as reported earlier [58].

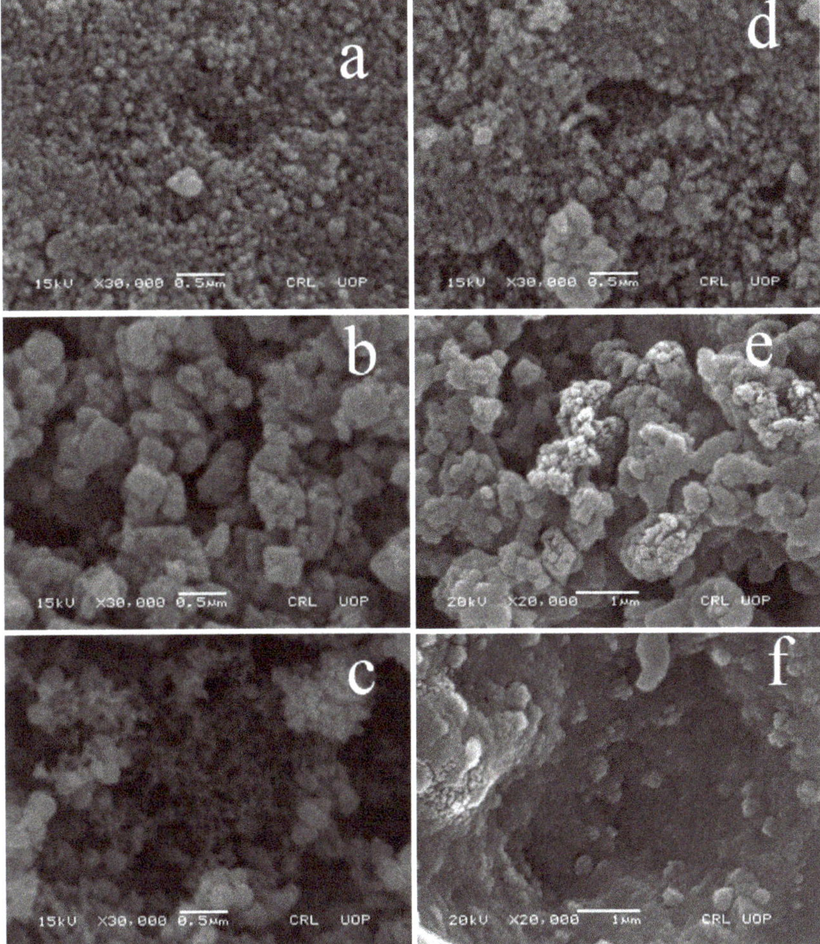

Figure 1. SEM images of Fe_2O_3, PANI, and PANI/Fe_2O_3 composites before (**a,c,e**) and after (**b,d,f**) adsorption of BB3.

In the spectrum of PANI, the band at 325–338 nm is due to π-π^* transitions of the benzenoid ring and the band at 660–680 nm is attributed to excitation of the quinoid ring [59]. The spectrum of PANI/Fe_3O_4 composites shows a small band at 441 nm due to doping of benzenoid amine with Fe_3O_4 particles, while the band at 770 nm is due to the change from polaron to bipolaron state, suggesting interactions between PANI and Fe_3O_4 materials, which is in close resemblance to the already reported results [60,61].

Figure 2B shows the UV-Vis spectra of Fe_3O_4, PANI, and PANI/Fe_3O_4 composites after adsorption of BB3, respectively. The appearance of absorption band at 647–651nm in all the spectra clearly indicates the adsorption of BB3 on Fe_3O_4, PANI, and PANI/Fe_3O_4 composites. As reported previously, BB3 gives a strong absorption band at 654 nm [62]. This absorption band is more intense in the spectrum of the composites as compared to the spectra of PANI and Fe_3O_4. The enhancement in the intensity of the absorption band of the composite around 650 nm shows strong interactions and adsorption capability of PANI/Fe_3O_4 composites towards BB3 as compared to pristine PANI and Fe_3O_4.

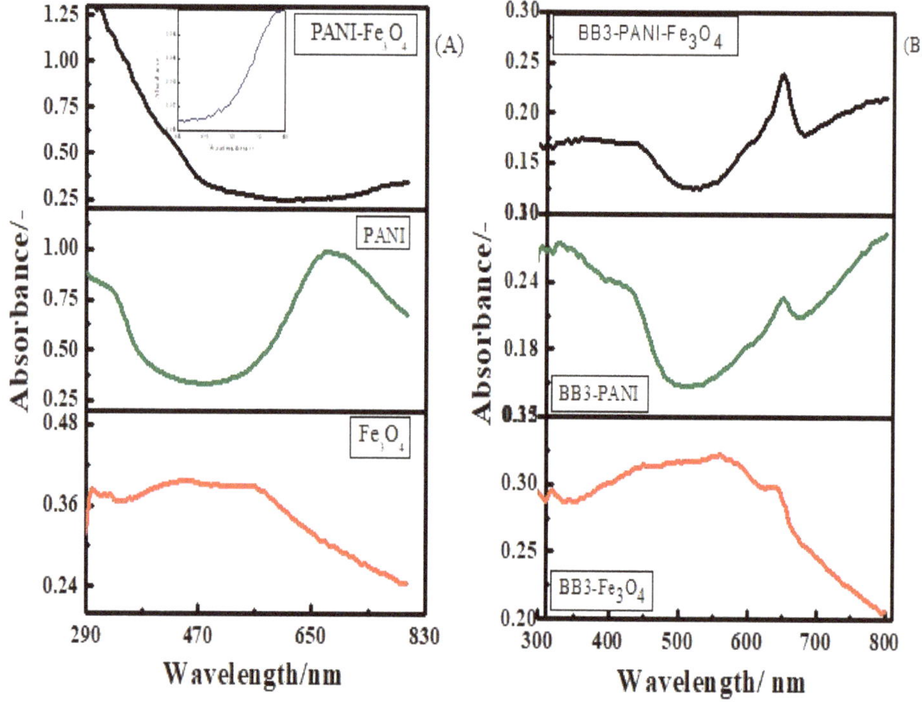

Figure 2. UV-Vis spectra of Fe_3O_4, PANI, and PANI/Fe_3O_4 composites (**A**) before and (**B**) after adsorption of BB3. The inset (A) shows the spectrum of PANI/Fe_3O_4 in the long wavelength region.

3.1.3. FTIR Spectroscopy

FTIR spectroscopy is used to study and identify organic, polymeric, and in some cases inorganic materials. Figure 3A shows FTIR spectra of Fe_3O_4, PANI, and PANI/Fe_3O_4 composites before adsorption of BB3. The details of FTIR signals associated with different types of vibrations are summarized in Table S2 of the Supplementary information.

A characteristic absorption band is observed at 554.8 cm^{-1} due to the stretching vibration of Fe–O bonds in the Fe_3O_4 spectrum. In an early study, stretching vibrations of Fe–O bonds were reported at 560 cm^{-1} [63]. This shift in the Fe–O band towards lower frequency in the present study may be due to the presence of DBSA in the Fe_3O_4 particles. Peaks at 1133.6 and 1534.6 cm^{-1} correspond to CH_2 bending modes of DBSA. Similarly, a weak peak at 3494.3 cm^{-1} is because of –OH stretching attached to the Fe_3O_4 surface and shows close resemblance to the already reported work [64]. Another weak band at 1734.7 cm^{-1} is assigned to residual NH_4OH, as already reported elsewhere [65]. The peak at 554.8 cm^{-1} is due to stretching vibrations of Fe–O disappearing and a new peak at 539.5 cm^{-1} appearing, showing BB3 dye adsorbtion onto Fe_3O_4, as shown in Figure 3B. This is because of the interaction of oxygen present in the dye structure with Fe of Fe_3O_4. The appearance of more intense peaks at 1224.6 and 1365.7 in Figure 3B is also attributed to the adsorption of BB3 [66].

FTIR spectrum of PANI is also shown in Figure 3A. Peaks at 1568 cm^{-1} and 1466 cm^{-1} are due to C=C and C=N stretching vibrations of benzoinoid and quinoid rings, respectively. Phang and Kuramoto have reported the C=C and C=N stretching vibrations of PANI at 1572 and 1497 cm^{-1}, respectively [54]. The bands at 1307.6 cm^{-1} are due to C–N•+ stretching of secondary aromatic amine of PANI doped with protic acid. The peak at 670.1 cm^{-1} shows out-of-plane bending vibrations of the C–H bond. The peak at 1017.9 cm^{-1} is assigned to –SO_3H group of DBSA bonded to nitrogen of PANI. The bands at 1133.7 and 829.2 cm^{-1} are assigned to the aromatic C–H bending in-plane and out-of-plane deformation of C–H. The peaks at 2844.6, 2931.6, and 3249.9 cm^{-1} are assigned to N–H stretching vibrations of secondary amines. In the early research, such peaks appeared in the range of 3000–3500 cm^{-1} [67]. The shifting towards the low frequency range in the present work may be due to the presence of DBSA. After adsorption of BB3 dye, all these peaks shift towards high frequency, with a decrease in the intensity of peaks at 2844.6 and 2931.6 cm^{-1}, as shown in the Figure 3B [50].

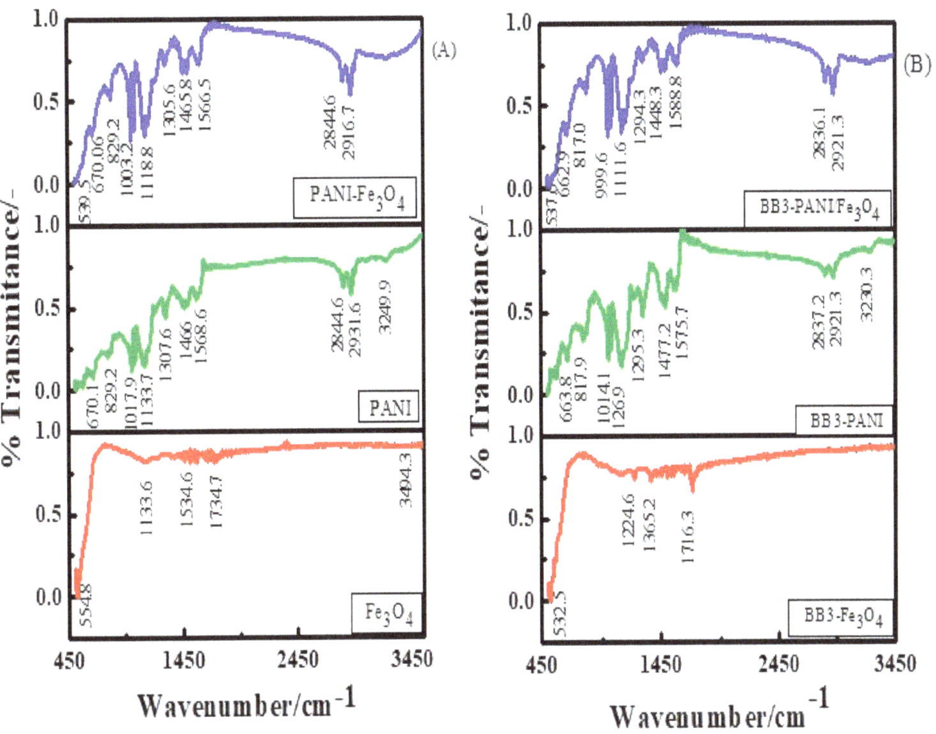

Figure 3. FTIR spectra of (**A**) Fe_3O_4, PANI, and PANI/Fe_3O_4 before and (**B**) after adsorption of BB3.

All these peaks appeared in the FTIR spectra of PANI/Fe_3O_4 composites, with a slight shift towards low frequency, as shown in Figure 3A. The shifting of absorption bands towards low frequency shows the existence of physical forces between PANI and Fe_3O_4. The band at 3249.9 cm^{-1} in the FTIR spectrum of PANI is replaced by a broad absorption plateau in the FTIR spectrum of PANI/Fe_3O_4 composites. The appearance of a very small peak at 539.5 cm^{-1}, due to Fe–O bond stretching, shows the presence of Fe_3O_4 in the composite [68]. The absorption bands in the FTIR spectrum of PANI/Fe_3O_4 shift towards low frequency after adsorption of BB3, as was also observed in the spectra of PANI and Fe_3O_4, but the peaks are more intense in the former case, as shown in Figure 3B.

3.1.4. EDX Spectroscopy

EDX study is very important to analyze elemental composition of materials. Figure 4 shows the EDX spectra of Fe_3O_4, PANI, and PANI/Fe_3O_4 composites before and after adsorption of BB3 dye, respectively. The highest percentages of Fe and O is present in Fe_3O_4, which are 53.23 and 46.77 by weight, respectively (Figure 4a). Elsewhere, Fe and O contents were reported to be 41.6 and 41.56% [69]. After BB3 adsorption, Fe content was decreased to 40.44%, while O was increased to 53.18%. The increase in oxygen and appearance of carbon in the EDX spectrum (Figure 4b) are evidence of the adsorption of dye onto Fe_3O_4.

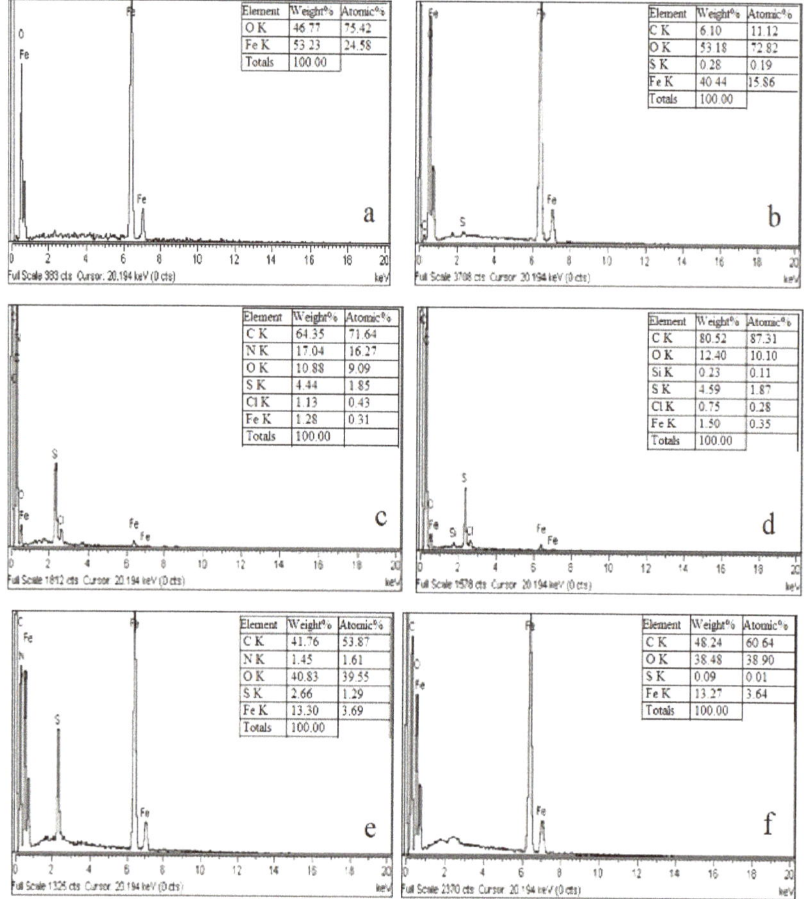

Figure 4. EDX of Fe_3O_4, PANI, and PANI/ Fe_3O_4 before (**a**,**c**,**e**) and after (**b**,**d**,**f**) adsorption of BB3.

The presence of C and N in the EDX spectra of PANI and PANI/ Fe_3O_4 composites indicates their formation. The weight percentages of C and N are 64.35 and 17.04 in PANI, respectively. Besides C and N, some other elements, such as Fe, O, S, and Cl, are also present in PANI texture. Their presence is attributed to the contribution from $FeCl_3$ and DBSA, which were used as oxidant and emulsifying agents, respectively. The increase in the percentage weights of C and O indicates adsorption of BB3 on PANI, shown in Figure 4d. Figure 4e shows the EDX spectrum of the PANI/Fe_3O_4 composite. It contains 41.76 and 1.45% C and N, respectively, in addition to 13.27% Fe, indicating formation of the PANI/Fe_3O_4 composite. Like PANI, PANI/Fe_3O_4 composites also contain 2.66% S due to DBSA. In the EDX spectrum of the composite, the contents of both C and O increase after interaction with BB3 (Figure 4f), which suggests the adsorption of BB3 on the composite [70]. These observations support the results obtained through UV-Vis and FTIR spectroscopies.

3.1.5. XRD Study

X-ray diffraction is an important technique used to determine the structure and composition of synthesized materials. Figure 5A shows XRD patterns of Fe_3O_4, PANI, and PANI/Fe_3O_4 composites before adsorption of BB3. The characteristic diffraction peaks appeared at $2\theta = 24.04°$, $33.06°$, $35.6°$, $49.3°$, $53.9°$, and $62.7°$ in the XRD spectrum of Fe_3O_4, which indicates spinel cubic crystals of Fe_3O_4.

The formation of a strong peak at 33.06° indicates the formation of Fe_3O_4. These peaks were matched with the standard cards on powder diffraction files-2 (PDF 89-598) and have close agreement [71]. After adsorption of BB3, the intensities of diffraction peaks decrease due to interactions between dye and Fe_3O_4 (Figure 5B) [72].

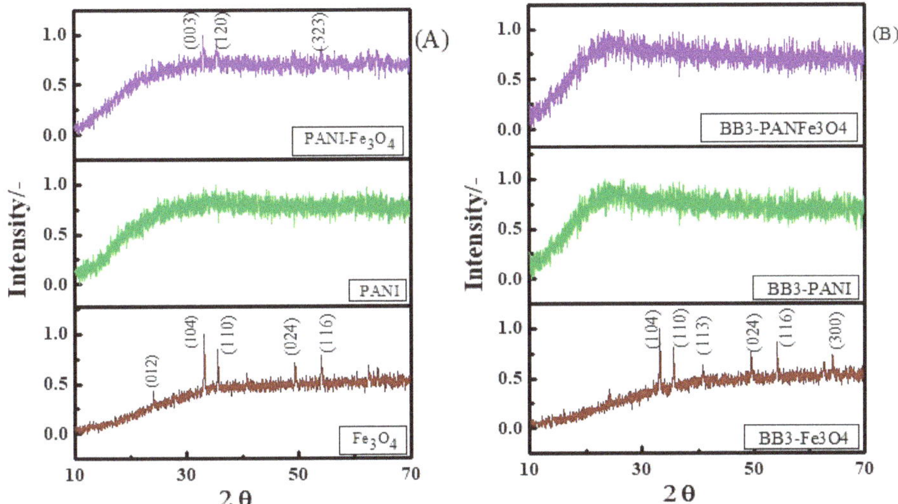

Figure 5. XRD of Fe_3O_4, PANI, and PANI/Fe_3O_4 (**A**) before and (**B**) after adsorption of BB3.

XRD spectrum (Figure 5A) of PANI shows its amorphous nature. No apparent change is observed in the spectrum of PANI after adsorption of BB3 (Figure 5B). Deshpande et al. [73] have reported a PANI film with amorphous shape. One can observe the presence of Fe_3O_4 in the PANI matrix due to diffraction peaks in the XRD spectrum of PANI/Fe_3O_4, but the intensities of these peaks are smaller than those in the spectrum of pure Fe_3O_4 particles, showing interaction between Fe_3O_4 and PANI. Obviously, the crystanality in the composites arises due to the presence of Fe_3O_4 particles. After adsorption of BB3 the peaks in the XRD spectrum of the composites simply disappeared. These observations indicate the strong overlaying layer of the dye on the surface of composites, thereby blunting the XRD peaks that were observed before adsorption of the dye [74].

3.1.6. Surface Area Analysis

Surface area analysis has a major role in the adsorption phenomenon. The surface areas of Fe_3O_4, PANI, and PANI/Fe_3O_4 composites before and after adsorption of BB3 were determined by adsorption–desorption of nitrogen gas through Brunauer–Emmett–Teller (BET) method (Figure 6) [75]. The obtained results are summarized in Table 1, which show that the surface areas of Fe_3O_4, PANI, and PANI/Fe_3O_4 composites before adsorption of BB3 are 65.818, 70.263, and 99.759 m^2/g, respectively (Figure 6A). After adsorption of BB3, the surface areas of Fe_3O_4, PANI, and PANI/Fe_3O_4 composites decreased to 46.608, 46.698, and 53.196 m^2/g, respectively (Figure 6B). The decrease in surface areas of Fe_3O_4, PANI, and PANI/Fe_3O_4 composites after adsorption of dye confirms that PANI/Fe_3O_4 composites can adsorb comparatively more dye than Fe_3O_4 and PANI. These results correlate to those obtained through SEM, XRD, EDX, and FTIR.

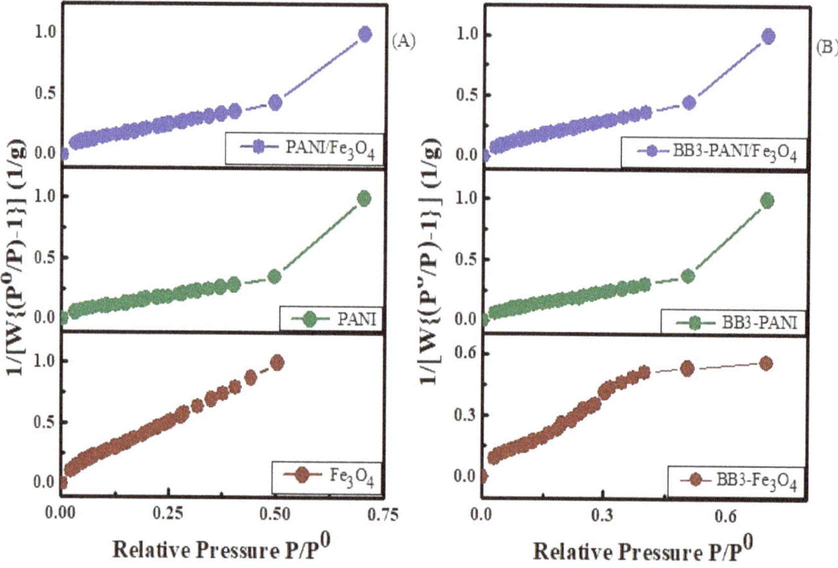

Figure 6. Surface area analysis of (**A**) Fe_3O_4, PANI, and PANI/ Fe_3O_4 before and (**B**) after adsorption of BB3.

Table 1. Surface area and Barrett, Joyner, and Halenda (BJH) para meters of Fe_3O_4, PANI, and PANI/Fe_3O_4 composites before and after adsorption of BB3 dye.

Observations	Sample	BJH Average Pore Radius (Å)	BJH Pore Volume (cc/g)	Surface Area (m²/g)
	Fe_3O_4	14.879	0.033	65.818
Before adsorption	PANI	15.500	0.021	70.263
	PANI/Fe_3O_4	14.951	0.062	99.759
	Fe_3O_4	14.864	0.023	46.608
After adsorption	PANI	14.822	0.020	46.698
	PANI/Fe_3O_4	14.944	0.046	53.196

Beside surface area, BET calculation can also be applied to determine the pore volume and average pore diameter, as shown in Table 1.

3.2. Equilibrium Study

An equilibrium study is very valuable for understanding the interaction of BB3 with Fe_3O_4, PANI, and PANI/Fe_3O_4 composites. The adsorption data are shown in Table 2, which shows that the adsorption capacity of the dye on these materials increases as the concentration of dye increases. BB3 is a cationic dye and gets adsorbed on Fe_3O_4, PANI, and PANI/Fe_3O_4 composites from aqueous solution due to interactions with negative sites on the surface of the adsorbent. In the literature it has been explained that these binding sites are present (electron pair) on oxygen of Fe_3O_4 and nitrogen of amine and imine PANI and PANI/Fe_3O_4, which are capable of interacting with oppositely charged ions present in the dye [76]. The data obtained from the equilibrium study were fitted into Freundlich, Langmuir, Tempkin, and D-R adsorption isotherms for estimation of various adsorption parameters.

Table 2. Parameters calculated from adsorption isotherm models applied for adsorption of BB3 on Fe_3O_4, PANI, and PANI/Fe_3O_4 composites.

Adsorbents	Freundlich		Langmuir		Tempkin		D-R	
Fe_3O_4	$1/n$	0.9593	q_{max}	7.474	β	−0.8096	q_s	0.888
	K_f	1.312	K_L	0.0911	K_T	8.565	E_{ads}	0.899
	R^2	0.9755	R_L	0.1210	R^2	0.5048	R^2	0.8036
	-	-	R^2	0.9788	-	-	-	-
PANI	$1/n$	0.8673	q_{max}	47.977	β	−5.626	q_s	9.183
	K_f	16.912	K_L	0.0141	K_T	22.26	E_{ads}	0.999
	R^2	0.9797	R_L	0.4710	R^2	0.6196	R^2	0.8620
	-	-	R^2	0.9849	-	-	-	-
PANI/Fe_3O_4	$1/n$	0.9112	q_{max}	78.13	β	−10.372	q_s	20.54
	K_f	44.719	K_L	0.0071	K_T	33.04	E_{ads}	0.897
	R^2	0.9911	R_L	0.6410	R^2	0.7514	R^2	0.9437
	-	-	R^2	0.9985	-	-	-	-

Freundlich adsorption equation is expressed by the following equation.

$$\ln\ q_e = \ln K_f + \frac{1}{n}\ln\ C_e \tag{2}$$

where q_e (mg/g) is the amount of dye adsorbed per gram of adsorbent, C_e (mg/L) is the concentration of dye at equilibrium, K_f is Freundlich isotherm constant, and n is the intensity of adsorbent. A plot of $\ln q_e$ vs. $\ln C_e$ is shown in Figure 7a.

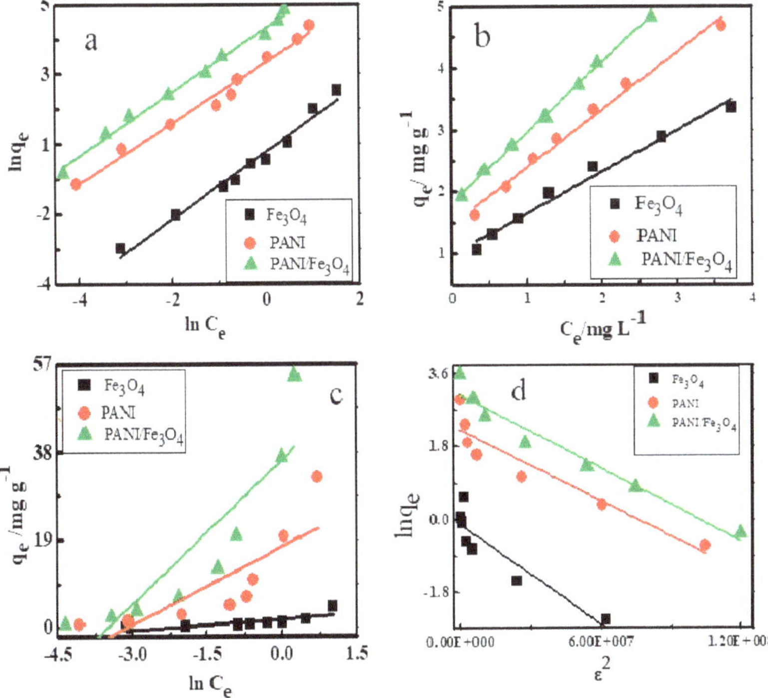

Figure 7. Adsorption of BB3 on Fe_3O_4, PANI, and PANI/Fe_3O_4. (**a**) Freundlich, (**b**) Langmuir, (**c**) Tempkin, and (**d**) D-R adsorption isotherms.

From the value of the slope obtained from the Freundlich adsorption isotherm, it can be demonstrated whether adsorption is favorable or unfavorable, reversible or irreversible. It also explains whether the system is heterogeneous or not [77]. If $1/n > 1$, adsorption is unfavorable at low concentration but favorable at high concentration; if $1/n < 1$, adsorption is favorable over the entire range of concentrations and the system is heterogeneous. However, if $1/n = 1$, then the system is homogenous [78]. The values of $1/n$ obtained from the Freundlich adsorption isotherm in the present study are 0.9593, 0.8673, and 0.9112 for adsorption of BB3 on Fe_3O_4, PANI, and PANI/Fe_3O_4 composites, respectively, as shown in Table 2. These values are in close resemblance with the literature showing that adsorption is favorable and heterogeneous. R^2 values show that the Freundlich adsorption isotherm fits to the adsorption data for Fe_3O_4, PANI, and PANI/Fe_3O_4 composites.

The adsorption data were also analyzed through the Langmuir adsorption isotherm, which is expressed in the following equation.

$$\frac{C_e}{q_e} = \frac{1}{q_{max}K_L} + \frac{1}{q_{max}}C_e \tag{3}$$

where q_{max} is the max adsorption capacity (mg/g), q_e is the amount of dye adsorbed at equilibrium (mg/g), C_e is the equilibrium adsorption concentration (mg/L), and K_L is the constant related to energy (Langmuir constant). From the Langmuir isotherm, R_L (dimensionless separating factor) is calculated by the following equation.

$$R_L = \frac{1}{(1 + K_L C_i)} \tag{4}$$

From R_L value it can be demonstrated whether adsorption is favorable, unfavorable, reversible, or irreversible. If R_L value is less than one but more than zero ($0 < R_L < 1$) adsorption is favorable, but if $1 < R_L$ adsorption is unfavorable. If $R_L = 0$ adsorption is irreversible and $R_L = 1$ indicates that adsorption is reversible [79]. The adsorption data obtained through the Langmuir isotherm are given in Table 2, which show that the maximum adsorption capacities (q_{max}) are 7.474, 47.977, and 78.13 mg/g for Fe_3O_4, PANI and PANI/Fe_3O_4 composites, respectively. The values of Langmuir constant (K_L) and dimensionless separating constant (R_L) for all the three types of adsorbents shows that adsorption of BB3 on Fe_3O_4, PANI, and PANI/Fe_3O_4 composites is monolayer and favorable. R^2 values show that the Langmuir adsorption isotherm fits more closely to the adsorption data than the other isotherms.

Tempkin adsorption isotherm, shown in the Equation (5), was also applied to explain the adsorption data.

$$q_e = \beta \ln K_T + \beta \ln C_e \tag{5}$$

R^2 values show that Tempkin isotherm does not fit very well to adsorption data as compared to Freundlich and Langmuir isotherms, but is still helpful in explaining the binding forces between adsorbents and adsorbate. K_T is the binding constant at equilibrium and corresponds to maximum binding energy [80]. Its values calculated from the intercept of plot q_e vs. $\ln C_e$ (Figure 7c) are 8.565, 22.26, and 33.04 L/g for Fe_3O_4, PANI, and PANI/Fe_3O_4 composites, respectively. These results show that there are strong binding forces between BB3 and PANI/Fe_3O_4 as compared to binding forces of dye with Fe_3O_4 and PANI, respectively. The value of constant β is related to the heat of adsorption in Equation (6)

$$\beta = \frac{RT}{b} \tag{6}$$

where b is Tempkin isotherm constant of binding energy (J/mol K). The negative sign of β values for all the three adsorbents shows that adsorption of BB3 on Fe_3O_4, PANI, and PANI/Fe_3O_4 composites is exothermic.

The Dubinin-Radushkevitch (D-R) adsorption equation has also been successfully applied to the data obtained by plotting lnq_e vs. ε^2, and is shown in Figure 7d. A linearized form of D.R adsorption equation is given below

$$lnq_e = lnq_s - B\varepsilon^2 \tag{7}$$

where q_s is the theoretical monolayer saturation capacity (mg/g), B is the constant, called D-R model constant, and ε^2 is the Polanyi potential, which is calculated by the Equation (8)

$$\varepsilon = RTlog(1 + \frac{1}{C_e}) \tag{8}$$

where R is the general gas constant and T is the absolute temperature. From the D-R model, energy of adsorption was calculated by Equation (9)

$$E_{ads} = \frac{1}{\sqrt{(1 - 2B)}} \tag{9}$$

In the literature it has been explained that for physical adsorption, the value of adsorption energy should be less than 40 kJ/mol [81]. Its value also tells about the route of adsorption through ion exchange process. In the early literature it has been explained that for ion exchange process the value of adsorption energy should be in the range of 8–16 kJ/mol. The values of q_s calculated from the linear plot of D-R isotherm are 0.888, 9.183, and 20.54 mg/g for Fe_3O_4, PANI, and PANI/Fe_3O_4 composites, respectively, showing that adsorption is physical. Similarly, values of (E_{ads}), shown in Table 2, demonstrate that adsorption does not follow ion exchange process [82]. A comparison of the adsorption efficiency of the synthesized materials with those reported earlier is also provided in Table 3.

Table 3. Comparative adsorption of BB3 on Fe_3O_4, PANI, and PANI/Fe_3O_4 with other adsorbents.

Adsorbents	pH	T (°C)	Maximum Adsorption (mg/g)	Refs.
Aleppo pine-tree sawdust	7	20	65.4	[83]
Ethylenediamine modified rice husk	4.7	25	3.29	[84]
Wood activated Charcoal	7	10–50	0.59–0.64	[85]
Quartinized sugarcane bagass	6–8	20	37.59	[86]
Activated sludge biomass	-	20	36.5	[87]
Palm fruit bunch particles	-	-	91.33	[88]
CM-60 weak acid acrylic resin	5.5	17–50	34.36–59.53	[89]
Activated Carbon	8	32	406	[90]
Fe_3O_4	12	30	7.474	Present study
PANI	8	30	47.97	Present study
PANI/Fe_3O_4	10	30	78.13	Present study

3.3. Effect of Ionic Strength

Electrostatic interactions, such as ionic strength, greatly affect the surface properties of the adsorbent [91]. The effect of ionic strength on adsorption of BB3 (dye concentration 80 mg/L in 20 mL volume) on Fe_3O_4, PANI, and PANI/Fe_3O_4 was observed by adding sodium sulphate solution in the range of 0.01–0.3 mol dm^{-3}. The obtained results (Figure 8) show that the adsorption capacities of Fe_3O_4, PANI, and PANI/Fe_3O_4 composites decrease as the concentration of salt (ionic strength) increases. The minimum dye adsorption on Fe_3O_4, PANI, and PANI/Fe_3O_4 was observed at 0.25, 0.21, and 0.25 ionic strengths, respectively. The competition of Na^+ or SO_4^{2-} ions with BB3 dye for active sites present on the surface of Fe_3O_4, PANI, and PANI/Fe_3O_4 might be a reason for the decrease in adsorption capability [71,92]. This competition is related to the interactions between hydrated ions and active sites of the adsorbent. Cations with a smaller hydrated radius occupy more active sites on the adsorbent, leading to stronger interaction with the adsorbent [93].

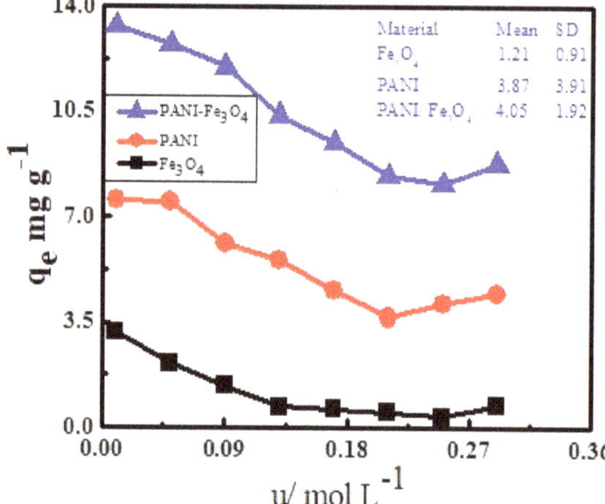

Figure 8. Effect of ionic strength on adsorption of BB3 on Fe$_3$O$_4$, PANI, and PANI/Fe$_3$O$_4$ composites.

3.4. Effect of pH

The pH of the solution plays a major role in the removal of adsorbates from aqueous solutions. Figure 9 shows the effect of pH on the adsorption of BB3 on Fe$_3$O$_4$, PANI, and PANI/Fe$_3$O$_4$. As BB3 is a cationic dye, at low pH the H$^+$ ions compete with dye for active sites present on the surface of the adsorbent and protonate them. These active sites are Fe–O and –C–N groups. Similarly, the nitrogen and oxygen in the dye are also protonated. This causes electrostatic repulsion between dye and adsorbent, hence reducing adsorption [94]. As the pH of dye solutions increases, the adsorption increases and reaches a maximum for Fe$_3$O$_4$, PANI, and PANI/Fe$_3$O$_4$ composites when the pH of the dye solution is 12, 8, and 10, respectively. At high pH de-protonation of Fe–OH and –C–N–H groups occurs, resulting in negatively charged sites, such as Fe–O$^-$ and –C–N$^-$, which have stronger interactions with dye. Figure 9 also indicates that after optimum pH, adsorption once again decreases. This may be due to hydroxylation of active sites of adsorbents [95].

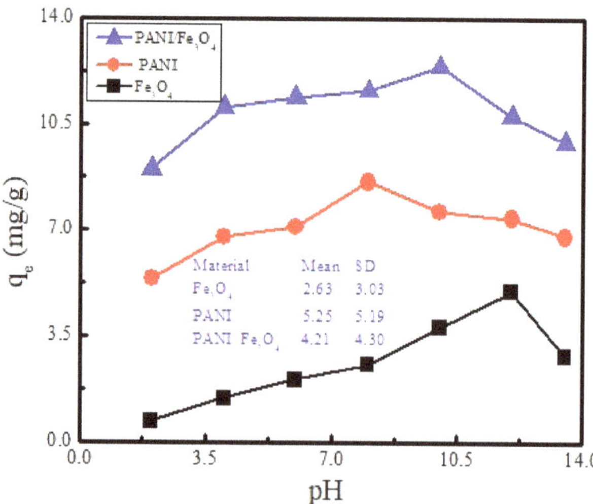

Figure 9. Effect of pH on adsorption of BB3 on Fe$_3$O$_4$, PANI, and PANI/Fe$_3$O$_4$ composite.

3.5. Effect of Contact Time and Temperature

Contact time and temperature are also important parameters for explaining the adsorption phenomenon. The adsorption of BB3 on Fe_3O_4, PANI, and $PANI/Fe_3O_4$ composites as a function of time is shown in Figure 10a, which shows that the adsorption increases with the passage of time. This figure also shows that initially adsorption is fast and contributes significantly to the equilibrium, but as the time passes, the adsorption slows down and its contribution to equilibrium decreases. This is due to filling of active sites on the surface of the adsorbent by the molecules of dye, and gradually adsorption becomes less effective. At this time, a dynamic equilibrium is established between the amount of dye adsorbed and desorbed from the adsorbent. This time is termed "equilibrium time" and the dye adsorbed at the equilibrium time is referred to as the maximum adsorption capacity of the adsorbent. It is evident from Figure 10a that the equilibrium time of adsorption is reached for Fe_3O_4, PANI, and $PANI/Fe_3O_4$ composites within 50 to 60 min [96]. Figure 10b shows that adsorption of BB3 on PANI and $PANI/Fe_3O_4$ composites is maximal at 30 °C and decreases beyond this temperature, indicating exothermic behavior.

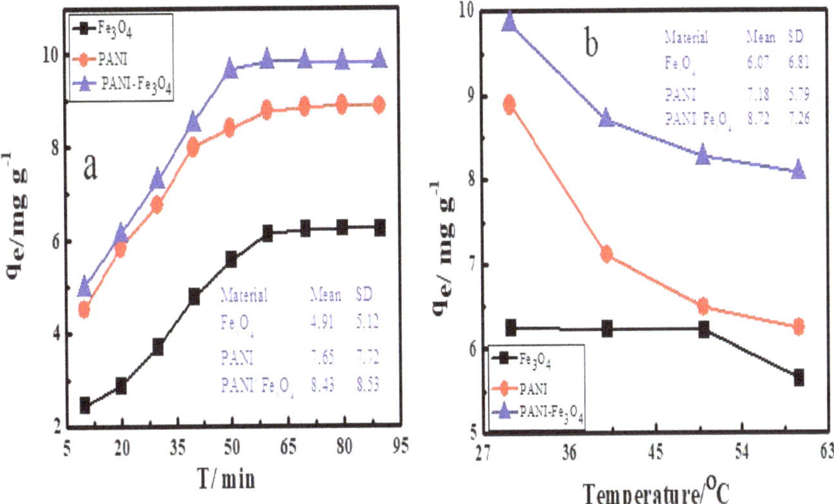

Figure 10. Adsorption of BB3 at (**a**) different time intervals and (**b**) temperature on Fe_3O_4, PANI, and $PANI/Fe_3O_4$ composites.

3.6. Effect of Adsorbent Dose

The effect of adsorbent dose on adsorption of BB3 (50 mg/L) is studied with different amounts (0.02 g, 0.06 g, and 0.1 g) of Fe_3O_4, PANI, and $PANI/Fe_3O_4$ composites, respectively. The results are shown in the Figure 11, which shows that amount of adsorption of BB3 increases as the amount of adsorbent increases. This shows that as the amount of adsorbent increases, more active sites are available for the adsorption of dye, which results in more interactions between dye and adsorbent. The figure shows that the adsorption capacity of $PANI/Fe_3O_4$ composites is more than Fe_3O_4 and PANI.

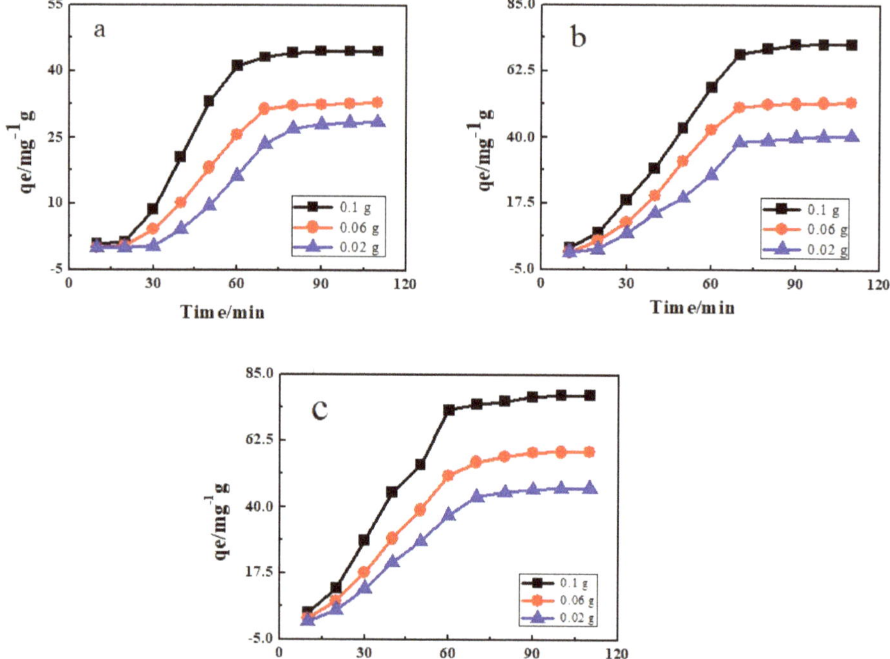

Figure 11. Effect of adsorbent dose for (**a**) Fe$_3$O$_4$, (**b**) PANI, and (**c**) PANI/ Fe$_3$O$_4$ composite on adsorption of BB3.

3.7. Kinetic Study

Kinetic study is very important to explain the adsorption phenomenon. The data obtained from adsorption of BB3 dye were analyzed through Lagergren's pseudo first order, Ho and McKay's pseudo second order, and Weber and Morris's intra particle diffusion models by using Equations (10)–(12).

$$\log(q_e - q_t) = \log q_e - \frac{K_1 t}{2.303} \tag{10}$$

$$\frac{t}{q_t} = \frac{1}{K_2 q_e^{\,2}} + \frac{t}{q_e} \tag{11}$$

$$q_t = k_d . t^{1/2} + C \tag{12}$$

where q_e and q_t are the amount of dye adsorbed (mg g^{-1}) at equilibrium and at time t, K_1, K_2, and K_d are rate constant of pseudo first order (min^{-1}), pseudo second order (g mg^{-1} min^{-1}), and intra-particle diffusion models (g mg^{-1} min$^{-1/2}$), respectively. C (mg g^{-1}) is the constant and t is the time in minutes. Figure 12a–c shows the fitted curves of pseudo first order, pseudo second order, and intra-particle diffusion models for BB3 adsorbed on Fe$_3$O$_4$, PANI, and PANI/Fe$_3$O$_4$ composites, respectively. The kinetics data obtained are shown in Table 4. The correlation factor (R^2) indicates that the pseudo second order kinetic model fits more closely to data as compared to the pseudo first order and intra-particle diffusion models. Values of rate constant indicate that as the temperature increases, rate of adsorption decreases [77,78].

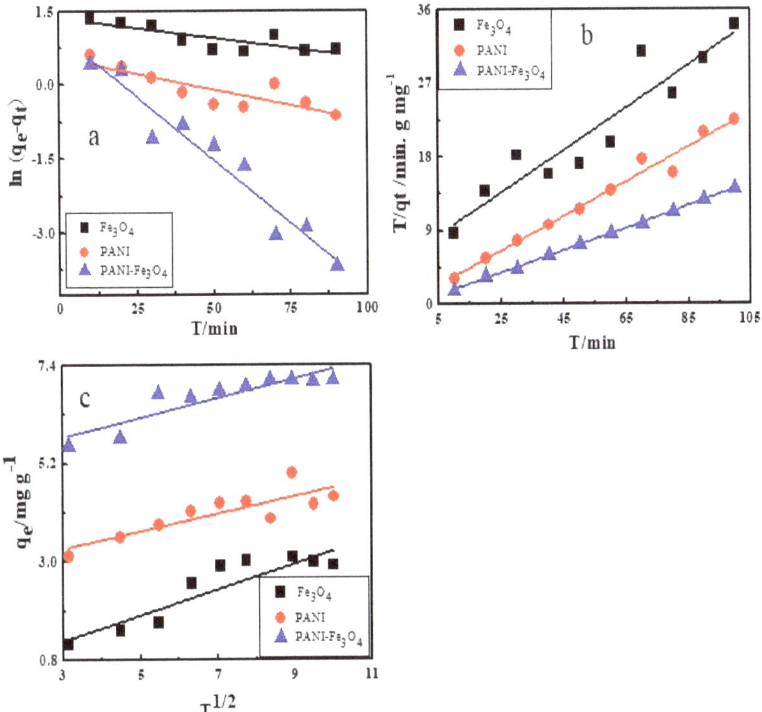

Figure 12. Kinetics models: (**a**) First order and (**b**) second order kinetics, (**c**) intra-particle diffusion model of adsorption of BB3 dye on Fe_3O_4, PANI, and PANI/Fe_3O_4 composite.

Table 4. Parameters of Kinetics models of BB3 adsorption onto Fe_3O_4, PANI, and PANI/Fe_3O_4 composite.

	Pseudo 1st Order			Pseudo 2nd Order			Intra Particle Diffusion		
Adsorbents	K_1 (min^{-1})	q_e (mg g^{-1})	R^2	K_2 (g mg^{-1} min^{-1})	q_e (mg g^{-1})	R^2	K_d (g mg^{-1} min$^{-1/2}$)	C (mg g^{-1})	R^2
Fe_3O_4	−0.0081	3.859	0.644	0.259	7.145	0.873	0.2887	0.353	0.745
PANI	−0.0128	1.733	0.688	0.211	44.50	0.979	0.1946	2.712	0.679
PANI/Fe_3O_4	−0.0513	2.869	0.932	0.136	76.71	0.999	0.2192	5.122	0.777

3.8. Mechanism of Adsorption

Actually, many factors, such as structure and charge on dye, surface of adsorbent, hydrophilic, and hydrophobic properties, electrostatic interaction, and physical forces, such as hydrogen bonding and dipole-dipole interaction, affect the adsorption of BB3 on PANI/Fe_3O_4 composites. Therefore, different mechanisms can be proposed for the adsorption of BB3 on Fe_3O_4, PANI, and PANI/ Fe_3O_4 composites. When BB3 is added to water, it dissociates in a positively-charged complex cation and negatively charged chloride ion as shown below (Scheme 3).

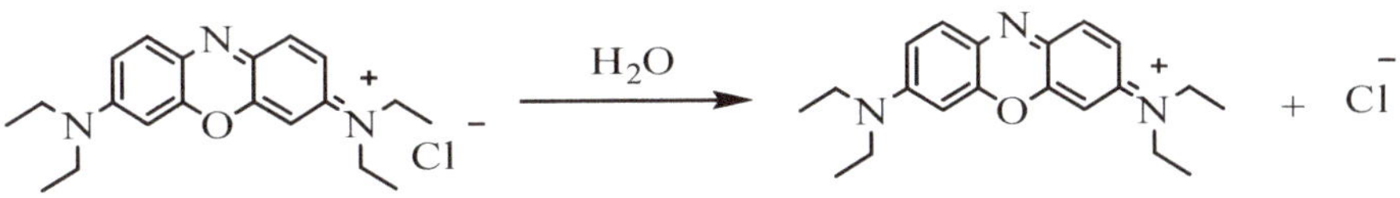

Scheme 3. Dissociation of BB3 in a positively-charged complex cation and negatively charged chloride ion into.

There is a possibility of H-bonding between amine and imine groups of PANI/Fe$_3$O$_4$ with nitrogen and oxygen present in the BB3 structure. Similarly, the surface hydroxyl groups of Fe$_3$O$_4$ may also form H-bonds with dye molecules [97].

There may exist Vander Waal's interaction between hydrophobic parts of the dye and hydrophobic parts of the PANI/Fe$_3$O$_4$ composite, because the nonpolar groups have a tendency to associate in aqueous solution. Another possibility is the existence of electrostatic interaction between positively-charged nitrogen present in the dye structure and a lone pair present on the nitrogen of amine and imine group of PANI and PANI/Fe$_3$O$_4$ [98]. The adsorption behavior of BB3 on PANI/Fe$_3$O$_4$ in basic medium is shown as the following (Scheme 4).

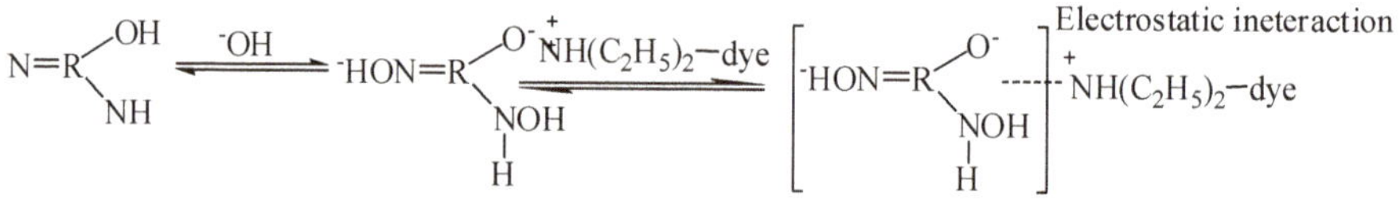

$$N{=}R\underset{OH,\ NH}{} \xrightleftharpoons{^{-}OH} {}^{-}HON{=}R\underset{NOH,\ H}{\overset{O^{-}\ \overset{+}{N}H(C_2H_5)_2{-}dye}{}} \rightleftharpoons \left[{}^{-}HON{=}R\underset{NOH,\ H}{\overset{O^{-}}{}} {-}{-}{-}{-}\ \overset{+}{N}H(C_2H_5)_2{-}dye \right]^{\text{Electrostatic interaction}}$$

Scheme 4. Adsorption behavior of BB3 on PANI/Fe$_3$O$_4$ in basic medium.

Where R represents the non-polar part of the PANI/Fe$_3$O$_4$ with = NH, $-$NH$_2$ of PANI, and $-$OH group of Fe$_3$O$_4$.

During the adsorption process the amount of energy released compensates for the entropy change of adsorbed molecules and depends upon the forces between adsorbent and adsorbate molecules; the stronger the force, the more energy will be released. The energy released during the adsorption process for H-bond is (2–40 kJ/mol), dipole-dipole interaction is (2–29 kJ/mol), Vander Waals forces is (4–10 kJ/mol), and is about 5 kJ/mol for hydrophobic forces, and more than 60 kJ/mol for electrostatic interaction [99]. In the present study the enthalpy change are −32.84, −62.93, and −74.26 kJ/mol when BB3 adsorbs on Fe$_3$O$_4$, PANI, and PANI/Fe$_3$O$_4$, respectively.

3.9. Calculation of Thermodynamic Parameters

Thermodynamic parameters, such as activation energy, Gibb's free energy change, enthalpy change, and entropy change, are helpful to explain the nature of adsorption. Activation energy is calculated by Arrhenius equation, shown below.

$$k = Aexp^{(-Ea/RT)} \tag{13}$$

where Ea is the activation energy, T is the absolute temperature, A is the Arrhenius constant, and k is the rate constant. Gibb's Free energy change is calculated by the following equation.

$$\Delta G = -RT\ln\frac{q_e}{C_e} \tag{14}$$

Enthalpy change and entropy change are calculated by Van't Hoff equation by plotting the $\ln q_e/C_e$ vs. 1/T, as given below.

$$\ln\frac{q_e}{C_e} = \frac{\Delta S}{R} - \frac{\Delta H}{RT} \tag{15}$$

where ΔH is the enthalpy change and ΔS is the change in entropy, and T is the absolute temperature. Figure 13a shows the Arrhenius plot, obtained by plotting $\ln K_2$ vs. 1/T after adsorption of BB3. From the slope the activation energy values of adsorption of BB3 were to found to be 11.14, 11.97, and 09.94 kJ/mol, respectively, which indicate that adsorption is physical and is a diffusion control process (Table 5) [100]. The value of enthalpy change is also helpful in explaining the adsorption phenomenon. It was reported that enthalpy change in the range of 84–420 kJ/mol suggests chemical interaction between dye and adsorbent (chemisorption), while its value below 84 kJ/mol indicates physical adsorption [95]. The values of enthalpy change in the present work, as shown in Table 5, are −32.84, −62.93, and −74.26 kJ/mol for the adsorption of BB3 on Fe_3O_4, PANI, and PANI/Fe_3O_4 composite, respectively, thereby confirming the physical process. The negative sign of ΔH indicates that adsorption is exothermic. The ΔG value is also helpful in explaining the adsorption phenomenon, it explains the spontaneity and non-spontaneity of adsorption. The negative sign for ΔG shown in Table 5 indicates that adsorption is exothermic and spontaneous. The ΔG values in the range of −20 to 0 kJ/mol show physiosorption, and from −400 to −80 kJ/mol show chemisorption [101,102]. The ΔG values for Fe_3O_4, PANI, and PANI/Fe_3O_4 composite used as adsorbents are −04.05, −07.78, and −10.63 kJ/mol, respectively, which suggest that adsorption of BB3 dye on all the three adsorbents is physical, exothermic, and spontaneous [103]. These observations strongly correlate with the data presented in Section 3.5 for temperature effect on the absorption phenomenon.

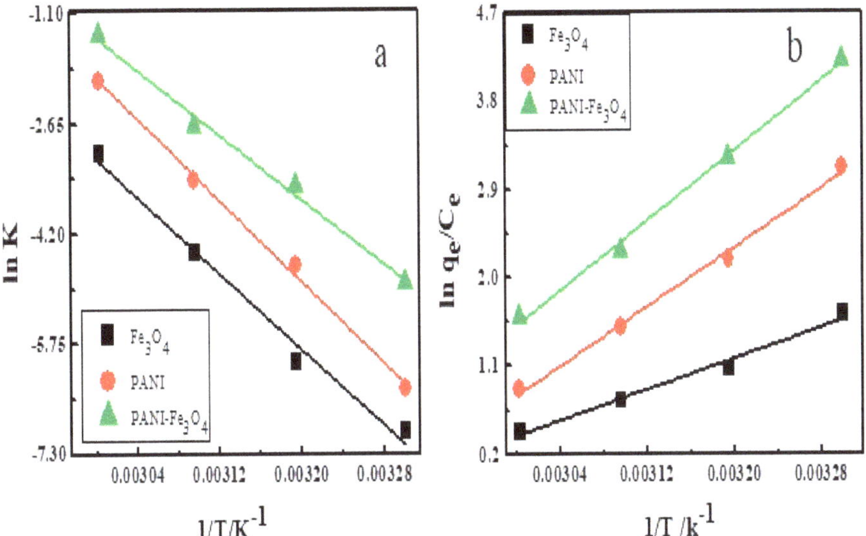

Figure 13. (a) Arrhenius plot for calculation of activation energy. (b) The van't Hoff plot for calculation of enthalpy and entropy.

Table 5. Thermodynamic parameters of adsorption of BB3 on Fe_3O_4, PANI, and PANI/Fe_3O_4 composite.

Adsorbents	(kJ/mol)	(kJ/mol)	(kJ/(mol·K))	Ea (kJ/mol)
Fe_3O_4	−04.05	−32.84	−0.095	11.14
PANI	−07.78	−62.93	−0.182	11.97
PANI/Fe_3O_4	−10.63	−74.26	−0.210	09.94

4. Conclusions

PANI/Fe_3O_4 composites, whose syntheses were confirmed through various spectroscopic techniques, such as SEM, FTIR, EDX, UV, and XRD, can effectively be utilized as adsorbents for removal of BB3 (cationic dye) from aqueous solution. It was envisaged that the synergy between PANI and magnetite would impart promising properties onto the composite material, as a high amount of dye (78.13 mg/g) was adsorbed on PANI/Fe_3O_4 composites in comparison to that adsorbed for Fe_3O_4 (7.474 mg/g) and PANI (47.977). The enhanced adsorption capability of the composites is attributed to the increase in surface area and pore volume of the hybrid materials. The adsorption followed pseudo second order kinetics, with R^2 values of 0.873, 0.979, and 0.999 for Fe_3O_4, PANI, and PANI/Fe_3O_4 composites, respectively. The activation energy, enthalpy, Gibbs free energy changes, and entropy changes were found to be 11.14, −32.84, −04.05, and −0.095 kJ/mol for Fe_3O_4, 11.97, −62.93, −07.78, and −0.18 kJ/mol for PANI, and 09.94, −74.26, −10.63, and −0.210 kJ/mol for PANI/Fe$_3$O$_4$, respectively, indicating the spontaneous and exothermic nature of the adsorption process. The Langmuir adsorption isotherm model fitted more closely to the data. The adsorption was greater in basic medium than in acidic medium. The adsorption was well-described by the pseudo second order kinetic model. Thermodynamically, adsorption is proven to be exothermic and spontaneous.

Supplementary Materials: The following are available online at http://www.mdpi.com/1996-1944/12/11/1764/s1, Table S1: Comparison of different synthesis methods for PANI/iron oxide and their use as adsorbent for removal of various dyes, Table S2: Summery of FTIR absorption bands.

Author Contributions: A.M. performed experimental work, formal analysis, and writing of main draft. A.-u.-H.A.S. supervised and contributed in writing and editing. S.B. and G.R. contributed to writing and formal analysis.

Funding: This research was funded by the Higher Education Commission Pakistan (project No. 20-1647 and 20–111/NRPU/R&D/HEC). The APC was funded by the German Research Foundation and the Open Access Publication Funds of the Technische Universität Braunschweig.

Acknowledgments: We acknowledge support from the German Research Foundation and the Open Access Publication Funds of the Technische Universität Braunschweig. S.B. acknowledges support from the Alexander von Humboldt Foundation Germany.

Conflicts of Interest: The authors declare no conflict of interest.

References

1. Ayad, M.M.; El-Nasr, A.A. Adsorption of Cationic Dye (Methylene Blue) from Water Using Polyaniline Nanotubes Base. *J. Phys. Chem. C* **2010**, *114*, 14377–14383. [CrossRef]

2. Gonga, J.L.; Wanga, B.; Zenga, G.M.; Yanga, C.P.; Niua, C.G.; Niua, Q.Y.; Zhoua, W.J.; Liangb, Y. Removal of cationic dyes from aqueous solution using magnetic multi-wall carbon nanotube nanocomposite as adsorbent. *J. Hazard. Mater.* **2009**, *164*, 1517–1522. [CrossRef]

3. Bukallah, S.B.; Rauf, M.A.; AlAli, S.S. Removal of Methylene Blue from aqueous solution by adsorption on sand. *Dyes Pigm.* **2007**, *74*, 85–87. [CrossRef]

4. Ghoreishi, S.M.; Haghighi, R. Chemical catalytic reaction and biological oxidation for treatment of non-biodegradable textile effluent. *J. Chem. Eng.* **2003**, *95*, 163–169. [CrossRef]

5. Blanco, M.; Martinez, A.; Marcaide, A.; Aranzabe, E.; Aranzabe, A. Heterogeneous Fenton Catalyst for the Efficient Removal of Azo Dyes in Water. *Am. J. Anal. Chem.* **2014**, *5*, 490–499. [CrossRef]

6. Amin, N.K. Removal of direct blue-106 dye from aqueous solution using new activated carbons developed from pomegranate peel: Adsorption equilibrium and kinetics. *J. Hazard. Mater.* **2009**, *165*, 52–62. [CrossRef] [PubMed]

7. Darvishmanesh, S.; Pethica, B.A.; Sundaresan, S. Forward osmosis using draw solutions manifesting liquid-liquid phase Separation. *Desalination* **2017**, *421*, 23–31. [CrossRef]

8. Konicki, W.; Aleksandrzak, M.; Mijowska, E. Equilibrium, kinetic and thermodynamic studies on adsorption of cationic dyes from aqueous solutions using graphene oxide. *Chem. Eng. Res. Des.* **2017**, *123*, 35–49. [CrossRef]

9. Zhao, B.; Xiao, W.; Shang, Y.; Zhu, H.; Han, R. Adsorption of light green anionic dye using cationic surfactant-modified peanut husk in batch mode. *Arab. J. Chem.* **2017**, *10*, S3595–S3602. [CrossRef]

10. Ali, A.F.; Kovo, A.S.; Adetunji, S.A. Methylene Blue and Brilliant Green Dyes Removal from Aqueous Solution Using Agricultural Wastes Activated Carbon. *J. Encaps. Adsorpt. Sci.* **2017**, *7*, 95–107. [CrossRef]

11. Bonou1, S.A.S.; Sagbo, E.; Charvillat, S.O.C.; Nissan, B.B.; Cazalbou, S. Adsorption of Textile Dyes on the Shells of Snails Achatina achatina and Lanistes varicus Acclimatized in Benin: Influence of Their Heating Treatment. *J. Environ. Prot.* **2018**, *9*, 158–174. [CrossRef]

12. Lin, L.; Zhai, S.R.; Xiao, Z.Y.; Song, Y.; Song, X.W. Dye adsorption of mesoporous activated carbons produced from NaOH-pretreated rice husks. *Bioresour. Technol.* **2013**, *136*, 437–443. [CrossRef] [PubMed]

13. Sahal, T.K.; Bhoumik, N.C.; Karmaker, S.; Ahmed, M.G.; Ichikawa, H.; Fukumori., Y. Adsorption of Methyl Orange onto Chitosan from Aqueous Solution. *J. Water Resour. Prot.* **2010**, *2*, 898–906. [CrossRef]

14. Muinde, V.M.; Onyari, J.M.; Wamalwa, B.; Wabomba, J.; Nthumbi, R.M. Adsorption of Malachite Green from Aqueous Solutions onto Rice Husks: Kinetic and Equilibrium Studies. *J. Environ. Prot.* **2017**, *8*, 215–230. [CrossRef]

15. Fernandez, M.E.; Nunell, G.V.; Bonelli, P.R.; Cukierman, A.L. Activated carbon developed from orange peels: Batch and dynamic competitive adsorption of basic dyes. *Ind. Crop. Prod.* **2014**, *62*, 437–445. [CrossRef]

16. Zhang, G.L.; Deng, H.; Sun, P. Dyes adsorption using a synthetic carboxymethyl cellulose-acrylic acid Adsorbent. *J. Environ. Sci.* **2014**, *26*, 1203–1211. [CrossRef]

17. Krishnani, K.K.; Srinives, S.; Mohapatra, B.C.; Boddu, V.M.; Hao, J.; Meng, X.; Mulchandani, A. Hexavalent chromium removal mechanism using conducting polymers. *J. Hazard. Mater.* **2013**, *252*, 99–106. [CrossRef] [PubMed]

18. Bilal, S.; Farooq, S.; Shah, H.A.A.; Holze, R. Improved solubility, conductivity, thermal stability and corrosion protection properties of poly(o-toluidine. synthesized via chemical polymerization. *Synth. Met.* **2014**, *197*, 144–153. [CrossRef]

19. Kaloni, T.P.; Giesbrecht, P.K.; Schreckenbach, G.; Freund, M.S. Polythiophene: From Fundamental Perspectives to Applications. *Chem. Mater.* **2017**, *29*, 10248–10283. [CrossRef]

20. Bai, H.; Chen, Q.; Li, C.; Lu, C.; Shi, G. Electrosynthesis of polypyrrole/sulfonated polyaniline composite films and their applications for ammonia gas sensing. *Polymer* **2007**, *48*, 4015–4020. [CrossRef]

21. Shih, Y.C.; Ke, C.Y.; Yu, C.J.; Lu, C.Y.; Tseng, W.L. Combined Tween 20-Stabilized Gold Nanoparticles and Reduced Graphite Oxide–Fe$_3$O$_4$ Nanoparticle Composites for Rapid and Efficient Removal of Mercury Species from a Complex Matrix. *ACS Appl. Mater. Interface* **2014**, *6*, 17437–17445. [CrossRef] [PubMed]

22. Li, J.; Zhang, Q.; Feng, J.; Yan, W. Synthesis of PPy-modified TiO$_2$ composite in H$_2$SO$_4$ solution and its novel adsorption characteristics for organic dyes. *Chem. Eng. J.* **2013**, *225*, 766–775. [CrossRef]

23. El-Naggar, I.M.; Zakaria, E.S.; Ali, M.; Khali, M.; El-Shahat, M.F. Removal of Cesium on Polyaniline Titanotungstate as Composite Ion Exchanger. *Adv. Chem. Eng. Sci.* **2012**, *2*, 166–179. [CrossRef]

24. Khuspe, G.D.; Navale, S.T.; Chougule, M.A.; Sen, S.; Agawane, G.L.; Kim, J.H.; Patil, V.B. Facile method of synthesis of polyaniline-SnO$_2$ hybrid nano composites: Microstructural, optical and electrical transport properties. *Synth. Met.* **2013**, *178*, 1–9. [CrossRef]

25. Vivekanandan, J.; Ponnusamy, V.; Mahudeswaraand, A.; Vijayanand, P.S. Synthesis, characterization and conductivity study of polyaniline prepared by chemical oxidative and electrochemical methods. *Arch. Appl. Sci. Res.* **2011**, *3*, 147–153.

26. Jamadade, V.S.; Dhawale, D.S.; Lokhande, C.D. Studies on electro synthesized leucoemeraldine, emeraldine and pernigraniline forms of polyaniline films and their super capacitive behavior. *Synth. Met.* **2010**, *160*, 955–960. [CrossRef]

27. Zaragoza, C.E.A.; Hernández, E.A.; Estrada, M.; Kobayashi, T. Synthesis of diphenylamine-co-aniline copolymers in emulsified systems using a reactive surfactant as the emulsifying agent and aniline monomer. *Synth. Met.* **2016**, *214*, 5–13. [CrossRef]

28. Bian, C.; Yu, A.; Wu, H. Fibriform polyaniline/nano-TiO$_2$ composite as an electrode material for aqueous redox supercapacitors. *Electrochem. Commun.* **2009**, *11*, 266–269. [CrossRef]

29. Gemeay, A.H.; El-Sharkawy, R.G.; Mansour, I.A.; Zaki, A.B. Catalytic activity of polyaniline/MnO$_2$ composites towards the oxidative decolorization of organic dyes. *Appl. Catal. B Environ.* **2008**, *80*, 106–115. [CrossRef]

30. Huang, X.; Wang, G.; Yang, M.; Guo, W.; Gao, H. Synthesis of polyaniline-modified Fe$_3$O$_4$/SiO$_2$/TiO$_2$ composite microspheres and their photocatalytic application. *Mater. Lett.* **2011**, *65*, 2887–2890. [CrossRef]

31. Jing, S.; Xing, S.; Yu, L.; Wu, Y.; Zhao, C. Synthesis and characterization of Ag/polyaniline core–shell nanocomposites based on silver nanoparticles colloid. *Mater. Lett.* **2007**, *61*, 2794–2797. [CrossRef]

32. Ji, Y.; Qin, C.; Niu, H.; Sun, L.; Jin, Z.; Bai, X. Electrochemical and electro chromic behaviors of polyaniline-graphene oxide composites on the glass substrate/Ag nano-film electrodes prepared by vertical target pulsed laser deposition. *Dyes Pigm.* **2015**, *117*, 72–82. [CrossRef]

33. Liu, P.; Huang, Y.; Zhang, X. Synthesis, characterization and excellent electromagnetic wave absorption properties of graphene@CoFe$_2$O$_4$@polyaniline nanocomposites. *Synth. Met.* **2015**, *201*, 76–81. [CrossRef]

34. Janaki, V.; Oh, B.; Shanthi, K.; Lee, K.; Ramasamy, A.K.; Kamala, S. Polyaniline/chitosan composite: An eco-friendly polymer for enhanced removal of dyes from aqueous solution. *Synth. Met.* **2012**, *162*, 974–980. [CrossRef]

35. Sultana, S.; Zain, K.M. Synthesis and characterization of copper ferrite nanoparticles doped polyaniline. *J. Alloy. Compd.* **2012**, *535*, 44–49. [CrossRef]

36. Neuberger, T.; Schöpf, B.; Hofmann, H.; Hofmann, M.; Rechenberg, B. Superparamagnetic nanoparticles for biomedical applications: Possibilities and limitations of a new drug delivery system. *J. Magn. Magn. Mater.* **2005**, *293*, 483–496. [CrossRef]

37. Khurshid, H.; Hadjipanayis, C.; Chen, H.W.; Li, H.; Mao, H.; Machaidze, R.; Tzitzios, V.; Hadjipanay, G.C. Core/shell structured iron/iron-oxide nanoparticles as excellent MRI contrast enhancement agents. *J. Magn. Magn. Mater.* **2013**, *331*, 17–20. [CrossRef]

38. Wang, G.; Chang, Y.; Wang, L.; Wei, Z.; Kang, J.; Sang, L.; Dong, X.; Chen, G.; Wang, H.; Qi, H. Preparation and characterization of PVPI-coated Fe_3O_4 nanoparticles as an MRI contrast agent. *J. Magn. Magn. Mater.* **2013**, *340*, 57–60. [CrossRef]

39. Pankhurst, Q.A.; Connolly, J.; Jones, S.K.; Dobson, J. Applications of magnetic nanoparticles in biomedicine. *J. Phys. D Appl. Phys.* **2003**, *36*, R167–R181. [CrossRef]

40. Ito, A.; Shinkai, M.; Honda, H.; Kobayashi, T. Medical application of functionalized magnetic nanoparticles. *J. Biosci. Bioeng.* **2005**, *100*, 1–11. [CrossRef]

41. Do, S.H.; Jo, Y.H.; Park, J.Y.; Hong, S.H. As^{3+} removal by $Ca–Mn–Fe_3O_4$ with and without H_2O_2: Effects ofcalcium oxide in $Ca–Mn–Fe_3O_4$. *J. Hazard. Mater.* **2014**, *280*, 322–330. [CrossRef] [PubMed]

42. Hu, J.; Irene, M.C.; Chen, G. Fast Removal and Recovery of Cr(VI) Using Surface-Modified Jacobsite ($MnFe_2O_4$) Nanoparticles. *Langmuir* **2005**, *21*, 11173–11179. [CrossRef] [PubMed]

43. Tavakoli, A.; Sohrabi, M.; Kargari, A. A review of methods for synthesis of nano structured metals with emphasis on iron compounds. *Chem. Pap.* **2007**, *61*, 151–170. [CrossRef]

44. Amer, M.A.; Meaz, T.M.; Attalah, S.S.; Ghoneim, A.I. Structural and magnetic characterization of the $Mg_{0.2-x}Sr_xMn_{0.8}Fe_2O_4$ nanoparticles. *J. Magn. Magn. Mater.* **2014**, *363*, 60–65. [CrossRef]

45. Lam, U.T.; Mammucari, R.; Suzuki, K.; Foster, N.R. Processing of iron oxide nanoparticles by supercritical fluids. *Ind. Eng. Chem. Res.* **2008**, *47*, 599–614. [CrossRef]

46. Teja, A.S.; Koh, P.Y. Synthesis, properties, and applications of magnetic iron oxide nanoparticles. *Prog. Cryst. Growth Charact. Mater.* **2009**, *55*, 22–45. [CrossRef]

47. Majewski, P.; Thierry, B. Functionalized magnetite nanoparticles-synthesis, properties, and bio-applications. *Solid State Mater. Sci.* **2007**, *32*, 203–215. [CrossRef]

48. Jia, Z.; Yujun, W.; Yangcheng, L.; Jingyu, M.; Guangsheng, L. In situ preparation of magnetic chitosan/Fe_3O_4 composite nanoparticles in tiny pools of water-in-oil microemulsion. *React. Funct. Polym.* **2006**, *66*, 1552–1558. [CrossRef]

49. Racuciu, M.; Creanga, D.E.; Airinei, A. Citric-acid coated magnetite nanoparticles for biological Applications. *Eur. Phys. J. E* **2006**, *21*, 117–121. [CrossRef] [PubMed]

50. Shena, J.; Shahida, S.; Amuraa, I.; Sarihana, A.; Tiana, M.; Emanuelsson, E.A. Enhanced adsorption of cationic and anionic dyes from aqueous solutions by polyacid doped polyaniline. *Synth. Met.* **2018**, *245*, 151–159. [CrossRef]

51. Ahmadi, S.; Chia, C.H.; Zakaria, S.; Saeedfar, K.; Asim, N. Synthesi of Fe_3O_4 nanocrystals using hydrothermal approach. *J. Magn. Magn. Mater.* **2012**, *324*, 4147–4150. [CrossRef]

52. Shreepathi, S.; Holze, R. Spectro electrochemical Investigations of Soluble Polyaniline Synthesized via New Inverse Emulsion Pathway. *Chem. Mater.* **2005**, *17*, 4078–4085. [CrossRef]

53. Samania, M.R.; Borghei, S.M.; Olad, A.; Chaichi, M. Removal of chromium from aqueous solution using polyaniline-Poly ethylene glycol composites. *Hazard. Mater.* **2010**, *184*, 248–254. [CrossRef]

54. Wai, P.B.S.; Kuramoto, N. Development and Investigation of Polyaniline Micro/nanocomposites that Possess Moderate Conductivity, Dielectric and Magnetic Properties. *Polymer* **2008**, *40*, 25–32. [CrossRef]

55. Liu, Y.; Drew, M.G.B.; Cao, F.L. A comparative study of Fe_3O_4/polyaniline composites with octahedral and microspherical inorganic kernels. *J. Mater. Sci.* **2014**, *49*, 3694–3704. [CrossRef]

56. Eskizeybek, V.; Sar, F.; Gülce, H.; Gülce, A.; Avc, A. Preparation of the new polyaniline/ZnO nano composite and its photocatalytic activity for degradation of methylene blue and malachite green dyes under UV and natural sun lights irradiations. *Appl. Catal. B* **2012**, *119*, 197–206. [CrossRef]

57. Tung, L.M.; Cong, N.X.; Huy, L.T.; Lan, N.T.; Phan, V.N.; Hoa, N.Q.; Vinh, L.K.; Thinh, N.V.; Tai, L.T.; Ngo, D.T.; Mølhave, K.; Tran Quang Huy, T.Q.; Le, A.T.; et al. Synthesis, Characterizations of Superparamagnetic Fe_3O_4–Ag Hybrid Nanoparticles and Their Application for Highly Effective Bacteria Inactivation. *J. Nanosci. Nanotechnol.* **2016**, *16*, 5902–5912. [CrossRef]

58. Bachan, N.; Asha, A.; Jeyarani, W.J.; Kumar, D.A.; Shyla, J.M. A Comparative Investigation on the Structural, Optical and Electrical Properties of SiO_2–Fe_3O_4 Core–Shell Nanostructures with Their Single Components. *Acta Metall. Sin. Engl. Lett.* **2015**, *28*, 1317–1325. [CrossRef]

59. Ullah, R.; Bowmaker, G.A.; Laslau, C.; Zujovic, Z.D.; Ali, K.; Shah, A.H.A.; Sejdic, J.T. Synthesis of polyaniline by using $CuCl_2$ as oxidizing agent. *Synth. Met.* **2014**, *198*, 203–211. [CrossRef]

60. Aphesteguy, J.C.; Jacobo, S.E. Synthesis of a soluble polyaniline ferrite composite: Magnetic and electric properties. *J. Mater. Sci.* **2007**, *42*, 7062–7068. [CrossRef]

61. Huang, J.; Li, Q.; Wang, Y.; Dong, L.; Xie, H.; Wang, J.; Xiong, C. Fluxible Nanoclusters of Fe_3O_4 Nanocrystals Embedded Polyaniline by Macromolecule-Induced Self-Assembly. *Langmuir* **2013**, *29*, 10223–10228. [CrossRef] [PubMed]

62. Khataee, A.R.; Mirzajani, O. UV/peroxydisulfate oxidation of C. I. Basic Blue 3: Modeling of key factors by artificial neural network. *Desalination* **2010**, *251*, 64–69. [CrossRef]

63. Roychowdhury, A.; Pati, S.P.; Mishra, A.K.; Kumar, S.; Das, D. Magnetically addressable fluorescent Fe_3O_4/ZnO nanocomposites: Structural, optical and magnetization studies. *J. Phys. Chem. Solids* **2013**, *74*, 811–818. [CrossRef]

64. Gholivand, M.B.; Yamini, Y.; Dayeni, M.; Seidi, S. Removal of Methylene Blue and Neutral Red from Aqueous Solutions by Surfactant-Modified Magnetic Nanoparticles as Highly Efficient Adsorbent. *Environ. Prog. Sustain. Energy* **2015**, *34*, 1683–1693. [CrossRef]

65. Khoshnevisan, K.; Barkhi, M.; Zare, D.; Davoodi, D.; Tabatabaei, M. Preparation and Characterization of CTAB-Coated Fe_3O_4 Nanoparticles. *Nano-Metal Chem.* **2012**, *42*, 644–648. [CrossRef]

66. Cao, C.; Xiao, L.; Chen, C.; Shi, X.; Cao, Q.; Gao, L. In situ preparation of magnetic Fe_3O_4/chitosan nanoparticles via a novel reduction–precipitation method and their application in adsorption of reactive azo dye. *Powder Technol.* **2014**, *260*, 90–97. [CrossRef]

67. Ayad, M.; Hefnawy, G.E.; Zaghlol, S. Facile synthesis of polyaniline nanoparticles; its adsorption behavior. *J. Chem. Eng.* **2013**, *217*, 460–465. [CrossRef]

68. Umare, S.S.; Shambharkar, B.H.; Ningthoujam, R.S. Synthesis and characterization of polyaniline–Fe_3O_4 nanocomposite: Electrical conductivity, magnetic, electrochemical studies. *Synth. Met.* **2010**, *160*, 1815–1821. [CrossRef]

69. Zhan, J.; Zhang, H.; Zhub, G. Magnetic photocatalysts of cenospheres coated with Fe_3O_4/TiO_2 core/shell nanoparticles decorated with Ag nanopartilces. *Ceram. Int.* **2014**, *40*, 8547–8559. [CrossRef]

70. Yang, Y.; Qi, S. Preparation of pyrrole with iron oxide precipitated on the surface of graphite nanosheet. *J. Magn. Magn. Mater.* **2012**, *324*, 2380–2387. [CrossRef]

71. Shariati, S.; Faraji, M.; Yamini, Y.; Rajabi, A.A. Fe_3O_4 magnetic nanoparticles modified with sodium dodecyl sulfate for removal of safranin O dye from aqueous solutions. *Desalination* **2011**, *270*, 160–165. [CrossRef]

72. Absalan, G.; Asadi, M.; Kamran, S.; Sheikhian, L.; Goltz, D.G. Removal of reactive red-120 and 4-(2-pyridylazo. resorcinol from aqueous samples by Fe_3O_4 magnetic nanoparticles using ionic liquid as modifier. *J. Hazard. Mater.* **2011**, *192*, 476–484. [CrossRef]

73. Deshpande, N.G.; Gudagea, Y.G.; Sharma, R.; Vyas, J.C.; Kim, J.B.; Lee, Y.P. Studies on tin oxide-intercalated polyaniline nanocomposite for ammonia gas sensing applications. *Sens. Actuators B* **2009**, *138*, 76–84. [CrossRef]

74. Jang, J.H.; Lim, H.B. Characterization and analytical application of surface modified magnetic nanoparticles. *Microchem. J.* **2010**, *94*, 148–158. [CrossRef]

75. Xuan, S.; Wang, F.; Lai, J.M.Y.; Sham, K.W.Y.; Xiang, J.Y.; Lee, W.S.F.; Yu, J.C.; Cheng, C.H.K.; Leung, K.C. Synthesis of Biocompatible, Mesoporous Fe_3O_4 Nano/Microspheres with Large Surface Area for Magnetic Resonance Imaging and Therapeutic Applications. *Appl. Mater. Interfaces* **2011**, *3*, 237–244. [CrossRef]

76. Etim, U.J.; Umoren, S.A.; Eduok, U.M. Coconut coir dust as a low cost adsorbent for the removal of cationic dye from aqueous solution. *J. Saudi Chem. Soc.* **2016**, *20*, S67–S76. [CrossRef]

77. Salem, M.A. The role of polyaniline salts in the removal of direct blue 78 from aqueous solution: A kinetic study. *React. Funct. Polym.* **2010**, *70*, 707–714. [CrossRef]

78. Rauf, M.A.; Bukallah, S.B.; Hamour, F.A.; Nasir, A.S. Adsorption of dyes from aqueous solutions onto sand and their kinetic behavior. *J. Chem. Eng.* **2008**, *137*, 238–243. [CrossRef]

79. Sharma, P.; Das, M.R. Removal of a Cationic Dye from Aqueous Solution Using Graphene Oxide Nanosheets: Investigation of Adsorption Parameters. *J. Chem. Eng. Data* **2013**, *58*, 151–158. [CrossRef]

80. Abdelwahab, O. Evaluation of the use of loofa activated carbons as potential adsorbents for aqueous solutions containing dye. *Desalination* **2008**, *222*, 357–367. [CrossRef]

81. Zreig, M.A.; Rudra, R.P.; Dickinson, W.T.; Evans, L. Effect of surfactants on sorption of atrazine by soil. *J. Contam. Hydrol.* **1999**, *36*, 249–263. [CrossRef]

82. Patil, M.R.; Khairnar, S.D.; Shrivastava, V.S. Synthesis, characterisation of polyaniline–Fe_3O_4 magnetic nanocomposite and its application for removal of an acid violet 19 dye. *Appl. Nanosci.* **2016**, *6*, 495–502. [CrossRef]

83. Ouazene, N.; Lounis, A. Adsorption characteristics of CI Basic Blue 3 from aqueous solution onto Alleppo pine-tree sawdust. *Color. Technol.* **2011**, *127*, 21–27.

84. Ong, S.T.; Lee, C.K.; Zainal, Z. A cmparison of sorption and photodegradation study in the removal of basic and reactive dyes. *Aust. J. Basic Appl. Sci.* **2009**, *3*, 3408–3416.

85. Bangash, F.K.; Manaf, A. Dyes removal from aqueous solution using wood activated charcoal of bombax cieba tree. *J. Chin. Chem. Soc.* **2005**, *52*, 489–494. [CrossRef]

86. Wong, S.Y.; Tan, Y.P.; Abdullah, A.H.; Ong, S.T. The removal of basic and reactive dyes using quartenised sugar cane bagasse. *J. Phys. Sci.* **2009**, *20*, 59–74.

87. Chu, H.C.; Chen, K.M. Reuse of activated sludge biomass: I. Removal of basic dyes from wastewater biomass. *Process Biochem.* **2002**, *37*, 595–600. [CrossRef]

88. Marungrueng, K.; Pavasant, P. Removal of basic dye (Astrazon Blue FGRL. using macroalga Caulerpa lentillifera. *J. Environ. Manag.* **2006**, *78*, 268–274. [CrossRef]

89. Barsa, A.; Buha, R.; Dulman, V. Removal of Basic Blue 3 by sorption onto weak acid acrylic resin. *J. Appl. Polym. Sci.* **2009**, *113*, 607–614. [CrossRef]

90. Vasanth, K.; Sivanesan, S. Isotherm parameters for basic dyes onto activated carbon: Comparison of linear and non-linear method. *J. Hazard. Mater.* **2006**, *129*, 147–150. [CrossRef]

91. Zhang, J.; Cai, D.; Zhang, G.; Cai, C.; Zhang, C.; Qiu, G.; Zheng, K.; Wu, Z. Adsorption of methylene blue from aqueous solution onto multiporous palygorskite modified by ion beam bombardment: Effect of contact time, temperature, pH and ionic strength. *Appl. Clay Sci.* **2013**, *83*, 137–143. [CrossRef]

92. Hu, Y.; Guo, T.; Ye, X.; Li, Q.; Guo, M.; Liu, H.; Wu, Z. Dye adsorption by resins: Effect of ionic strength on hydrophobic and electrostatic interactions. *Chem. Eng. J.* **2013**, *228*, 392–397. [CrossRef]

93. Xu, D.; Tan, X.L.; Chen, C.L.; Wang, X.K. Adsorption of Pb(II) from aqueous solution to MX-80 bentonite: Effect of pH, ionic strength, foreign ions and temperature. *Appl. Clay Sci.* **2008**, *41*, 37–46. [CrossRef]

94. Lian, L.; Guo, L.; Guo, C. Adsorption of Congo red from aqueous solutions onto Ca-bentonite. *J. Hazard. Mater.* **2009**, *16*, 126–131. [CrossRef]

95. Mahanta, D.; Madras, G.; Radhakrishnan, S.; Patil, S. Adsorption of Sulfonated Dyes by Polyaniline Emeraldine Salt and Its Kinetics. *J. Phys. Chem. B* **2008**, *112*, 10153–10157. [CrossRef]

96. Ai, L.; Jianga, J.; Zhang, R. Uniform polyaniline microspheres: A novel adsorbent for dye removal from aqueous solution. *Synth. Met.* **2010**, *160*, 762–767. [CrossRef]

97. Muller-Dethlefs, K.; Hobza, P. Noncovalent interactions: A challenge for experiment and theory. *Chem. Rev.* **2000**, *100*, 143–167. [CrossRef]

98. Wang, J.; Deng, B.; Chen, H.; Wang, X.O.; Zheng, J.Z. Removal of aqueous Hg(II. by polyaniline: Sorption characteristics and mechanisms. *Environ. Sci. Technol.* **2009**, *43*, 5223–5228. [CrossRef]

99. Ahmad, R.; Kumar, R. Conducting Polyaniline/Iron Oxide Composite: A Novel Adsorbent for the Removal of Amido Black 10B. *J. Chem. Eng. Data* **2010**, *55*, 3489–3493. [CrossRef]

100. Yang, C.; Du, J.; Peng, Q.; Qiao, R.; Chen, W.; Xu, C.; Shuai, Z.; Gao, M. Polyaniline/Fe_3O_4 Nanoparticle Composite: Synthesis and Reaction Mechanism. *J. Phys. Chem. B* **2009**, *113*, 5052–5058. [CrossRef]

101. Fernandes, A.N.; Almedia, C.A.P.; Debacher, N.A.; Sierra, M.D.S. Isotherm and thermodynamic data of adsorption of methylene blue from aqueous solution onto peat. *J. Mol. Struct.* **2010**, *982*, 62–65. [CrossRef]
102. Gupta, V.K.; Pathania, D.; Kothiyal, N.C.; Sharma, G. Polyaniline zirconium (IV) silicophosphate nanocomposite for remediation of methylene blue dye from waste water. *J. Mol. Liq.* **2014**, *190*, 139–145. [CrossRef]
103. Ai, L.; Li, M.; Li, L. Adsorption of Methylene Blue from Aqueous Solution with Activated Carbon/Cobalt Ferrite/Alginate Composite Beads: Kinetics, Isotherms, and Thermodynamics. *J. Chem. Eng. Data* **2011**, *56*, 3475–3483. [CrossRef]

Article

Comparative Study of the Adsorption of Acid Blue 40 on Polyaniline, Magnetic Oxide and Their Composites: Synthesis, Characterization and Application

Amir Muhammad [1], Anwar ul Haq Ali Shah [1,*] and Salma Bilal [2,3,*]

[1] Institute of Chemical Sciences, University of Peshawar, Peshawar 25120, Pakistan
[2] National Centre of Excellence in Physical Chemistry, University of Peshawar, Peshawar 25120, Pakistan
[3] TU Braunschweig Institute of Energy and Process Systems Engineering, Franz-Liszt-Straße 35, 38106 Braunschweig, Germany
* Correspondence: anwarulhaqalishah@uop.edu.pk (A.u.H.A.S.); s.bilal@tu-braunschweig.de or salmabilal@uop.edu.pk (S.B.); Tel.: +92-919216652 (A.u.H.A.S.); +49-531-39163651 or +92-919216766 (S.B.)

Received: 24 July 2019; Accepted: 31 August 2019; Published: 4 September 2019

Abstract: Conducting polymers (CPs), especially polyaniline (PANI) based hybrid materials have emerged as very interesting materials for the adsorption of heavy metals and dyes from an aqueous environment due to their electrical transport properties, fascinating doping/de-doping chemistry and porous surface texture. Acid Blue 40 (AB40) is one of the common dyes present in the industrial effluents. We have performed a comparative study on the removal of AB40 from water through the application of PANI, magnetic oxide (Fe_3O_4) and their composites. Prior to this study, PANI and its composites with magnetic oxide were synthesized through our previously reported chemical oxidative synthesis route. The adsorption of AB40 on the synthesized materials was investigated with UV-Vis spectroscopy and resulting data were analyzed by fitting into Tempkin, Freundlich, Dubinin–Radushkevich (D–R) and Langmuir isotherm models. The Freundlich isotherm model fits more closely to the adsorptions data with R^2 values of 0.933, 0.971 and 0.941 for Fe_3O_4, PANI and composites, respectively. The maximum adsorption capacity of Fe_3O_4, PANI and composites was, respectively, 130.5, 264.9 and 216.9 mg g^{-1}. Comparatively good adsorption capability of PANI in the present case is attributed to electrostatic interactions and a greater number of H-bonding. Effect of pH of solution, temperature, initial concentration of AB40, contact time, ionic strength and dose of adsorbent were also investigated. Adsorption followed pseudo-second-order kinetics. The activation energy of adsorption of AB40 on Fe_3O_4, PANI and composites were 30.12, 22.09 and 26.13 kJmol^{-1} respectively. Enthalpy change, entropy change and Gibbs free energy changes are −6.077, −0.026 and −11.93 kJ mol^{-1} for adsorption of AB40 on Fe_3O_4. These values are −8.993, −0.032 and −19.87 kJ mol^{-1} for PANI and −10.62, −0.054 and −19.75 kJ mol^{-1} for adsorption of AB40 on PANI/Fe_3O_4 composites. The negative sign of entropy, enthalpy and Gibbs free energy changes indicate spontaneous and exothermic nature of adsorption.

Keywords: Acid blue 40 dye; adsorption isotherms; kinetics and thermodynamic study

---

1. Introduction

The discovery of conducting polymers in 1977 initiated an interesting field of research. These polymers showcased fascinating physico-chemical properties which made them suitable for numerous applications [1]. Polyaniline, polythiophene, polypyrrole and their derivatives are the most studied conducting polymers [2–5] and show optical as well as conducting properties due to the presence of π conjugated electrons in their skeleton [6]. Polyaniline (PANI) has gained a lot interest among the conducting polymers family because it can be synthesized easily from low-cost materials. It is highly conductive and possesses good environmental stability [3,7,8].

A number of methods including chemical oxidation, electro-chemical oxidation, enzymatic, interfacial, self-assembling and seeding methods have been applied to synthesize PANI [9–13]. Chemical and electro-chemical oxidation methods are the most common methods which involve the polymerization of aniline in an acidic or basic medium. However, the conducting emeraldine form of PANI is usually synthesized in an acidic environment [14]. PANI has been effectively applied in corrosion protection, batteries, solar cells, supercapacitors and adsorption of heavy metals and dyes from an aqueous solution [3,15–18]. The suitability of PANI as an adsorbent to remove dyes from an aqueous environment is due to the presence of a large number of amine and imine functional groups which are expected to interact with dyes. The charge transfer induced by doping enables PANI to interact with ionic species through electrostatic interactions [19]. Although PANI has been used widely as an adsorbent for the removal of dyes from water, its performance is restricted due to two main challenges. Firstly, its particles aggregate due to intermolecular interactions, resulting in the decrease of surface area and hence the adsorption capacities [20]. Secondly, acid doped PANI is prone to de-doping due to the evaporation of the small acid molecules at room temperature. This causes a reduction in the surface charge of PANI which ultimately affect the electrostatic interaction between PANI and dye [21].

To overcome these challenges, considerable work has been done in recent years to synthesize composites of PANI with inorganic substances such as Ag, Cd, SiO_2, TiO_2, ZnO, MnO_2 and magnetic oxide (Fe_3O_4) [22–25]. These composites exhibit characteristics electrical, optical, catalytic and mechanical properties that are better than single components in some cases. The composites of PANI and Fe_3O_4 have attracted much attention because of easy synthesis and numerous applications in areas such as in biosensors, sensors, solar cells and purification of water [26–29].

Just like PANI, magnetic oxide also finds applications in drug delivery systems [30], clinical diagnosis [31], efficient hyperthermia for the removal of cancer [32], microwave devices, magnetic resonance imaging (MRI) [33,34] and the removal of heavy metals from an aqueous solution [35,36]. Electric explosion of wire, laser target evaporation and biomineralization are commonly used for controlled size and morphology of Fe_3O_4 [37], but the wet chemical methods, like the chemical co-precipitation method [38], sol-gel [39], hydrothermal method [40], gas phase [41], liquid phase [42] and two-phase methods such as microemulsion methods [43] are also used for the preparation of Fe_3O_4.

In general, composites of PANI and Fe_3O_4 have been synthesized either through in situ formations of magnetic oxide composites in the presence of PANI [44] or polymerization of aniline monomers in the presence of iron oxide. In comparison with the former, the latter strategy gives better results because of the magnetic properties of the resulting hybrid materials [45].

Bhaumik et al. [46] prepared nanofibers composites from metallic nanoparticles and PANI and applied these composites to remove arsenic (V), chromium (VI) and Congo red from an aqueous solution. Different polymer salts (PANI–HNO_3, PANI–H_2SO_4 and PANI–H_3PO_4) are reported to use as adsorbents to remove Direct Blue 78 (DB78) from water [47]. The dye uptake was in the order PANI–H_3PO_4 > PANI–H_2SO_4 > PANI–HNO_3. The rate of adsorption was decreased as the concentration of DB78 and pH of dye solution increased. The adsorption followed pseudo-second-order kinetics. Cui and co-workers [48], studied the adsorption of Hg (II) onto polyaniline/attapulgite (PANI/ATP) composites. (PANI/ATP) composites were synthesized by the chemical oxidation method. The maximum amount of dye adsorbed was 800 mg/g when the pH of Hg (II) solution was 5.9 and followed pseudo-second-order kinetics.

In the present study, PANI/Fe_3O_4 is used as an adsorbent to remove Acid Blue 40 (AB40) from water. The adsorption behaviors of PANI/Fe_3O_4 were compared with PANI and Fe_3O_4 which were synthesized and tested according to our previous work [49]. The chemical oxidation method was used to synthesize PANI and PANI/Fe_3O_4 composites using $FeCl_3·6H_2O$ as an oxidant in an acidic medium, while the chemical co-precipitation method was adopted to synthesize Fe_3O_4 materials in the basic medium at a temperature of 85–90 °C. All these synthesized materials were characterized through UV-Vis, SEM, FTIR, EDX and surface area measurements. Adsorption study was carried out to determine the effect of pH, initial concentration, temperature, contact time, adsorbent dosage and ionic strength on adsorption phenomenon using UV-Vis spectroscopy. Freundlich, Langmuir, D–R and Tempkin adsorption isotherm models were applied to analyze the adsorption data. The adsorption mechanism was determined on the basis of kinetic study. Thermodynamic aspects of adsorption of AB40 on these materials were also investigated.

2. Experiment

2.1. Materials

Aniline was purchased from Across and distilled under vacuum. Acid Blue 40 dye, $FeCl_3 \cdot 6H_2O$ and Na_2SO_4 (Sigma-Aldrich, St. Louis, MO, USA), Dodecyl benzene sulfonic acid (DBSA) (Across) and $FeSO_4 \cdot 7H_2O$ (Merck, Kenilworth, NJ, USA) were used without further purification.

2.2. Synthesis of PANI

PANI was synthesized via our previously reported chemical oxidation method [49]. Typically, 0.02 M (1.182 mL) aniline was suspended in 50 mL of 0.01 M H_2SO_4 solution. To this suspension 0.01 M (0.15 mL) DBSA was added as an emulsifying agent. Then 50 mL of 0.01 M $FeCl_3 \cdot H_2O$ prepared in 0.01 M H_2SO_4 was added drop by drop as an oxidant with constant stirring. After 20 min of continuous stirring, a milky white color suspension turned to light green and then dark green in one hour. The final product was thoroughly washed with acetone and then with double-distilled water until the filtrate became clear. The obtained powder was dried in an oven for 24 h at 60 °C.

2.3. Synthesis of Fe_3O_4

Fe_3O_4 was synthesized by the chemical co-precipitation method by adding 0.15 mL DBSA and 2 M of $FeCl_3 \cdot 6H_2O$ dissolved in 50 mL of 0.1 M NaOH to 0.5 M $FeSO_4 \cdot 7H_2O$ solution. The whole mixture was stirred continuously at 85–90 °C. After 20 min of stirring, 30 mL of 5 M ammonia solution was added at once which turned the color of the reacting mixture to black. The pH of the reacting mixture was kept at 10 during the whole experiment. After two hours of continuous stirring at 85–90 °C, the precipitate was washed with ethanol and double distilled water until the effluent became clear. The black precipitate was dried at 80 °C for 10 h and then annealed at 600 °C for 5 h in a furnace (NEYCRAFT JFF 2000 Fiber Furnace, France).

2.4. Synthesis of PANI/Fe_3O_4 Composites

PANI/Fe_3O_4 composites were synthesized by suspending 0.15 g Fe_3O_4 particles in 30 mL of 0.01 M H_2SO_4 solution followed by addition of 50 mL (0.02 M) of aniline solution prepared in 0.01 M H_2SO_4 and 0.5 mL (0.01) DBSA, respectively. After 30 min of continuous stirring, 50 mL of 0.01 M $FeCl_3 \cdot 6H_2O$ was added as an oxidizing agent. A light green color appeared within the stirring mixture after 20 min of the oxidant's addition. The color of this mixture turned dark black after about one hour. After continuously stirring for 6 h, the product was separated and washed with acetone and double-distilled water. The clean precipitate was dried in oven at 60 °C for 24 h.

2.5. Batch Adsorption Study for Removal of AB40 Dye

Twenty-milliliter solutions of different concentrations between 5–120 mgL^{-1} were prepared from the stock solution of AB40 dye. To these solutions, PANI was added and shacked for about 120 min. These solutions were then filtered to determine the concentration of dye in the filtrate using UV-Visible spectrophotometer and applying Equation (1) [50].

$$q_e = \frac{(C_i - C_e)V}{m} \tag{1}$$

where q_e (mg g^{-1}) refers to adsorption at equilibrium, C_i (mg L^{-1}) and C_e (mg L^{-1}) show initial and equilibrium concentration of dye, respectively, m (g) is the mass of adsorbent while V represents the volume of solution in mL. The effect of temperature, contact time, ionic strength, pH and initial concentration of dye solution on the adsorption behavior was studied. The adsorption data were utilized to calculate the thermodynamic and kinetics parameters. The same procedure was employed for studying adsorption of AB40 on Fe_3O_4 and PANI/Fe_3O_4 composite.

Before the adsorption process, standard solutions of AB40 dye were prepared was in the range 0.005–2 mg L^{-1} and their maximum absorption was determined via UV-Visible spectrophotometer (Perkin Elmer, Waltham, MA, USA). The absorption values were plotted against concentration of the standard dye solutions and calibration curve was obtained according to the Beer–Lambert law. The slope so obtained was used as reference for the determination of concentration in rest of the experiments. The calibration curve is also shown in the supporting files (Figure S1).

After the adsorption of AB40 dye, the adsorbents were put onto the filter paper and washed several times with double-distilled water to run out the adsorbed dye. Then it was washed with 0.1 M NaOH to remove the remaining dye. This process enables the reutilization of the adsorbent.

2.6. Characterization

FTIR spectra of the synthesized materials were registered in the spectral range of 400 to 4000 cm^{-1} through a Fourier transmission infrared spectrophotometer (Shimadzu, Tokyo, Japan). X-ray diffraction (XRD) patterns were recorded by using Cu Kα radiations of wavelength 1.5405 Å with the help of a JEOL JDX-3532 (JEOL, Tokyo, Japan). The concentration of dye in the solution and its adsorbed amount onto the synthesized materials were checked through UV-visible spectrophotometer (Perkin Elmer, Buckinghamshire, UK). An energy-dispersive X-ray (EDX) spectrophotometer (Inca 200, Oxford, UK) was utilized to determine the percentage of different elements. Brunauer–Emmett–Teller (BET) surface areas of the composite as well as PANI and Fe_3O_4, were determined in the N_2 atmosphere through adsorption–desorption method with a surface area analyzer model 2200 e Quanta Chrome (Quanta Chrome, Boynton Beach, FL, USA). The surface morphologies were studied through scanning electron microscopy (SEM) (JSM-6490, JEOL, Tokyo, Japan).

3. Results and Discussion

3.1. Scanning Electron Microscopy (SEM)

SEM images provide interesting information about surface morphology and size of adsorbent materials under investigation. Figure 1a,b shows SEM images of Fe_3O_4 particles before and after adsorption of AB40 dye. The Fe_3O_4 particles are round in shape with an average size of 0.15 μm. After adsorption of AB40, the porosity decreases in the agglomerated surface of Fe_3O_4 but the average particles size increases to 0.23 μm (Figure 1b). Keyhanian et al. [50] have reported the agglomeration of magnetic particles of Fe_3O_4 after adhering of methyl violet dye from an aqueous solution.

Rods- or wires-like porous structure can be seen in the SEM image of PANI with 0.21 μm average diameter of the rods (Figure 1c). These rods are aggregated to each other like fibers. After the adsorption of AB40 (Figure 1d), the morphology of PANI changes to a cauliflower shape with some needle-like structures present on the surface. Such a change in morphology was also reported during the adsorption of anionic dyes on PANI doped with Potash Alum [51]. Figure 1e shows an SEM image of PANI/Fe_3O_4 composites. It shows a porous morphology where Fe_3O_4 particles have adhered with PANI interconnected rods. Similar morphology was depicted by nanocomposites of PANI/ Fe_3O_4 coated on $MnFe_2O_4$ [52]. Just like PANI, morphology of PANI/Fe_3O_4 composites also changes after the adsorption of AB40. The dye distributes homogeneously over the surface of composite imparting a broccoli-like appearance to it as shown in Figure 1f.

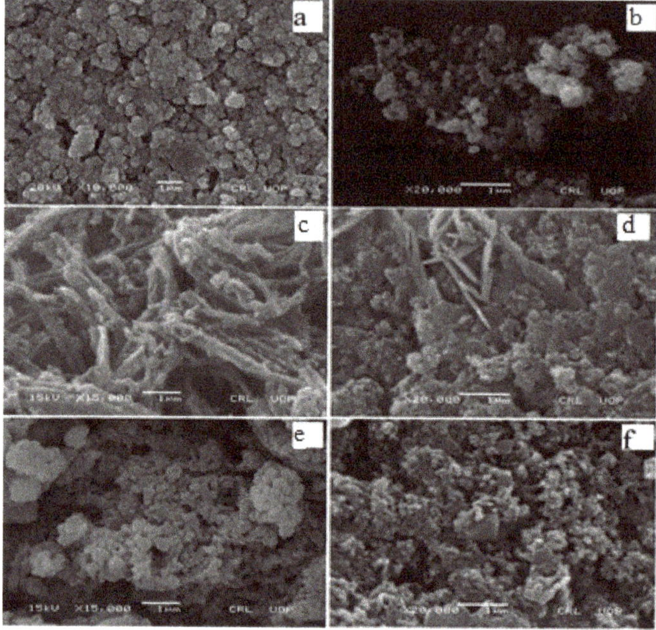

Figure 1. SEM images of magnetic oxide (Fe_3O_4), polyaniline (PANI) and PANI/Fe_3O_4 composite before (**a,c,e**) and after (**b,d,f**) adsorption of acid blue 40 (AB40) dye.

3.1.1. Optical Studies

Figure 2A represents the UV-Vis spectra of the synthesized materials before adsorption of the dye. A weak band in the region of 450 nm arises due to the interaction of electromagnetic radiations with the valence electrons of iron in the Fe_3O_4. As a result, the valence electron of the metal atom starts to oscillate with the frequency of the electromagnetic source [53]. This phenomenon is known as surface plasmon resonance (SPR). Another band at 485.85 nm is due to the presence of DBSA moiety with Fe_3O_4 and closely resembles already reported work [54]. The two characteristic bands of PANI can be observed in its spectrum at 333.91 and 633.42 nm. The band at 633.42 nm is due to charge transfer from the benzenoid ring to the quinoid ring and the band at 333.91 nm is attributed to π-π^* transitions of the benzenoid ring [55]. In the spectrum of PANI/Fe_3O_4 composites, the band at 333.91 nm shows a redshift due to the doping of the benzenoid amine with Fe_3O_4 particles. Moreover, the bipolaron band at 633.42 nm is shifted to 773.14 nm suggesting that some physical interactions between PANI and Fe_3O_4 particles may exist [56].

Figure 2B represent UV-visible spectra of Fe_3O_4, PANI and composite of Fe_3O_4 with PANI after adsorption of AB40. One can observe a band in the region of 618–620 nm in all the spectra of Fe_3O_4, PANI and composite of Fe_3O_4 and PANI which indicates the adsorption of AB40. This band has been demonstrated that AB40 shows strong absorption at 620 nm [57]. The intensity of this band is higher for PANI, which is different from our previous work where more intense peaks, due to adsorption of Basic Blue 3 dye, was observed in the spectrum of PANI/Fe_3O_4 composite [49]. The reason can be explained by the fact that in the PANI/Fe_3O_4 composite, the positively charged active sites of PANI are covered by Fe_3O_4. Moreover, the oxygen of Fe_3O_4 behave as negatively charged sites, which may cause repulsion to the negative charge of the anionic dye and hence reduces its adsorption.

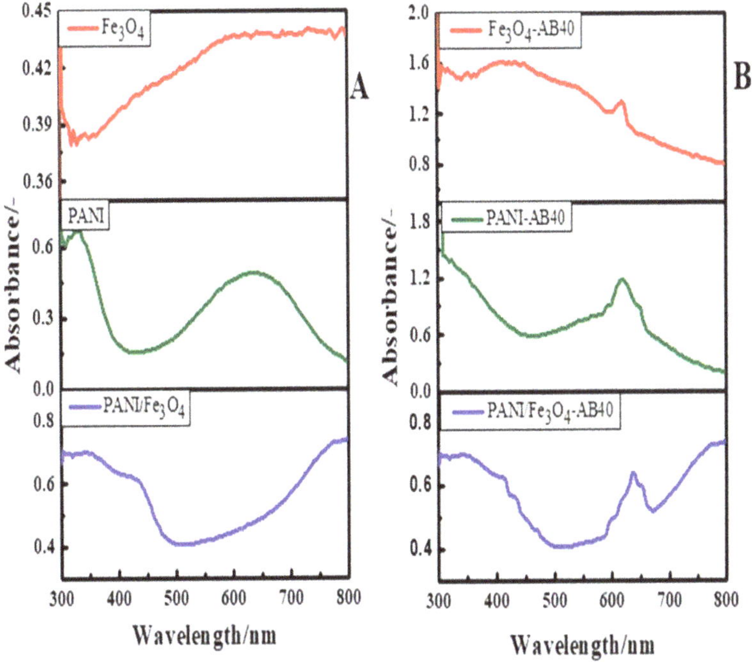

Figure 2. UV-visible spectrum of Fe$_3$O$_4$, PANI and PANI/ Fe$_3$O$_4$ composite (**A**) before and (**B**) after adsorption of AB40.

3.1.2. Energy Dispersive X-ray (EDX) Study

Figure 3 shows the EDX analysis of PANI, Fe$_3$O$_4$ and PANI/Fe$_3$O$_4$ composites before and after adsorption of AB40. The weight percent of Fe and O in Fe$_3$O$_4$ is 68.74 and 29.15, respectively. After the adsorption of AB40, the weight percent of Fe decreases from 68.74 to 62.09, while the percent weight of O and C increases due to the presence of oxygen and carbon in the AB40 texture. Similarly, the appearance of nitrogen and Sulphur in spectrum 3b is more evidence of the adsorption of AB40 onto Fe$_3$O$_4$, as these elements are present in the dye texture [58]. Figure 3c shows the EDX spectrum of PANI before adsorption of AB40. One can observe a 9.54 percent nitrogen and 68.06 percent carbon by weight in this spectrum. The presence of sulfur and oxygen may be due to the presence of DBSA while Fe and Cl may be due to the presence of FeCl$_3$·H$_2$O which was used as oxidant. After adsorption of AB40, although weight percent of carbon decreases but weight percent of nitrogen and oxygen increases which shows that AB40 adsorb on PANI (Figure 3d) [59]. In the EDX spectrum of PANI/Fe$_3$O$_4$ composite, peaks for nitrogen, oxygen, carbon and iron can clearly be observed in Figure 3e, confirming the formation of composites. The sulfur percent by weight is 2.66 and is due to the presence of some moiety of DBSA. After the adsorption of AB40, the weight percent of carbon, nitrogen and sulfur is increased (Figure 3f) [60].

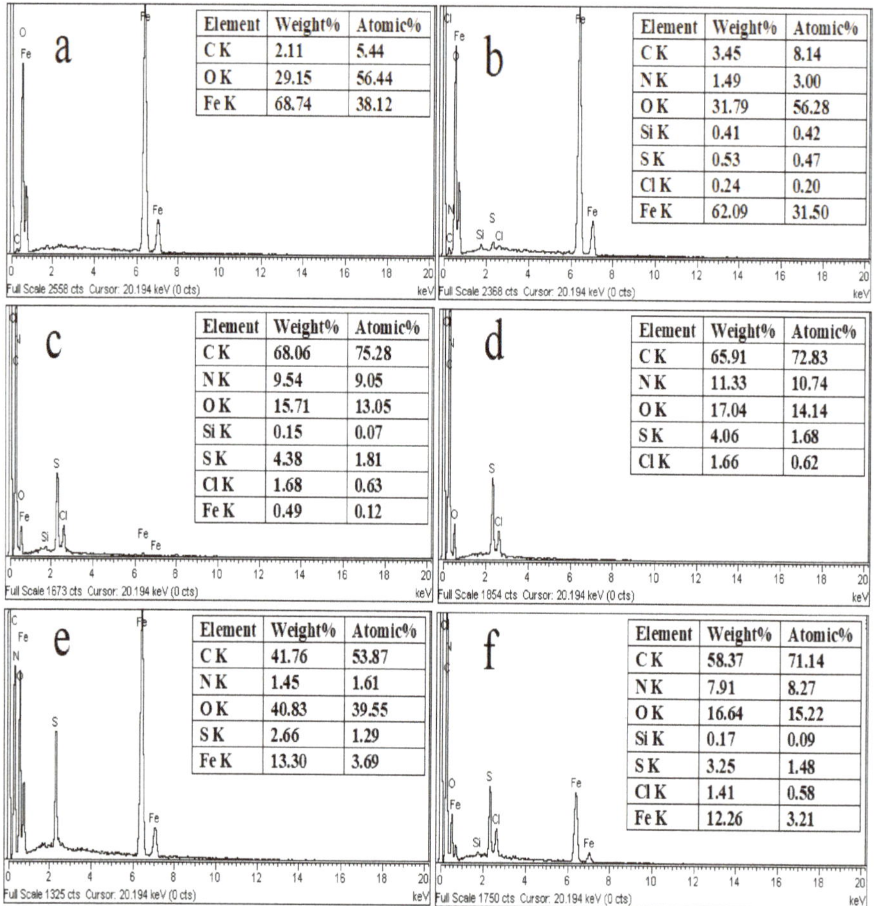

Figure 3. EDX spectra of Fe_3O_4, PANI and PANI/Fe_3O_4 composites before (**a,c,e**) and after (**b,d,f**) adsorption of AB40.

3.1.3. FTIR Study

Figure 4A,B represent, respectively, FTIR spectra of Fe_3O_4, PANI and PANI/Fe_3O_4 composite before and after adsorption of AB40. The peak located at 543.1 cm^{-1} is due to the stretching vibration of the Fe–O bond in the Fe_3O_4 spectrum [61]. A wide peak at 3427.34 cm^{-1} shows stretching vibrations of –OH group attached to Fe_3O_4 surface [62]. The shifting of all peaks towards a lower frequency and the appearance of a very small peak at 2343.2 cm^{-1} in Figure 4B indicates that the AB40 dye comes in contact with Fe_3O_4 after adsorption [9,63].

FTIR spectrum of PANI shows –N–H group of secondary amine at 3231.5 cm^{-1}. Similarly, the peaks at 2842.8 and 2932.8 cm^{-1} can be attributed to the symmetric and asymmetric stretching vibrations of the C–H bond, respectively. Vivekanandan et al. have reported such asymmetric and symmetric C–H stretching vibrations at 2923.62 and 2825.55 cm^{-1}, respectively [9]. Peaks at 1602.8 and 1469.3 cm^{-1} attribute to C=C and C=N stretching vibrations of the benzenoid and quinoid rings. The band at 1304.2 cm^{-1} corresponds to the –C–N$^+$ stretching vibrations of the secondary aromatic amine. Similarly, the peaks at 1140.3 and 826.5 cm^{-1} represent the bending vibrations of the aromatic C–H bond in plane and out of plane deformation [64]. The peak at 1020.4 cm^{-1} shows the S=O stretching vibrations of the –SO_3H group, confirming the presence of DBSA moiety in the PANI texture [65,66]. The peak at 677.2 cm^{-1} shows the out of plane bending vibrations of the C–H bond.

In the spectrum of PANI/Fe_3O_4 composites, all peaks are shifted to the low-frequency range in comparison with PANI, indicating a presence of some physical forces between PANI and Fe_3O_4 particles. The appearance of the small peak at 542.7 cm^{-1} shows Fe-O stretching, which confirms the formation of PANI/Fe_3O_4 composites [67]. After the adsorption of AB40, there is a slight shift of peaks towards a low frequency, both in the spectrum of PANI and PANI/Fe_3O_4 composites. Moreover, the appearance of the peak at 2356.7 cm^{-1} shows the adsorption of AB40 dye on PANI and PANI/Fe_3O_4 composites [68]. This peak is more intense in the spectrum of AB40 adsorbed on PANI as compared to the PANI/Fe_3O_4 composite which is in agreement with the UV-visible study.

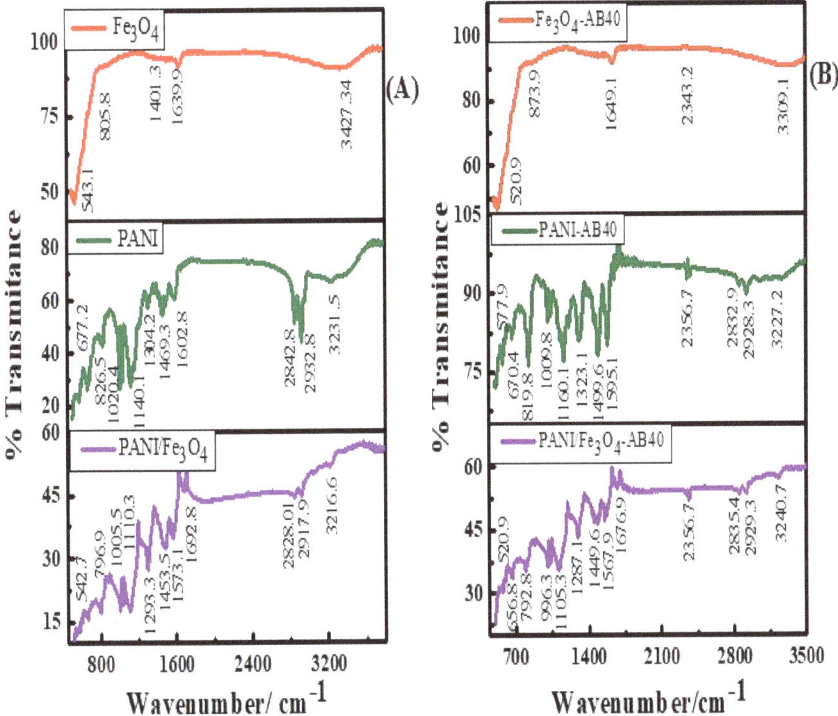

Figure 4. FTIR spectra of Fe_3O_4, PANI and PANI/ Fe_3O_4 composites before (**A**) and after (**B**) the adsorption of AB40.

3.1.4. Surface Area Study

The surface area of adsorbent plays a unique role in the adsorption study. The Brunauer–Emmett–Teller (BET) technique was employed to determine the average pore size radius, pore volume and specific surface area of PANI, Fe_3O_4 and PANI/Fe_3O_4 composite via nitrogen adsorption–desorption analysis (Figure 5). The results obtained are summarized in Table 1. The data shows that specific surface area of PANI/Fe_3O_4 composite is greater than PANI and Fe_3O_4 particles [69]. After the adsorption of AB40, the surface area of Fe_3O_4, PANI and PANI/Fe_3O_4 composite decreases [70]. However, the extent of reduction is more for PANI as compared to Fe_3O_4 and PANI/Fe_3O_4 composites, showing a greater adsorption of the dye on PANI.

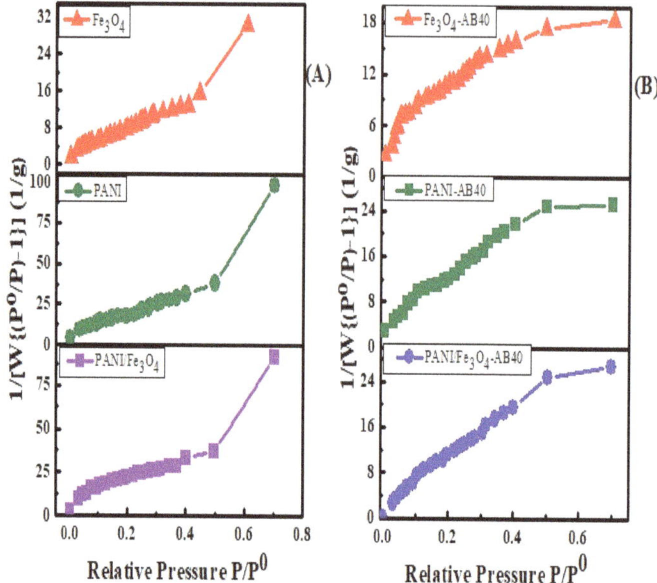

Figure 5. Brunauer–Emmett–Teller (BET) surface area of Fe$_3$O$_4$, PANI and PANI/Fe$_3$O$_4$ composites before (**A**) and after (**B**) the adsorption of AB40.

Table 1. Surface area, average pore volume and pore radius of PANI, Fe$_3$O$_4$ and PANI/Fe$_3$O$_4$ composites.

Status	Materials	Surface Area (m^2/g)	BJH Average Pore Radius (A^0)	BJH Pore Volume (cc/g)
Before adsorption	Fe$_3$O$_4$	71.314	15.749	0.053
	PANI	95.423	16.565	0.049
	PANI/Fe$_3$O$_4$	98.184	15.501	0.069
After adsorption	Fe$_3$O$_4$-AB40	53.707	13.334	0.033
	PANI-AB40	43.938	11.743	0.043
	PANI/Fe$_3$O$_4$-AB40	65.269	12.804	0.036

3.2. Isotherms Study

The most important step in the adsorption study is the fitting of adsorption isotherm models to adsorption data in order to describe how interaction occurs between adsorbent and dye. A number of adsorption isotherms are available and have been successfully applied by the earlier researcher to analyze the adsorption data [71]. In this study, four adsorption isotherms models, namely Freundlich, Tempkin, Langmuir and Dubinin–Radushkevich (D–R) were tested. Adsorption parameters so calculated have been summarized in Table 2. The correlation factor R^2, indicates that Freundlich adsorption isotherm equation fit more closely to the adsorption data. The linearized form of Freundlich adsorption equation is expressed in Equation (2);

$$\ln q_e = \ln k_f + \frac{1}{n}\ln c_e \tag{2}$$

where q_e (mg g^{-1}) and C_e (mg L^{-1}) are the solid and liquid phase equilibrium concentration of dye. K_f is constant, and is known as the Freundlich constant and $1/n$ is the slope obtained by plotting $\ln q_e$ vs. C_e shown in Figure 6A. The values of $1/n$ vary due to heterogeneity of the adsorbing materials.

The values of $1/n$ shows favorable $(0 < 1/n < 1)$, unfavorable $(1/n > 1)$ or irreversible $(1/n = 0)$ adsorption. However, if its value is unity, the system is at equilibrium and will show heterogeneity [72]. In the present work, the values of $1/n$ calculated from Freundlich for Fe_3O_4, PANI and PANI/Fe_3O_4 composites are 0.126, 0.504 and 0.723 respectively showing favorable physical adsorption [73].

Table 2. Summary of parameters calculated from adsorption isotherms models.

Adsorbents	Adsorption Isotherms												
	Freundlich			Langmuir				Tempkin			D-R		
	$1/n$	K_f	R^2	q_{max}	K_L	R_L	R^2	β_T	K_T	R^2	q_S	E_{ads}	R^2
Fe_3O_4	0.126	22.88	0.933	130.5	0.195	0.059	0.499	26.38	3.050	0.917	98.83	3.199	0.933
PANI	0.504	98.21	0.971	264.9	1.579	0.339	0.773	14.24	153.6	0.957	166.7	23.63	0.971
PANI/Fe_3O_4	0.723	58.99	0.946	216.9	0.499	0.167	0.859	22.15	14.17	0.902	134.5	11.29	0.909

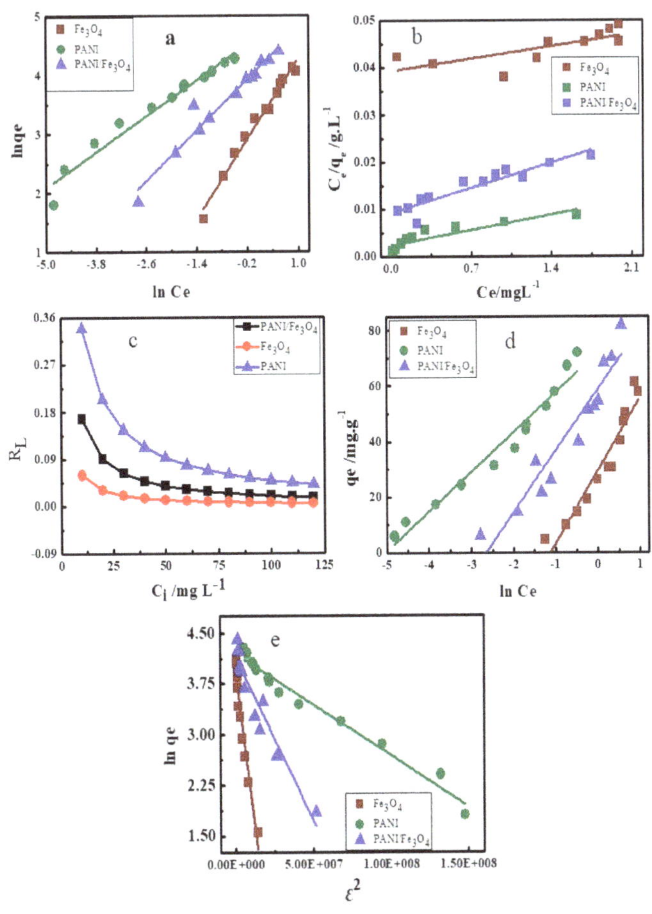

Figure 6. The Isotherm plots (**a**) Freundlich, (**b**) Langmuir, (**c**) Separation factor, (**d**) Tempkin and (**e**) D–R of adsorption of AB40 on Fe_3O_4, PANI and PANI/Fe_3O_4 composite.

The data were also fitted in the Langmuir adsorption isotherm equation (Equation (3)) as shown below;

$$\frac{c_e}{q_e} = \frac{1}{q_{max}k_L} + \frac{1}{q_{max}}c_e \tag{3}$$

where C_e (mg L^{-1}) and q_e (mg g^{-1}) indicates the concentration of dye and amount of dye adsorbed per gram of adsorbent at equilibrium, respectively. Similarly, K_L (mg L^{-1}) represent the Langmuir constant related to adsorption energy and q_{max} (mg g^{-1}) is the maximum adsorption capacity of adsorbing materials which can be calculated from the slope. The maximum adsorption capacity of AB40 onto Fe$_3$O$_4$, PANI and PANI/Fe$_3$O$_4$ composites are 130.5, 264.9 and 216.9 mg g^{-1}, respectively, as compared in Table 3A. The dimensionless constant (R_L) also called separation factor, expresses essential features of the Langmuir isotherm and is represented by Equation (3a).

$$R_L = \frac{1}{(1 + K_L C_i)} \tag{3a}$$

where C_i (mg L^{-1}) is the initial concentration of AB40. Values of R_L indicate that isotherm is either favorable ($1 > R_L > 0$), linear ($R_L = 1$), irreversible ($R_L = 0$) or unfavorable ($1 < R_L$) [74]. In the present study, the values of R_L range from 0.00525 to 0.34988 as depicted in Figure 6c, which shows that adsorption of AB40 onto Fe$_3$O$_4$, PANI and PANI/Fe$_3$O$_4$ composites is favorable at low concentration [75].

Table 3. Kinetics parameters for adsorption of AB40 on Fe$_3$O$_4$, PANI and PANI/Fe$_3$O$_4$ composite based on pseudo-first-order and pseudo-second-order equations.

Adsorbents	Pseudo 1st Order			Pseudo 2nd Order		
	K_1 (min^{-1})	qe (mg g^{-1})	R^2	K_2 (g mg^{-1} min^{-1})	qe (mg g^{-1})	R^2
Fe$_3$O$_4$	-0.015	1.765	0.812	0.0665	126.3	0.999
PANI	-0.033	4.823	0.885	0.0028	258.8	0.983
PANI/ Fe$_3$O$_4$	-0.047	2.495	0.881	0.0213	207.3	0.994

The Tempkin isotherm is also an important isotherm model and has been used by researchers to analyze their adsorption data [76,77]. The Tempkin isotherm assumes that due to interactions of the dye with the adsorbent, the adsorption decreases linearly and is characterized by binding energies. It is represented by the following equation (Equation (4));

$$q_e = \beta \ln k_T + \beta \ln c_e \tag{4}$$

where C_e (mg L^{-1}), q_e (mg g^{-1}) and K_T (L g^{-1}) are equilibrium concentration, equilibrium adsorption and binding constant at equilibrium. It is obtained by plotting q_e vs. $\ln C_e$ (Figure 6d). The constant β, considers the interaction between adsorbent and dye (Equation (4a)).

$$\beta = \frac{RT}{b} \tag{4a}$$

where b is the Tempkin isotherm constant of binding energy (J mol^{-1}K^{-1}). The correlation factors (R^2) given in Table 2 show that the Tempkin isotherm also fit the adsorption data. The values of K_T show that there is strong interaction between AB40 and PANI as compared to Fe$_3$O$_4$ and PANI/Fe$_3$O$_4$ composites (Table 2).

Dubinin–Radushkevich (D–R) as expressed in Equation (5) was also fitted to the adsorption data.

$$\ln q_e = \ln q_s - \beta \varepsilon^2 \tag{5}$$

where q_e is the amount of dye in mg adsorbed per gram of adsorbent (mg g^{-1}), β (mol^2 K^{-1}J^{-2}) is the activity coefficient useful in obtaining the mean adsorption energy E_{ad} (kJ mol^{-1}), q_s is the adsorption maximum, and ε is Polanyi potential. ε and E_{ad} are expressed by Equations (5a) and (5b) respectively.

$$\varepsilon = RT\ln\left(1 + \frac{1}{C_e}\right) \tag{5a}$$

$$E_{ad} = \frac{1}{\sqrt{1 - 2\beta}} \tag{5b}$$

where R is the gas constant which has a value of 8.314 (J mol^{-1} K^{-1}) and T is the kelvin temperature.

D–R adsorption model is a unique model used to differentiate between the chemical and physical adsorption on the basis of adsorption energy. In early research, it has been demonstrated that if the value of adsorption energy is less than 40 kJ mol^{-1}, the adsorption is physical [78]. In the present work, the values of E_{ad}, calculated by the Equation (5b) are less than 40 kJ mol^{-1} for adsorption of AB40 and PANI as compared to Fe$_3$O$_4$ and PANI/Fe$_3$O$_4$ composites showing physical adsorption as shown in Table 2.

3.3. Effect of Contact Time and Temperature on Adsorption

The contact time between adsorbent and dye is of great interest in the adsorption process. The optimum time of equilibrium was determined by adding 0.0340 ± 0.0001 g of each Fe$_3$O$_4$, PANI and PANI/Fe$_3$O$_4$ composite to 20 mL of AB40 (50 mg L^{-1}) in a series of experiments and was shacked at 150 rpm at 30 °C. The adsorption data so obtained was plotted as a function of time (Figure 7a). The graph shows that adsorption is very fast in the initial 10–15 min. The initial fast adsorption is due to a strong interaction between active sites of adsorbents and dye molecules. After 40–50 min, adsorption rate of dye become constant due to filling of active sites on the surface of adsorbents. This time period is defined as the dynamic equilibrium time. At the equilibrium time, rate of adsorption and desorption occurs simultaneously with the same speed [79]. Maximum adsorption of AB40 on Fe$_3$O$_4$, PANI and PANI/Fe$_3$O$_4$ composite is observed at 30 °C indicating exothermic nature (Figure 7b).

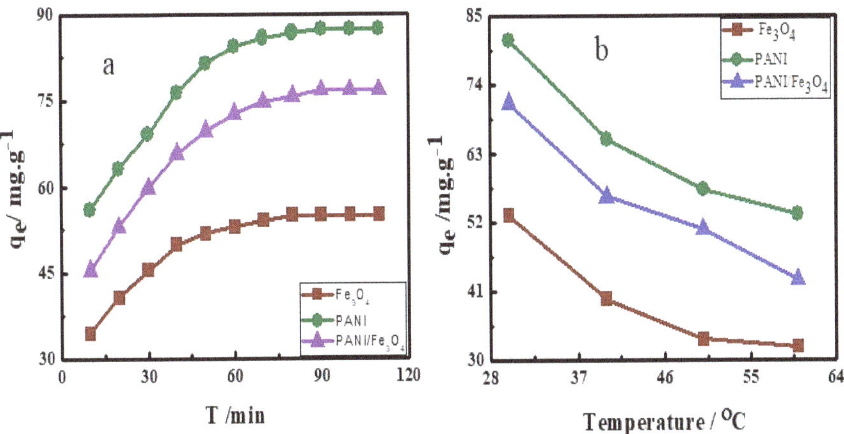

Figure 7. Effect of (**a**) time and (**b**) temperature on adsorption of AB40 onto Fe$_3$O$_4$, PANI and PANI/Fe$_3$O$_4$ composite.

3.4. Effect of pH on Adsorption

The pH of the dye solution plays a unique role in adsorption process. In the present work, the effect of pH on adsorption was investigated between 2–12. Results so obtained are plotted as adsorption versus pH (Figure 8). The plot indicates that adsorption of AB40 is high in acidic medium on all three adsorbents. When at a low pH, the backbone of adsorbents is positively charged and the active sites like Fe–O and –C=N are protonated. These positively charged sites have a strong interaction with the negatively charged sites of AB40 dye and hence enhance the adsorption. On the other hand, in a basic medium, the deprotonation of Fe–O–H and –C–N–H will create a negative charge in these groups which will repel the negatively charged sites of dye electrostatically, thus adsorption reduces [80].

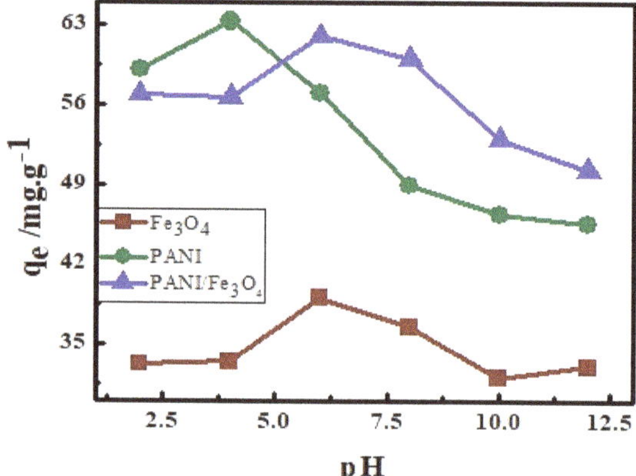

Figure 8. The effect of solution pH on adsorption of AB40 on PANI, Fe_3O_4 and PANI/Fe_3O_4 composite.

3.5. Effect of Ionic Strength on Adsorption

The effluent of industrial water also contains several ions. Therefore, the presence of these ions will also affect the adsorption process. In the present study, ionic strength effect of sodium sulfate and calcium chloride on adsorption has been studied in the pH between 5–6. The adsorption data so obtained are plotted against ionic strength (Figure 9). The plots (Figure 9a,b) show that adsorption of AB40 on PANI, Fe_3O_4 and PANI/Fe_3O_4 composite increases with an increase in ionic strength. This can be attributed to the fact that when both dye and adsorbent have similar charges, an increase in ionic strength will increase adsorption. This effect is more prominent in the adsorption of AB40 on PANI/Fe_3O_4 composite as compared to pristine PANI, because PANI/Fe_3O_4 composite contains a greater number of sites with a lone pair of electrons which behave as negatively charged groups [81]. Moreover, the significant increase in the adsorption of AB40 by increasing the ionic strength can be attributed to the dimerization of dye. A number of intermolecular forces like dipole-dipole, ion-dipole and Van der Waals forces have been suggested as the cause of the dimerization. Alberghina and co-workers have observed such type of dimerization while studying salts and temperature effect on adsorption of reactive dyes onto activated carbon [51].

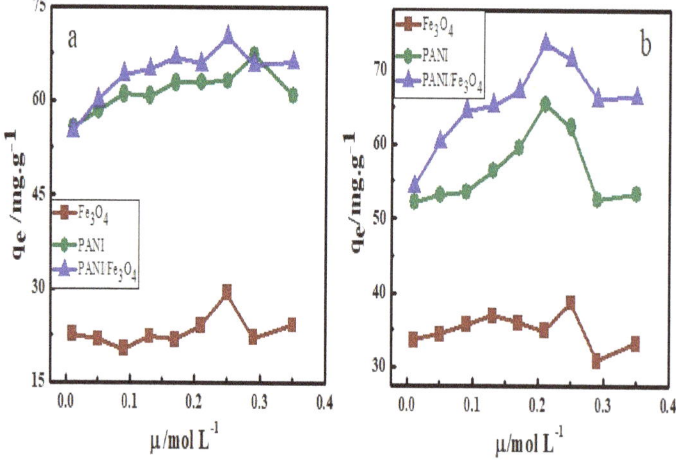

Figure 9. The effect of ionic strength of (**a**) $Na_2SO_4.7H_2O$ and (**b**) $CaCl_2.2H_2O$ on adsorption of AB40 onto Fe_3O_4, PANI and PANI/Fe_3O_4 composite.

3.6. Effect of Adsorbent Dosage on Adsorption

To investigate the effect of adsorbent dosage on adsorption, 0.034, 0.045, 0.075 and 0.1 g of each Fe_3O_4, PANI and PANI/Fe_3O_4 composite were added to 100 mg L^{-1} of AB40 separately and shook at 150 rpm at 30 °C and the amount adsorbed was noted (Figure 10). An increase in the adsorption of dye was observed by increasing the adsorbent dose. Initially the rate of adsorption is fast due to greater number of active site and splitting effect of the flux between adsorbents and dye [82].

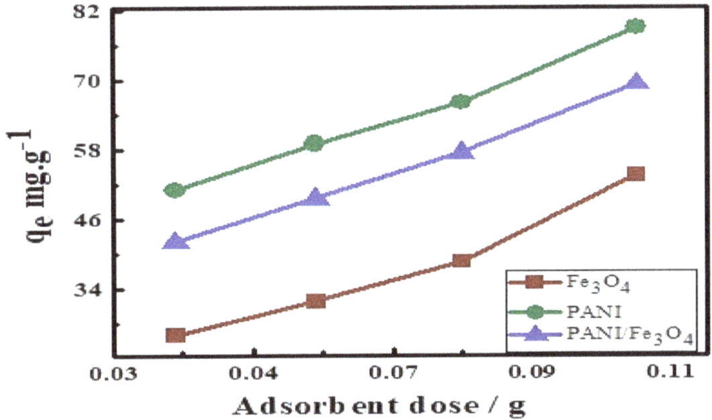

Figure 10. Effect of adsorbent dosage on adsorption of AB40 onto Fe_3O_4, PANI and PANI/Fe_3O_4 composite.

3.7. Adsorption Kinetics

Four kinetic equations namely pseudo 1st order, pseudo 2nd order, Elovich model and intra-particle diffusion models were used to analyze the adsorption data. The relationship between the amount of dye adsorbed on adsorbents and adsorption time was determined. Pseudo-first-order and pseudo-second-order equations are expressed in Equations (6) and (7) respectively as below;

$$\ln(q_e - q_t) = \ln q_e - k_1 t \tag{6}$$

$$\frac{t}{q_t} = \frac{1}{k_2 q_e^2} + \frac{t}{q_e} \tag{7}$$

where q_e (mg g^{-1}) is the equilibrium adsorption and q_t (mg g^{-1}) is the amount of dye adsorbed after time t (min). K_1 (min^{-1}) and K_2 (g mg^{-1} min^{-1}) are the rate constants of pseudo-first-order and pseudo-second-order equations respectively.

The Elovich kinetic model can be expressed as shown below in Equation (8);

$$q_t = \frac{1}{\beta} \ln(\alpha\beta) + \frac{1}{\beta} \ln t \tag{8}$$

where α (mg g^{-1} min^{-1}) shows an initial rate of adsorption and β (mg g^{-1}) is the desorption constant relating to the activation energy and the extent of surface coverage.

Intra-particle diffusion model is expressed in Equation (9);

$$q_t = k_d t^{1/2} + c \tag{9}$$

where K_d (g mg^{-1} $min^{-1/2}$) represent the rate of diffusion constant and C (mg g^{-1}) is the constant of boundary layer thickness.

The fitted curves of adsorption of AB40 onto Fe_3O_4, PANI and PANI/Fe_3O_4 composites are shown in Figure 11. The parameters of kinetics are summarized in Tables 3 and 4. The correlation factor of pseudo-first-order (R^2), are 0.812, 0.885 and 0.881 for adsorption of AB40 onto Fe_3O_4, PANI and PANI/Fe_3O_4 composites. These values indicate that adsorption of AB40 does not follow pseudo-first-order kinetics [83]. Similarly R^2 of Elovich model for PANI/Fe_3O_4 composites is 0.707 and intra-particle diffusion model for Fe_3O_4 is 0.864 indicating that these models also do not fit well for the adsorption data of AB40 on all of the three adsorbents [84]. R^2 values of pseudo-second-order equation show that the adsorption kinetics are more accurately described by this model (Table 3). Moreover, the q_e values calculated by the pseudo-second-order equation agree more closely with the adsorption isotherm values [85].

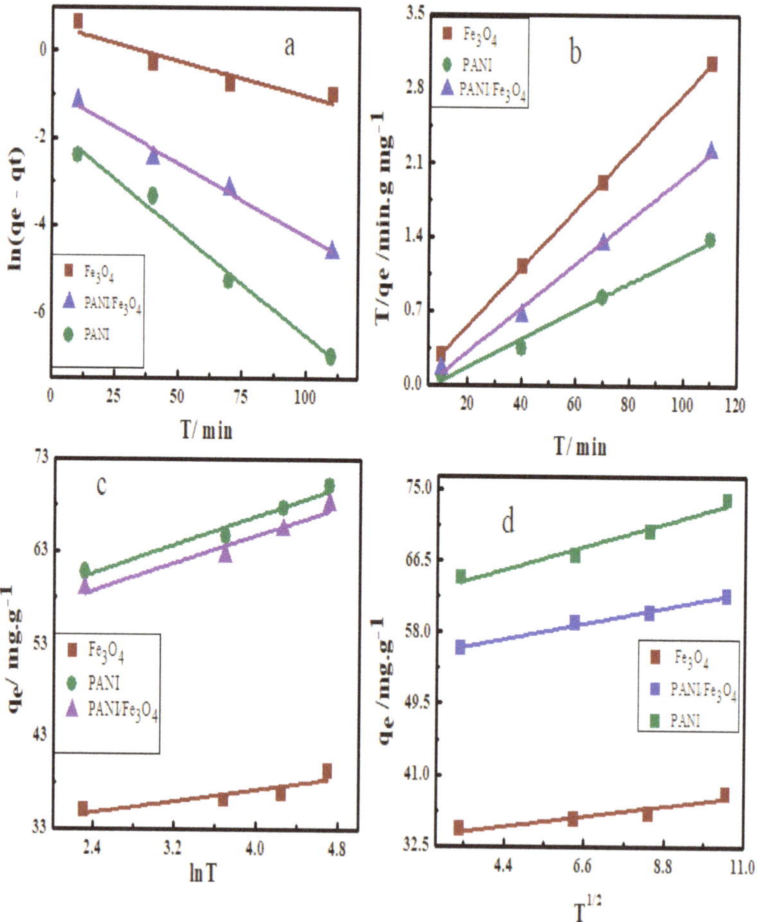

Figure 11. The Kinetics model (**a**) pseudo-first-order, (**b**) pseudo-second-order, (**c**) Elovich model and (**d**) intra-particle diffusion model for adsorption of AB40 on Fe_3O_4, PANI and PANI/Fe_3O_4 composite.

Table 4. Kinetics parameters for adsorption of AB40 on Fe_3O_4, PANI and PANI/Fe$_3$O$_4$ composite with Elovich model and intra-particle diffusion model.

	Elovich Model			Intra Particle Diffusion Model		
Adsorbents	α (mg g^{-1}min^{-1})	β (g mg^{-1})	R^2	k_d (g mg^{-1}min$^{-1/2}$)	C (mg g^{-1})	R^2
Fe_3O_4	131.9	0.258	0.911	0.556	32.45	0.864
PANI	439.8	0.637	0.929	1.257	61.89	0.916
PANI/ Fe$_3$O$_4$	378.7	0.267	0.707	0.847	53.47	0.917

3.8. Adsorption Mechanism

Two routes can be proposed for the adsorption of AB40 on the surface of PANI salt and PANI/Fe$_3$O$_4$ composite. In the first one, electrostatic interaction may occur between the molecules of AB40 and PANI. The second one involves the formation of an H-bond between the dye and –NH group of PANI. H-bond formation is also possible between AB40 and –OH group present on the surface of Fe$_3$O$_4$.

The electrostatic interactions are based on the fact that when dye is dissolved in water it splits into positively and negatively charged ions (Dye-SO_3^-). These negatively charged anions (Dye-SO_3^-) interact with positively charged sites ($-^+$NH–) on PANI surface. The enhancement of dye adsorption in acidic medium is good evidence of electrostatic interaction expressed in Section 3.4. Existence of physical forces (H-bond) is also supported by FTIR spectra shown in Figure 4B. After adsorption of AB40, all peaks in the spectra of PANI and PANI/Fe$_3$O$_4$ composite are shifted towards low-frequency values. Moreover, appearance of peak at 2356.7 cm^{-1} shows existence of AB40 adheres to the surface of PANI and PANI/Fe$_3$O$_4$ composites [72].

3.9. Thermodynamics of Adsorption

The nature of adsorption can be described well with thermodynamic parameters like Gibbs free energy, change in enthalpy and change in entropy. Values of Gibbs free energy were calculated by the equation shown below (Equation (10));

$$\Delta G = -RT\ln K_e \tag{10}$$

where K_e is the equilibrium constant, R is the gas constant having the value of 8.314 J K^{-1} mol^{-1} and T represents the Kelvin temperature. The negative sign of ΔG values shows that the adsorption of AB40 onto Fe$_3$O$_4$, PANI and PANI/ Fe$_3$O$_4$ composites are spontaneous (Table 5). The values of ΔG which range from −20 to zero kJ mol^{-1} show physical adsorption [47]. The values of ΔH and ΔS were calculated from the slope and intercept of van't Hoff equation respectively by plotting lnk_e vs. 1/T (Figure 12b). The van't Hoff equation is expressed as below;

$$\ln K_e = -\frac{\Delta H}{RT} + \frac{\Delta S}{R} \tag{11}$$

$$K_e = \frac{q_e}{c_e} \tag{11a}$$

where q_e (mg g^{-1}) is the adsorption maximum and C_e (mg L^{-1}) is the concentration of dye at equilibrium. The negative values of ΔH and ΔS shown in Table 4 show that adsorption is exothermic and correlate to the effect of temperature on adsorption expressed in Section 3.3. Activation energy also expresses the nature of adsorption. Its values are calculated from the slope of Arrhenius equation by plotting lnK vs. 1/T shown in Figure 12b. The Arrhenius equation is expressed as below;

$$\ln k = \ln A - \frac{Ea}{RT} \tag{12}$$

where K is the rate constant, A is Arrhenius constant, Ea is the activation energy, R is the general gas constant and T is kelvin temperature. The activation energy of adsorption of AB40 onto Fe_3O_4, PANI and PANI/Fe_3O_4 composites are 30.12, 22.09 and 26.13 kJ mol^{-1} showing physical adsorption. Ozcan and co-workers have demonstrated that physical adsorption is characterized by the activation energy values range from 5 to 40 kJ mol^{-1} and its higher values (40–800) kJ mol^{-1} express chemical adsorption [85].

Table 5. Activation energy and thermodynamic parameters of AB40 adsorption.

Adsorbents	ΔH (kJ mol^{-1})	ΔS (kJ mol^{-1})	ΔG (kJ mol^{-1})	Ea (kJ mol^{-1})
Fe_3O_4	−6.077	−0.026	−11.93	30.12
PANI	−8.993	−0.032	−19.87	22.09
PANI/Fe_3O_4	−10.62	−0.054	−19.75	26.13

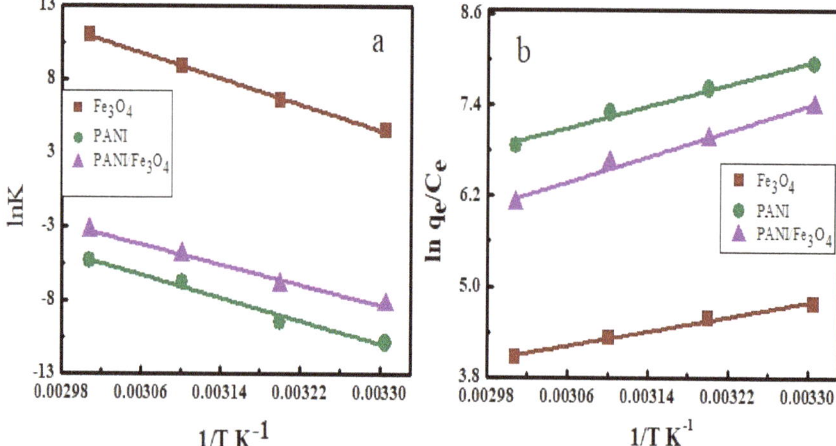

Figure 12. (a) Arrhenius plot and (b) van't Hoff plot for calculation of activation energy and thermodynamic parameters.

4. Conclusions

PANI, Fe_3O_4 and their composite can effectively be utilized for the removal of AB40 dye from aqueous environment. The comparison of adsorption behavior of the three materials for the uptake of AB40 reveals that the dye interaction with PANI was higher than both Fe_3O_4 and composites materials. This enhancement in adsorption on PANI can be attributed to the electrostatic interaction between oppositely charged sites of PANI and AB40. Greater number of active sites leading to physical forces also enhanced the adsorption of dye on PANI. In case of PANI/Fe_3O_4 composites the lone pair electrons present on the oxygen causes repulsive interaction with the negatively charged dye and reduces the adsorption. This fact was confirmed in the effect of ionic strength on adsorption where PANI/Fe_3O_4 composites showed higher adsorption than pristine PANI. The maximum amount of dye adsorbed on PANI, Fe_3O_4 and PANI/Fe_3O_4 composites were 264.9, 130.5 and 216.9 mg g^{-1}, respectively. The enhancement of adsorption on PANI was also supported by its smaller value of activation energy than Fe_3O_4 and PANI/Fe_3O_4 composites. Freundlich adsorption isotherm model fitted more closely with the adsorption data. The adsorption was high in acidic conditions and followed pseudo-second-order kinetics. The negative sign of the values of enthalpy changes, entropy changes and Gibbs free energy changes confirmed spontaneous and exothermic nature of adsorption.

Supplementary Materials: The following are available online at http://www.mdpi.com/1996-1944/12/18/2854/s1, Figure S1: UV-Visible calibration curve of AB40.

Author Contributions: A.M. wrote the original draft and executed all the experiments. A.u.H.A.S. supervised. A.u.H.A.S. and S.B. contributed to writing and corrected and edited the manuscript.

Funding: This research was funded by the Higher Education Commission Pakistan (project No. 20-1647 and 20–111/NRPU/R&D/HEC). The APC was funded by the German Research Foundation and the Open Access Publication Funds of the Technische Universität Braunschweig.

Acknowledgments: We acknowledge support from the German Research Foundation and the Open Access Publication Funds of the Technische Universität Braunschweig. S.B. wants to thank the Alexander von Humboldt Foundation Germany for support.

Conflicts of Interest: The authors declare no conflict of interest.

References

1. Bhadraa, S.; Khastgir, D.; Singhaa, N.K.; Leeb, J.H. Progress in preparation, processing and applications of polyaniline. *Prog. Polym. Sci.* **2009**, *34*, 783–810. [CrossRef]
2. Kiskan, B.; Yagci, Y. Synthesis and characterization of thermally curable polyacetylenes by polymerization of propargyl benzoxazine using rhodium catalyst. *Polymer* **2008**, *49*, 2455–2460. [CrossRef]
3. Shah, A.-U.-H.A.; Kamran, M.; Bilal, S.; Ullah, R. Cost Effective Chemical Oxidative Synthesis of Soluble and Electroactive Polyaniline Salt and Its Application as Anticorrosive Agent for Steel. *Materials* **2019**, *12*, 1527. [CrossRef] [PubMed]
4. Del Valle, M.A.; Diaz, F.R.; Armijo, F.; Soto, P. Electro-synthesis and characterization of polythiophene nano-wires/platinum nano-particles composite electrodes. Study of formic acid electro-catalytic oxidation. *Electrochim. Acta* **2012**, *71*, 277–282. [CrossRef]
5. Shen, C.; Sun, Y.; Yao, W.; Lu, Y. Facile synthesis of polypyrrole nanospheres and their carbonized products for potential application in high-performance supercapacitors. *Polymer* **2014**, *55*, 2817–2824. [CrossRef]
6. Fahim, M.; Shah, A.H.A.; Bilal, S. Highly stable and efficient performance of binder free symmetric supercapacitor fabricated with electroactive polymer synthesized via interfacial polymerization. *Materials* **2019**, *12*, 1626. [CrossRef] [PubMed]
7. Ganesan, R.; Shanmugam, S.; Gedanken, A. Pulsed sonoelectrochemical synthesis of polyaniline nanoparticles and their capacitance properties. *Synth. Met.* **2008**, *158*, 848–853. [CrossRef]
8. Kulkarni, S.B.; Joshi, S.S.; Lokhande, C.D. Facile and efficient route for preparation of nanostructured polyaniline thin films: Schematic model for simplest oxidative chemical polymerization. *Chem. Eng. J.* **2011**, *166*, 1179–1185. [CrossRef]
9. Vivekanandan, J.; Ponnusamy, V.; Mahudeswaran, A.; Vijayanand, P.S. Synthesis, characterization and conductivity study of polyaniline prepared by chemical oxidative and electrochemical methods. *Arch. Appl. Sci. Res.* **2011**, *3*, 147–153.
10. Xu, P.; Singh, A.; Kaplan, D.L. Enzymatic Catalysis in the Synthesis of Polyanilines and Derivatives of Polyanilines. *Adv. Polym. Sci.* **2006**, *194*, 69–94.
11. Abdolahi, A.; Hamzah, E.; Ibrahim, Z.; Hashim, S. Synthesis of Uniform Polyaniline Nanofibers through Interfacial Polymerization. *Materials* **2012**, *5*, 1487–1494. [CrossRef]
12. Guo, X.; Fei, G.T.; Su, H.; Zhang, L.D. Synthesis of polyaniline micro/nanospheres by a copper(II)-catalyzed self-assembly method with superior adsorption capacity of organic dye from aqueous solution. *J. Mater. Chem.* **2011**, *21*, 8618. [CrossRef]
13. Lee, S.; Hong, J.Y.; Jang, J. Synthesis and electrical response of polyaniline/poly(styrene sulfonate)-coated silica spheres prepared by seed-coating method. *J. Colloid. Inter. Sci.* **2013**, *398*, 33–38. [CrossRef] [PubMed]
14. Zhang, Y.; Shao, Y.; Zhang, T.; Meng, G.; Wang, F. The effect of epoxy coating containing emeraldine base and hydrofluoric acid doped polyaniline on the corrosion protection of AZ91D magnesium alloy. *Corros. Sci.* **2011**, *53*, 3747–3755. [CrossRef]
15. Xu, W.; Zhao, K.; Niu, C.; Zhang, L.; Cai, Z.; Han, C.; He, L.; Shen, T.; Yan, M.; Qu, L. Heterogeneous branched core–shell SnO2–PANI nanorod arrays with mechanical integrity and three dimensional electron transport for lithium batteries. *Nano Energy* **2014**, *8*, 196–204. [CrossRef]
16. Farooq, S.; Tahir, A.; Krewer, U.; Shah, A.A.; Bilal, S. Efficient photocatalysis through conductive polymer coated FTO counter electrode in platinum free dye sensitized solar cells. *Electrochim. Acta* **2019**, *320*, 134544. [CrossRef]

17. Mujawar, S.H.; Ambade, S.B.; Battumur, T.; Ambade, R.B.; Lee, S.H. Electropolymerization of polyaniline on titanium oxide nanotubes for supercapacitor application. *Electrochim. Acta* **2011**, *56*, 4462–4466. [CrossRef]
18. Li, R.; Liu, L.; Yang, F. Preparation of polyaniline/reduced graphene oxide nanocomposite and its application in adsorption of aqueous Hg(II). *Chem. Eng. J.* **2013**, *229*, 460–468. [CrossRef]
19. Shen, J.; Shahid, S.; Amura, I.; Sarihan, A.; Mi Tian, M.; Emanuelsson, E.A. Enhanced adsorption of cationic and anionic dyes from aqueous solutions by polyacid doped polyaniline. *Synth. Met.* **2018**, *245*, 151–159. [CrossRef]
20. Huang, Y.; Li, J.; Chen, X.; Wang, X. Applications of conjugated polymer based composites in wastewater purification. *RSC Adv.* **2014**, *4*, 62160–62178. [CrossRef]
21. Lee, Y.M.; Nam, S.Y.; Ha, S.Y. Pervaporation of water/isopropanol mixtures through polyaniline membranes doped with poly(acrylic acid). *J. Membr. Sci.* **1999**, *159*, 41–46.
22. Bober, P.; Stejskal, J.; Trchováa, M.; Prokes, J. In-situ prepared polyaniline–silver composites: Single-and two-step strategies. *Electrochim. Acta* **2014**, *122*, 259–266. [CrossRef]
23. He, K.; Li, M.; Guo, L. Preparation and photocatalytic activity of PANI-CdS composites for hydrogen evolution. *Int. J. Hydrogen Energy* **2012**, *37*, 755–759. [CrossRef]
24. Yilmaz, H.; Zengin, H.S.; Unal, H.I. Synthesis and electrorheological properties of polyaniline/silicon dioxide composites. *J. Mater. Sci.* **2012**, *47*, 5276–5286. [CrossRef]
25. Wanga, J.G.; Yang, Y.; Huanga, Z.H.; Kang, F. Interfacial synthesis of mesoporous MnO2/polyaniline hollow spheres and their application in electrochemical capacitors. *J. Power Sources* **2012**, *204*, 236–243. [CrossRef]
26. Chandraa, S.; Lang, H.; Bahadur, D. Polyaniline-iron oxide nanohybrid film as multi-functional label-free electrochemical and biomagnetic sensor for catechol. *Anal. Chim. Acta* **2013**, *795*, 8–14. [CrossRef] [PubMed]
27. Vellakkat, M.; Kamath, A.; Raghu, S.; Chapi, S.; Hundekal, D. Dielectric Constant and Transport Mechanism of Percolated Polyaniline Nanoclay Composites. *Ind. Eng. Chem. Res.* **2014**, *53*, 16873–16882. [CrossRef]
28. Salem, M.A.; Salem, I.A.; Hanfy, M.G.; Ahmed, B.; Zak, A.B. Removal of titan yellow dye from aqueous solution by polyaniline/Fe3O4 nanocomposite. *Eur. Chem. Bull.* **2016**, *5*, 113–118.
29. Das, S.; Chakraborty, P.; Ghosh, R.; Paul, S.; Mondal, S.; Panja, A.; Nand, A.K. Folic Acid-Polyaniline Hybrid Hydrogel for Adsorption/Reduction of Chromium(VI) and Selective Adsorption of Anionic Dye from Water. *Chem. Eng.* **2017**, *5*, 9325–9337. [CrossRef]
30. Neuberger, T.; Schöpf, B.; Hofmann, H.; Hofmann, M.; Rechenberg, B. Superparamagnetic nanoparticles for biomedical applications: Possibilities and limitations of a new drug delivery system. *J. Magn. Magn. Mater.* **2005**, *293*, 483–496. [CrossRef]
31. Ito, A.; Shinkai, M.; Honda, H.; Kobayashi, T. Medical application of functionalized magnetic nanoparticles. *J. Biosci. Bioeng.* **2005**, *100*, 1–11. [CrossRef] [PubMed]
32. Pankhurst, Q.A.; Connolly, J.; Jones, S.K.; Dobson, J. Applications of magnetic nanoparticles in biomedicine. *J. Phys. D Appl. Phys.* **2003**, *36*, 167–181. [CrossRef]
33. Khurshid, H.; Hadjipanayis, C.; Chen, H.W.; Li, H.; Mao, H.; Machaidze, R.; Tzitzios, V.; Hadjipanay, G.C. Core/shell structured iron/iron-oxide nanoparticles as excellent MRI contrast enhancement agents. *J. Magn. Magn. Mater.* **2013**, *331*, 17–20. [CrossRef]
34. Wang, G.; Chang, Y.; Wang, L.; Wei, Z.; Kang, J.; Sang, L.; Dong, X.; Chen, G.; Wang, H.; Qi, H. Preparation and characterization of PVPI-coated Fe3O4 nanoparticles as an MRI contrast agent. *J. Magn. Magn. Mater.* **2013**, *340*, 57–60. [CrossRef]
35. Hyun Do, S.I.; Hoon Jo, Y.; Park, J.Y.; Hong, S.H. As3+removal by Ca–Mn–Fe3O4with and without H2O2: Effects ofcalcium oxide in Ca–Mn–Fe3O4. *J. Hazard. Mater.* **2014**, *280*, 322–330.
36. Hu, J.; Irene, M.C.; Chen, G. Fast Removal and Recovery of Cr(VI) Using Surface-Modified Jacobsite (MnFe2O4) Nanoparticles. *Langmuir* **2005**, *21*, 11173–11179. [CrossRef]
37. Aphesteguy, J.C.; Kurlyandskaya, G.V.; de Celis, J.P.; Safronov, A.P.; Schegoleva, N.N. Magnetite nanoparticles prepared by co-precipitation method in different conditions. *Mater. Chem. Phys.* **2015**, *161*, 243–249. [CrossRef]
38. Amer, M.A.; Meaz, T.M.; Attalah, S.S.; Ghoneim, A.I. Structural and magnetic characterization of the Mg0.2_xSrxMn0.8Fe2O4 nanoparticles. *J. Magn. Magn. Mater.* **2014**, *363*, 60–65. [CrossRef]
39. Lam, U.T.; Mammucari, R.; Suzuki, K.; Foster, N.R. Processing of iron oxide nanoparticles by supercritical fluids. *Ind. Eng. Chem. Res.* **2008**, *47*, 599–614. [CrossRef]
40. Tavakoli, A.; Sohrabi, M.; Kargari, A. A review of methods for synthesis of nano structured metals with emphasis on iron compounds. *Chem. Pap.* **2007**, *61*, 151–170. [CrossRef]

41. Teja, A.S.; Koh, P. Y Synthesis, properties, and applications of magnetic iron oxide nanoparticles. *Prog. Cryst. Growth Charact. Mater.* **2009**, *55*, 22–45. [CrossRef]

42. Majewski, P.; Thierry, B. Functionalized magnetite nanoparticles synthesis, properties, and bio-applications. *Solid State Mater. Sci.* **2007**, *32*, 203–215. [CrossRef]

43. Jia, Z.; Yujun, W.; Yangcheng, L.; Jingyu, M.; Guangsheng, L. In situ preparation of magnetic chitosan/Fe3O4 composite nanoparticles in tiny pools of water-in-oil microemulsion. *React. Funct. Polym.* **2006**, *66*, 1552–1558.

44. Khan, A.; Aldwayyan, A.S.; Alhoshan, M.; Alsalhi, M. Synthesis by in situ chemical oxidative polymerization and characterization of polyaniline/iron oxide nanoparticle composite. *Polym. Int.* **2010**, *59*, 1690–1694. [CrossRef]

45. Rasha, M.K. Synthesis, characterization, magnetic and electrical properties of the novel conductive and magnetic Polyaniline/MgFe2O4 nanocomposite having the core–shell structure. *J. Alloys Compounds* **2011**, *509*, 9849–9857.

46. Bhaumik, M.; Choi, H.J.; McCrindle, R.I.; Maity, A. Composite nanofibers prepared from metallic iron nanoparticles and polyaniline: High performance for water treatment applications. *J. Colloid. Interface Sci.* **2014**, *425*, 75–82. [CrossRef] [PubMed]

47. Mohamed, A.S. The role of polyaniline salts in the removal of direct blue 78 from aqueous solution: A kinetic study. *React. Funct. Polym.* **2010**, *70*, 707–714.

48. Cui, H.; Qian, Y.; Li, Q.; Zhang, Q.; Zhai, J. Adsorption of aqueous Hg(II) by a polyaniline/attapulgite composite. *Chem. Eng. J.* **2012**, *211*, 216–223.

49. Muhammad, A.; Shah, A.H.A.; Bilal, S.; Rahman, G. Basic Blue Dye Adsorption from Water using Polyaniline/Magnetite(Fe3O4) Composites: Kinetic and Thermodynamic Aspects. *Materials* **2019**, *12*, e1764. [CrossRef]

50. Keyhanian, F.; Shariati, S.; Faraji, M.; Hesabi, M. Magnetite nanoparticles with surface modification for removal of methyl violet from aqueous solutions. *Arab. J. Chem.* **2016**, *9*, 348–354. [CrossRef]

51. Patra, B.N.; Majhi, D. Removal of Anionic Dyes from Water by Potash Alum Doped Polyaniline: Investigation of Kinetics and Thermodynamic Parameters of Adsorption. *J. Phys. Chem. B* **2015**, *119*, 8154–8164. [CrossRef] [PubMed]

52. Hosseini, S.H.; Asadnia, A. Polyaniline/Fe3O4 coated on MnFe2O4 nanocomposite: Preparation, characterization, and applications in microwave absorption. *Int. J. Phys. Sci.* **2013**, *8*, 1209–1217.

53. Tung, L.M.; Cong, N.X.; Huy, L.T.; Lan, N.T.; Phan, V.N.; Hoa, N.Q.; Vinh, L.K.; Thinh, N.V.; Tai, L.T.; Ngo, D.T.; et al. Synthesis, Characterizations of Superparamagnetic Fe3O4–Ag Hybrid Nanoparticles and Their Application for Highly Eective Bacteria Inactivation. *J. Nanosci. Nanotechnol.* **2016**, *16*, 5902–5912. [CrossRef]

54. Bachan, N.; Asha, A.; Jeyarani, W.J.; Kumar, D.A.; Shyla, J.M. A Comparative Investigation on the Structural, Optical and Electrical Properties of SiO2–Fe3O4 Core–Shell Nanostructures with Their Single Components. *Acta Metall. Sin. Engl. Lett.* **2015**, *28*, 1317–1325. [CrossRef]

55. Bilal, S.; Gul, S.; Holze, R.; Shah, A.A. An impressive emulsion polymerization route for the synthesis of highly soluble and conducting polyaniline salt. *Synth. Met.* **2015**, *206*, 131–144. [CrossRef]

56. Hatamzadeh, M.; Ahar, M.J.; Jaymand, M. In Situ Chemical Oxidative Graft Polymerization of Aniline from Fe3O4 Nanoparticles. *Int. J. Nanosci. Nanotechnol.* **2012**, *8*, 51–60.

57. Akar, T.; Ozcan, A.S.; Tunali, S.; Ozcan, A. Biosorption of a textile dye (Acid Blue 40) by cone biomass of Thuja orientalis: Estimation of equilibrium, thermodynamic and kinetic parameters. *Bioresour. Technol.* **2008**, *99*, 3057–3065. [CrossRef] [PubMed]

58. Khoshsang, H.; Ghaffarinejad, A.; Kazemi, H.; Jabarian, S. Synthesis of Mesoporous Fe3O4 and Fe3O4/C Nanocomposite for Removal of Hazardous Dye from Aqueous Media. *J. Water Environ. Nanotechnol.* **2018**, *3*, 191–206.

59. Ballav, N.; Debnath, S.; Pillay, K.; Maity, A. Efficient removal of Reactive Black from aqueous solution usingpolyaniline coated ligno-cellulose composite as a potential adsorbent. *J. Mol. Liq.* **2015**, *209*, 387–396. [CrossRef]

60. Konicki, W.; Pełech, I.; Mijowska, E.; Jasinska, I. Adsorption of anionic dye Direct Red 23 onto magnetic multi-walled carbonnanotubes-Fe3C nanocomposite: Kinetics, equilibrium and thermodynamics. *Chem. Eng. J.* **2012**, *210*, 87–95. [CrossRef]

61. Sun, M.; Zhu, A.; Zhang, Q.; Liu, Q. A facile strategyto synthesize mono disperse super paramagnetic OA-modified Fe3O4 nanoparticles with PEG assistant. *J. Magn. Magn. Mater.* **2014**, *369*, 49–54. [CrossRef]
62. Asgari, S.; Fakhari, Z.; Berijanic, S. Synthesis and Characterization of Fe3O4 Magnetic Nanoparticles Coated with Carboxymethyl Chitosan Grafted Sodium Methacrylate. *J. Nanostruct.* **2014**, *4*, 55–63.
63. Ömeroglu Ay, C.; Özcan, A.S.; Erdogan, Y.; Özcan, A. Characterization of Punica granatum L. peels and quantitatively determination of its biosorption behavior towards lead(II) ions and Acid Blue 40. *Colloids Surf. B* **2012**, *100*, 197–204. [CrossRef] [PubMed]
64. Gul, H.; Shah, A.A.; Bilal, S. Fabrication of Eco-Friendly Solid-State Symmetric Ultracapacitor Device Based on Co-Doped PANI/GO Composite. *Polymers* **2019**, *11*, 1315. [CrossRef] [PubMed]
65. Kellenberger, A.; Dmitrieva, E.; Dunsch, L. Structure Dependence of Charged States in Linear Polyaniline as Studied by In Situ ATR-FTIR Spectroelectrochemistry. *J. Phys. Chem. B* **2012**, *116*, 4377–4385. [CrossRef] [PubMed]
66. Ding, S.; Mao, H.; Zhang, W. Fabrication of DBSA-Doped Polyaniline Nanorods by Interfacial Polymerization. *J. Appl. Polym. Sci.* **2008**, *109*, 2842–2847. [CrossRef]
67. Umare, S.S.; Shambharkar, B.H.; Ningthoujam, R.S. Synthesis and characterization of polyaniline–Fe3O4 nanocomposite: Electrical conductivity, magnetic, electrochemical studies. *Synth. Met.* **2010**, *160*, 1815–1821. [CrossRef]
68. Oppong, S.O.B.; Anku, W.; Shukla, S.K.; Govender, P. Lanthanum doped–TiO2 decorated on graphene oxide nanocomposite: A photocatalyst for enhanced degradation of acid blue 40 under simulated solar light. *Adv. Mater. Lett.* **2017**, *8*, 295–302. [CrossRef]
69. Ayad, M.; Zaghlol, S. Nanostructured crosslinked polyaniline with high surface area: Synthesis, characterization and adsorption for organic dye. *Chem. Eng. J.* **2012**, *204*, 79–86. [CrossRef]
70. Germain, J.; Frechet, J.M.; Svec, F. Hypercrosslinked polyanilines with nanoporous structure and high surface area: Potential adsorbents for hydrogen storage. *J. Mater. Chem.* **2007**, *17*, 4989–4997. [CrossRef]
71. Kegl, T.; Ban, I.; Lobnik, A.; Košak, A. Synthesis and characterization of novel γ-Fe2O3-NH4OH@SiO2(APTMS) nanoparticles for dysprosium adsorption. *J. Hazard. Mater.* **2019**, *378*, 120764. [CrossRef] [PubMed]
72. Javadian, H.; Angaji, M.T.; Naushad, M. Synthesis and characterization of polyaniline/g-alumina nanocomposite: A comparative study for the adsorption of three different anionic dyes. *J. Ind. Eng. Chem.* **2014**, *20*, 3890–3900. [CrossRef]
73. Crini, G. Kinetic and equilibrium studies on the removal of cationic dyes from aqueous solution by adsorption onto a cyclodextrin polymer. *Dyes Pigm.* **2008**, *77*, 415–426. [CrossRef]
74. Sharma, P.; Das, M.R. Removal of a Cationic Dye from Aqueous Solution Using Graphene Oxide Nanosheets: Investigation of Adsorption Parameters. *J. Chem. Eng. Data* **2013**, *58*, 151–158. [CrossRef]
75. Bhatt, A.S.; Sakaria, P.L.; Vasudevan, M.; Radheshyam, R.; Sudheesh, P.N.; Bajaj, H.C.; Mody, H.M. Adsorption of an anionic dye from aqueous medium by organoclays: Equilibrium modeling, kinetic and thermodynamic exploration. *RSC Adv.* **2012**, *2*, 8663–8671. [CrossRef]
76. Song, W.; Gao, B.; Xu, X.; Xing, L.; Han, S.; Duan, P.; Song, W.; Jia, R. Adsorption–desorption behavior of magnetic amine/Fe3O4functionalized biopolymer resin towards anionic dyes from wastewater. *Bioresour. Technol.* **2016**, *210*, 123–130. [CrossRef] [PubMed]
77. Mittal, A.; Mittal, J.; Malviya, A.; Gupta, V.K. Adsorptive removal of hazardous anionic dye "Congo red" from wastewaterusing waste materials and recovery by desorption. *J. Colloid Interface Sci.* **2009**, *340*, 16–26. [CrossRef]
78. Patil, M.R.; Khairnar, S.D.; Shrivastava, V.S. Synthesis, characterisation of polyaniline–Fe3O4 magnetic nanocomposite and its application for removal of an acid violet 19 dye. *Appl. Nanosci.* **2016**, *6*, 495–502. [CrossRef]
79. Abramian, L.; El-Rassy, H. Adsorption kinetics and thermodynamics of azo-dye Orange II onto highly porous titania aerogel. *Chem. Eng. J.* **2009**, *150*, 403–410. [CrossRef]
80. German-Heins, J.; Flury, M. Sorption of Brilliant Blue FCF in soils as affected by pH and ionic strength. *Geoderma* **2000**, *97*, 87–101. [CrossRef]
81. Alberghina, G.; Bianchini, R.; Fichera, M.; Fisichella, S. Dimerization of Cibacron Blue F3GA and other dyes: Influence of salts and temperature. *Dyes. Pigm.* **2000**, *46*, 129–137. [CrossRef]
82. Mahanta, D.; Madras, G.; Radhakrishnan, S.; Patil, S. Adsorption and Desorption Kinetics of Anionic Dyes on Doped Polyaniline. *J. Phys. Chem. B* **2009**, *113*, 2293–2299. [CrossRef] [PubMed]

83. Cao, J.S.; Lin, J.X.; Fang, F.; Zhang, M.T.; Hu, Z.R. A new absorbent by modifying walnut shell for the removal of anionic dye: Kinetic and thermodynamic studies. *Bioresour. Technol.* **2014**, *163*, 199–205. [CrossRef] [PubMed]
84. Weng, C.H.; Lin, Y.T.; Tzeng, T.W. Removal of Methylene Blue from Aqueous Solution by Adsorption onto Pineapple Leaf Powder. *J. Hazard. Mater.* **2009**, *170*, 417–424. [CrossRef]
85. Özcan, S.; Erdem, B.; Özcan, A. Adsorption of Acid Blue 193 from Aqueous Solutions Onto Na-Bentonite and DTMA-Bentonite. *J. Colloid Interface Sci.* **2004**, *280*, 44–54. [CrossRef] [PubMed]

 *materials*

Article

Enhanced Adsorptive Properties and Pseudocapacitance of Flexible Polyaniline-Activated Carbon Cloth Composites Synthesized Electrochemically in a Filter-Press Cell

César Quijada [1,*], Larissa Leite-Rosa [1], Raúl Berenguer [2] and Eva Bou-Belda [1]

[1] Departamento de Ingeniería Textil y Papelera, Universitat Politècnica de València. Pza Ferrándiz y Carbonell, Alcoy, E-03801 Alicante, Spain
[2] Departamento de Química Física e Instituto Universitario de Materiales, Universidad de Alicante, Apartado 99, E-03080 Alicante, Spain
* Correspondence: cquijada@txp.upv.es; Tel.: +34-966-528-419

Received: 30 June 2019; Accepted: 1 August 2019; Published: 7 August 2019

Abstract: Electrochemical polymerization is known to be a suitable route to obtain conducting polymer-carbon composites uniformly covering the carbon support. In this work, we report the application of a filter-press electrochemical cell to polymerize polyaniline (PAni) on the surface of large-sized activated carbon cloth (ACC) by simple galvanostatic electropolymerization of an aniline-containing H_2SO_4 electrolyte. Flexible composites with different PAni loadings were synthesized by controlling the treatment time and characterized by means of Scanning Electron microscopy (SEM), X-Ray Photoelectron Spectroscopy (XPS), physical adsorption of gases, thermogravimetric analysis (TGA), cyclic voltammetry and direct current (DC) conductivity measurements. PAni grows first as a thin film mostly deposited inside ACC micro- and mesoporosity. At prolonged electropolymerization time, the amount of deposited PAni rises sharply to form a brittle and porous, thick coating of nanofibrous or nanowire-shaped structures. Composites with low-loading PAni thin films show enhanced specific capacitance, lower sheet resistance and faster adsorption kinetics of Acid Red 27. Instead, thick nanofibrous coatings have a deleterious effect, which is attributed to a dramatic decrease in the specific surface area caused by strong pore blockage and to the occurrence of contact electrical resistance. Our results demonstrate that mass-production restrictions often claimed for electropolymerization can be easily overcome.

Keywords: conducting polymer; emeraldine salt state; valence band; flexible composite electrode; dye adsorption kinetics; pseudo-second order model; capacitance; electrical conductivity

---

1. Introduction

Conducting polymers (CPs) are a fascinating family of organic materials that can be easily synthesized with a large diversity of chemical structures and a wide variety of micro- and nano-morphologies, in order to get tailored macroscopic physical and chemical properties [1]. Further, facile and reversible doping enables a set of unique and tunable optical, electronic and redox properties, particularly an electrical conductivity ranging from insulating to metallic. For these reasons, CPs have found promising application as flexible and lightweight functional materials in opto-electronic devices (e.g., light-emitting diodes, thin-film transistors, electrochromic displays, etc.), energy conversion and storage systems (rechargeable batteries, supercapacitors, solar cells or thermoelectric devices), as well as sensor and actuator devices [1–4]. Owing to the ease of synthesis, low cost, good environmental and chemical stability, high electrical conductivity and capacitance, together with its electrochromic character and ion-exchange properties [5], polyaniline (PAni) has emerged as one of the most industrially important CPs today. These virtues have made PAni an attractive material for a broad spectrum of technologically important applications, such as Li-ion batteries, supercapacitors, electromagnetic interference shielding, electrochemical sensors, and anti-corrosion coatings [1,6]. Also, it is worth mentioning the advent of recent new applications in solid-phase micro-extraction [7] or as (electro)adsorbent/ion-exchange materials for environmental issues [8–10], where the affinity of target pollutants for PAni active groups (charged and neutral amine and imine groups) is exploited to an advantage.

However, PAni has the main shortcomings of a relatively low porosity, specific surface area, slow degradation, and poor mechanical stability because of volume changes produced upon repeated charge/discharge process [2,4,11,12]. Composites of PAni with carbon materials, metal oxides, natural and modified clay minerals, zeolites or mesoporous silica [2,4,9,10], have been extensively investigated with the goal of overcoming these disadvantages and further improving existing properties via synergistic effects. Carbons are outstanding versatile materials with regards to their use as composite supports because of the wide availability of allotropes, microtextures, 0 to 3D dimensionality, and macroscopic forms. In addition, their excellent chemical and thermal stability, good electrical conductivity, large specific surface area, wide range of pore structures, and mechanical strength [11, 12] make them particularly suitable as composite constituents for electrochemical applications or environmental adsorbents. To date, much work has been undertaken on PAni composites with carbon nanotubes [13,14], graphene/reduced graphene oxide nanosheets [4,13], porous carbon foams or rods [15,16] and activated carbon fibers [11,17] or powders [18,19]. Only recently, activated carbon fiber cloths (ACC) have gained popularity as inexpensive, highly porous and flexible mechanical supports for conducting polymers in electrode materials for wearable power microelectronics or roll-up electrochemical systems [8,20–25]. Furthermore, the highly porous 3D conductive framework of the carbon fabric allowed PAni-ACC composites to be directly used as electrodes without the use of insulating binders and conductive additives. Thus, the unique combination of wide accessibility of the fabric pore nanoarchitecture [24]; the binderless CP–carbon interface [25]; and simple, fast and reversible surface redox reactions in PAni [24] has shown to provide the composite with shortened path lengths for direct electron transfer and fast ion transport [20,22,24,25].

PAni-carbon composites have been synthesized by conventional in situ chemical, emulsion/interfacial, vapor chemical or electrochemical polymerization, just to cite a few [6,26,27]. Electrochemical methods have proven to be simple and powerful tools to produce uniform, adherent thin PAni films over a number of different conductive substrates [6], including porous carbon materials, which have served as hard templates capable of transcribing their nanostructure to the growing polymer. However, it is often claimed that electrochemical routes for the preparation of PAni are limited in terms of mass production [23]. In fact, the vast majority of examples of electrosynthesized carbon-PAni composites deal with small size (typically 1 to 2 cm^2) samples, while research treating significantly larger areas is scarce [8,28] and generally does not focus on a systematic study of the effect of synthesis variables on the structure and properties of the resulting composites.

In this work, we show the feasibility of producing flexible PAni-ACC composites of large size (~20 cm^2) by simple galvanostatic (i.e., constant current) electrolpolymerization in a filter-press electrolyzer, a type of cell design advantageous for industrial scaling-up [29]. The polymer loading density was controlled by changing the electropolymerization time (i.e., the amount of passed charge), and the surface microstructure, chemical composition, porous texture, and thermal stability of the fabricated composites were studied by Scanning Electron microscopy (SEM), X-Ray Photoelectron Spectroscopy (XPS), N_2/CO_2 adsorption experiments, and thermogravimetric analysis (TGA). Some important properties for practical application as electrodes in supercapacitors, secondary batteries or as adsorbent materials in dyestuff effluent treatment were examined and correlated with their structural and chemical features. For this purpose, the capacitance and electrical conductivity were derived from cyclic voltammetry (CV) and four-strip probe conductance experiments, and the liquid-phase adsorptive capability was studied by using aqueous solutions of Acid Red 27, a model dye of anionic azo dyes used in the food and textile industries.

2. Materials and Methods

2.1. Materials

A viscose-based activated carbon cloth (CTex-20, knitted fabric, areal density: ~18 mg cm^{-2}, mean fiber diameter: ~10 μm) was purchased from Mast Carbon International Ltd. (Hampshire, UK). The carbon fabric was cut in sheets of 45 mm × 55 mm. The untreated carbon samples were repeatedly washed with distilled water until constant pH. Then, they were dried in an oven at 80 °C overnight and finally allowed to cool down in a desiccator. Analytical reagent-grade aniline (≥99.0%) and sulfuric acid (98%) were supplied by Merck. Prior to use, aniline was distilled under vacuum and stored in the dark at 5 °C. Solutions of aniline (0.1 M) in aqueous H_2SO_4 electrolyte (0.5 M) were made up with distilled water. Background electrolyte solutions (0.5 M H_2SO_4) for cyclic voltammetry were prepared from Millipore Milli-Q water. The anionic monoazo dye Acid Red 27 (Colour Index no. 16185, empirical formula $C_{20}H_{11}N_2Na_3O_{10}S_3$) was purchased from Sigma-Aldrich (91 wt.% dye content) and used as received.

2.2. Preparation of Hybrid PAni-ACC Composites

Polyaniline was polymerized on the carbon fabric substrate by galvanostatic treatment in an undivided home-made filter-press electrolyzer, designed for the electrochemical processing of textile-structured materials (Figure 1a) [30,31]. The cell was assembled in a flow-through parallel plate-and-frame configuration, with two identical stainless steel (SS) mesh electrodes separated by a 5-mm-thick plastic spacer, providing an interelectrode rectangular channel of 20 cm^2. A dry ACC sheet (typically 0.4 g) was pressed against the anode SS mesh with a silicone rubber sealing gasket to leave an exposed geometric area of 20 cm^2. The aniline-containing electrolyte was fed into the cell with the aid of a peristaltic pump (Dinko Instruments D-21V, Barcelona, Spain), until excess solution was collected at the outlet. As a pre-conditioning step, the system was left at open circuit for 30 min to allow carbon pore filling and facilitate aniline adsorption. Then, an input current of 100 mA (~14 mA cm^{-2} g^{-1}) was passed through the cell for different processing times (10–120 min) at room temperature. The anode potential was measured against an Ag/AgCl reference electrode connected through a Luggin capillary drilled in the plastic spacer, and it was found to lie within the range 0.6–0.8 V. A schematic view of the electrochemical set-up is shown in Figure 1b. After the prescribed electropolymerization time, the modified carbon cloth was washed repeatedly with 0.5 M H_2SO_4 and subsequently with distilled water. The obtained PAni-ACC composite was dried at 40 °C under dynamic vacuum for 24 h and stored in a desiccator until further characterization studies.

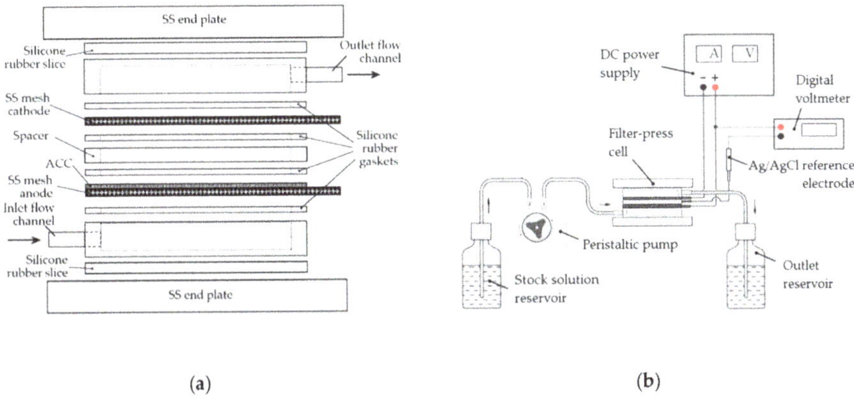

(**a**) (**b**)

Figure 1. (**a**) Cut-away view of the electrochemical filter-press cell; (**b**) Schematic diagram of the electropolymerization system.

2.3. Materials Characterization

The morphology and micro-structure of the unmodified ACC and the PAni-ACC composites were examined by SEM. Secondary electron micrographs were obtained with a Phenom Microscope (FEI Co., Hillsboro, USA). X-Ray Photoelectron spectroscopy (XPS) was conducted in a K-ALPHA spectrometer (ThermoFisher Scientific) by using a microfocused monochromatized Al Kα radiation (1486.6 eV) of 400 μm spot size at 173 K and a base pressure below 5×10^{-10} kPa. Photoelectrons were collected into a hemispherical analyzer operated in the constant energy mode at pass energy of 50 eV for narrow core-level and valence band spectra. Peak binding energies (BE) were referenced to the principal C1s line at 284.6 eV and given to an accuracy of ±0.2 eV. XPS data were analyzed with Thermo Scientific™ Avantage software. A smart correction function was used for background correction. Peak synthesis was done with mixed Gaussian (70%)/Lorentzian (30%) function lineshapes. Surface charging was compensated with a flood electron gun.

The porous texture was determined by physical adsorption of N_2 (at 77 K) and CO_2 (at 273 K) by using an automatic adsorption system (Autosorb-6, Quantachrome Instruments, Boynton Beach, USA). In order to remove moisture and adsorbed gases while avoiding thermal degradation of PAni, the samples were previously out-gassed under vacuum at 423 K for 4 h. The apparent specific surface area (S_{BET}) and total micropore volume (d < 2 nm, V_μ) were calculated from the N_2 isotherm by applying the Brunauer-Emmett-Teller (BET) and the Dubinin-Radushkevich (DR) equations, respectively [30]. The DR theory was also applied to obtain the ultramicropore volume (d < 0.7 nm, $V_{ultra\ \mu}$) from the CO_2 isotherm [30]. The good fitting of the N_2 and CO_2 adsorption data to the DR equation ($R^2 > 0.99$) validated the application of this method for the studied materials. The N_2 uptake at a relative pressure near to capillary condensation (~0.97) was used to calculate the total pore volume (V_{tp}) [32]. The mesopore volume (2 nm < d < 50 nm, V_{meso}) was estimated as the difference between total and micropore volumes.

The thermal stability of the samples was studied by thermogravimetric analysis (Mettler Toledo 851E/1600/LG) under a He stream at a flow rate of 100 mL min^{-1}. About 10 mg of sample was placed in a 70 μL aluminum crucible and submitted to a two-stage heating protocol at a rate of 20 °C min^{-1}. First, the samples were heated from 25 °C to 120 °C and kept at the latter temperature for 30 min for drying. Then, they were allowed to cool down to thermal equilibrium at 25 °C. In the second stage, the samples were heated up to 1000 °C.

2.4. Electrochemical and Sheet Resistance Measurements

Cyclic voltammetry (CV) experiments were conducted in an all-Pyrex glass cell with provision for a standard three-electrode assembly. Round-shaped cut, dry ACC and PAni-ACC samples of approximately 1 cm^2 (10–13 mg) were weighed to an accuracy of $\pm$0.001 mg. The samples were pressed at 20–25 kg cm^{-2} for 40 s in between a folded SS mesh, used as a current collector for the working electrode assembly. Prior to characterization, these electrodes were immersed in 0.5 M H_2SO_4 overnight to promote the material impregnation. A Pt wire was employed as a counter electrode. The working electrode potential was given with reference to a commercial Ag/AgCl/Cl$^-$ (3.5 M KCl) electrode. The measurements were carried out in a potentiostat system (VSP model, Bio-logic Science Instruments). The working solution was previously de-aerated by bubbling N_2 and blanketed throughout all the experiments by a N_2 stream flowing over it. Cyclic voltammograms were recorded at room temperature at a scan rate of 1 mV s^{-1} and presented as mass-normalized current (mA g^{-1}) vs. potential plots. The gravimetric specific capacitance was evaluated from CV (Appendix A, Equation (A1)).

The conductivity of ACC and PAni-ACC composite fabrics was measured by the four-strip probe method [33]. Dry sample sheets (11 mm $\times$ 40 mm) were sputter coated with four Pd-Au thin strip probes (5 mm $\times$ 11 mm) in an EMITECH SC7620 Sputter coater (Quorum Technologies Ltd, Laughton, UK). The sputtered probes were 5 mm equally spaced across the length of the sample sheet. Copper wires were glued to the probes by conducting silver epoxy resin, and the contacts were secured with thin aluminum foils. A constant current, I, from a DC power supply (EP-613, Silver electronics) was passed through the two outermost probes and measured with a digital multimeter in series. The voltage drop, V, across the two inner probes was measured in a FLUKE 45 dual Display digital voltmeter. The electrical resistance, R (Ω), was obtained from the linear slope of V vs. I plots in the range 0–10 mA. The surface sheet resistance, R_s ($\Omega\ \square^{-1}$), was derived from the calculated electrical resistance according to Equation (A2) (see Appendix A) [34].

2.5. Dye Adsorption Measurements

A synthetic stock amaranth solution was prepared by dissolving 1 g of dye in 1 L of distilled water. Working solutions of concentration in the range 25–200 mg L^{-1} were obtained by proper dilution with distilled water. Dry ACC or PAni-ACC composites were cut in pieces of about 2 cm^2 and accurately weighed to $\pm$0.1 mg (typical weights ~0.035–0.04 g). In a typical adsorption experiment, 50 mL of dye solution of known initial concentration were contacted with the adsorbent in 100 mL glass bottles with a screw cap, that were further sealed with Parafilm$^®$. The contact was made in a thermostatic water bath shaker (model WNE22, Memmert, Schwabach, Germany) at a constant temperature of 25 °C and at an agitation speed of 120 rpm, for the time necessary to reach equilibrium. After prescribed time intervals, the liquid-phase dye concentration was analyzed in a Thermo Scientific Helios γ UV-vis spectrophotometer at λ_{max} = 520 nm.

3. Results

3.1. Characterization of Surface Morphology, Surface Chemistry and Porous Structure

3.1.1. Analysis of the Surface Morphology by SEM Imaging

Figure 2a (low magnification) shows the typical morphology of a yarn in the knitted AC fabric. Each yarn is a twisted bundle of loose carbon fibers of about 10 μm in average diameter. A higher magnification image (Figure 2b) reveals that each fiber has a ridge surface with grooves parallel to the fiber axis direction. This shape is typical of wet spun viscose fibers [35] used as precursor material in the activated carbon fabric manufacture. At an electropolymerization time of 10 min, PAni can already be discerned on the surface of ACC (Figure 2c–d). The polymer appears to be distributed over the carbon fabric in a scattered fashion and in the form of smooth and compact deposits on the surface of the fibers. These deposits are preferentially localized around the grooves in a carbon fiber. The number of such PAni deposits increases slightly with the electropolymerization time, their size seems also to grow along the fiber grooves, and some degree of roughening appears (Figure 2e–f). In all these examples, the polymeric material was only distinguished on the ACC face in contact with the stainless steel anode inside the filter-press cell, whereas no evidence of PAni was found on the opposite face. However, the surface N/C quotient (as measured by XPS, see Section 3.1.2) increases with the increasing time of treatment at both sides of the ACC. The evolution of this ratio is a diagnostic signal that the polymer grows throughout the whole fabric surface, but whenever it does as a very smooth and thin film, it may appear morphologically featureless and barely discernable by SEM at the magnification reached in Figure 2 [36].

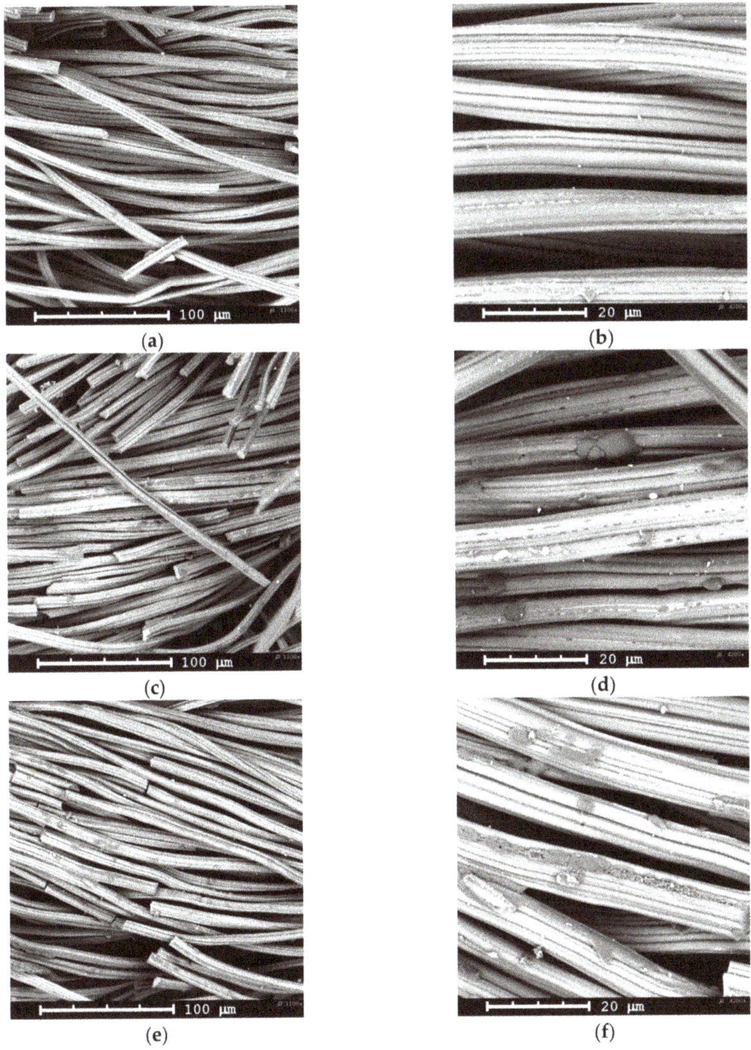

Figure 2. Low- and high-magnification Scanning Electron microscopy (SEM) micrographs of (**a**), (**b**) untreated activated carbon cloth (ACC) and PAni-ACC composites synthesized after (**c**), (**d**) 20 min and (**e**), (**f**) 40 min of electropolymerization.

After 120 min of galvanostatic treatment, a remarkable increase in the amount of electrodeposited conducting polymer is observed (Figure 3). At this stage, a thick PAni layer grows on the ACC face in contact with the anode surface, fully covering most of the carbon fibers and even filling most of the void space among fiber bundles in a yarn (Figure 3a). This thick PAni coating appears brittle and easily peels off upon slight fabric bending. Also, some polymer material detached from the fabric during manipulation for withdrawal from the cell and during subsequent rinsing. High magnification micrographs show that the PAni layer is rather porous (Figure 3b) and can be properly described as a nanofibrous mat (see inset to Figure 3b). Localized open networks of either nanofibrous polyaniline or aggregates of short and randomly aligned nanowires can also be discerned on the face exposed to the electrolyte (Figure 3c–d). These supramolecular structures of polymeric material are unevenly distributed among the individual carbon fibers forming a yarn and their proportion to the carbon fabric is much lower than that in the opposite face of the cloth. Accordingly, the XPS N/C atomic ratio is also lower than that for the side fully coated by a thick polymer layer (see below).

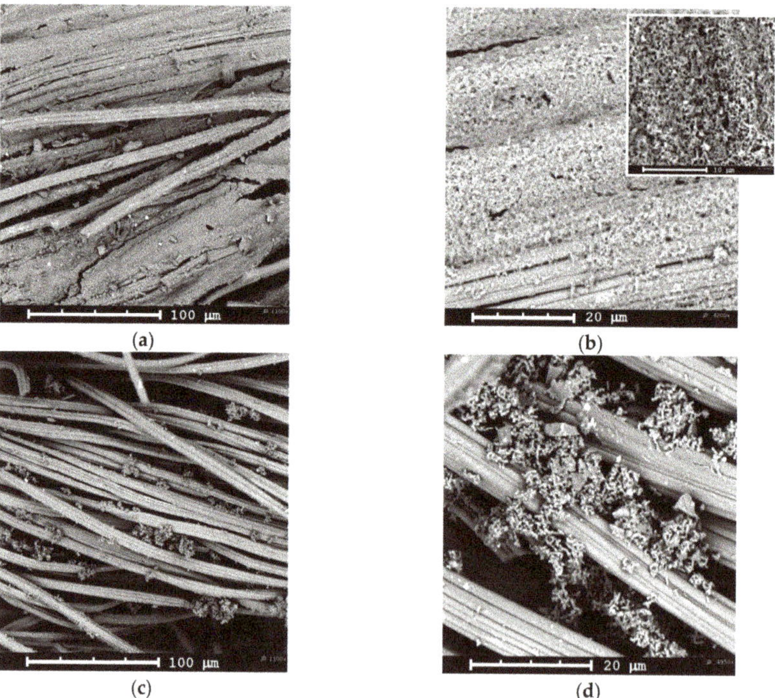

Figure 3. Low- and high-magnification SEM micrographs of PAni-ACC composites synthesized after 120 min of electropolymerization, (**a**), (**b**) electrode side; (**c**), (**d**) electrolyte side.

3.1.2. Surface Chemical Composition by XPS.

(1) Core-Level Spectra

The surface elemental composition, expressed as atomic percentage, of fresh and PAni-modified ACC at different electropolymerization times, was computed from integrated photoelectron areas and is summarized in Table S1. The small amount of nitrogen-containing surface complexes present in fresh ACC (1.2 at.%) most likely stems from nitrogen existing in the precursor carbon source [37]. Sulfur can also be present in the precursor source as elemental sulfur impurities or in the form of inorganic or organosulfur compounds at low proportion [37]. The oxygen content mostly results from carbon-oxygen surface complexes that develop as chemical defects at the edges of graphene layers.

The nitrogen and sulfur content on the surface increases with the increasing electropolymerization time in connection with the growth of polyaniline (Table S1). Nitrogen is a characteristic constituent of PAni polymeric chains either in the form of amine or imine links between benzenoid or quinonoid moieties in the linear polymer backbone or as charged (protonated) nitrogen connected to semiquinone segments. The increased amount of sulfur is due to electrolyte (bi) sulfate anions incorporated into the polymer structure as the dopant, although some residue from inefficient electrolyte removal by rinsing cannot be ruled out. The analysis of the different bonding states of these elements will be treated in detail further below.

The plot of the N/C and S/C atomic ratios vs. time (Figure 4) shows that the N and S content in the hybrid PAni-ACC composites increases moderately at short times, it seems to level off at intermediate electropolymerization time and eventually rises sharply at prolonged process time (120 min). The value of the N/C ratio at short electropolymerization time is lower than the theoretical one for pure linear polyaniline chains (N/C$_{PAni}$ = 1/6), which means that most of the C photoemission signal arises from carbon atoms belonging to either uncoated regions of the carbon fabric or covered by ultrathin, i.e. <~10-nm-thick, smooth PAni films looking featureless in SEM [36]. On the contrary, the N/C ratio is rather close to 1/6 at long electropolymerization time, indicating that XPS is mainly probing the surface chemistry of the polymeric fraction in the hybrid composite (Figure 4). Therefore, the evolution of the N/C ratio is in close agreement with SEM micrographs in Figures 2 and 3a,b, which showed scattered compact PAni deposits at short electropolymerization time and thick PAni layers almost completely covering the carbon cloth substrate at long treatment time (120 min). At short to moderate electropolymerization time, the N/C values corresponding to analyzed areas on the fabric side exposed to the electrolyte were only slightly lower than those described above and changed in a similar fashion (open square circles in inset to Figure 4). This suggests that a smooth thin layer of PAni also forms on this face within this time interval, although it is indiscernible by SEM imaging under our experimental conditions. The long-term N/C atomic ratio (inset to Figure 4) is high but noticeably lower than at the electrode side, in line with the surface topography shown in Figure 3c,d. Finally, the S/C ratio parallels the N/C tendency (open circles in Figure 4), which strongly suggests that most of the sulfur present on the surface of the PAni-ACC composite plays a role as a dopant ion.

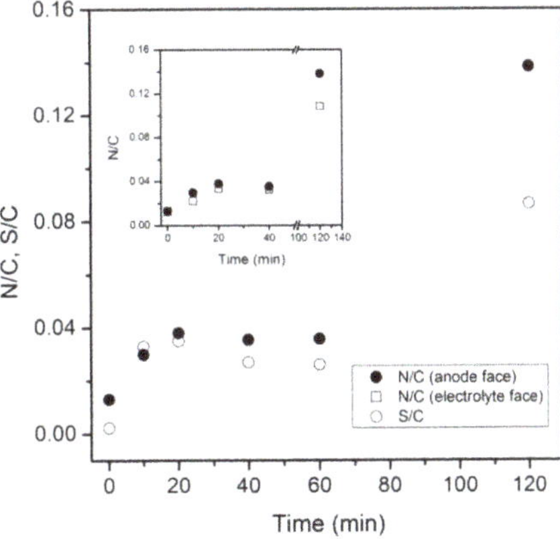

Figure 4. Evolution of surface N/C and S/C ratios for PAni-ACC composites as a function of the electropolymerization time. Inset: Comparison of N/C ratios at both sides of the fabric.

Figure 5 shows the high-resolution photoelectron spectra of C1s, N1s and S2p core levels of fresh (Figure 5a) and PAni-modified ACC after electropolymerization at different times (Figure 5b,c). The

C1s peak was satisfactorily fitted with photoelectron contributions at average values of 284.6 ± 0.1 (C1), 285.9 ± 0.3 (C2) and 287.9 ± 0.2 eV (C3) respectively (Table S2). The C1 subpeak is the major component and corresponds to aromatic carbons in graphene layers of the carbon fabric [32,38–40] and also to those belonging to carbon rings in the polymer backbone [41]. The C2 subpeak can be assigned to carbon atoms singly bound to oxygen groups (i.e., C–OH/C–O–C functionalities) [38,40], but to some extent it can be contributed to by C–N and C=N/C=N$^+$ groups from the polymeric fraction in the hybrid composite [41,42]. The chemical shift of the C3 subpeak is characteristic of double-bonded carbon–oxygen complexes (e.g., C=O functional groups) [38]. Also, C atoms singly bound to positively charged N atoms in PAni have been associated with this energy region [43].

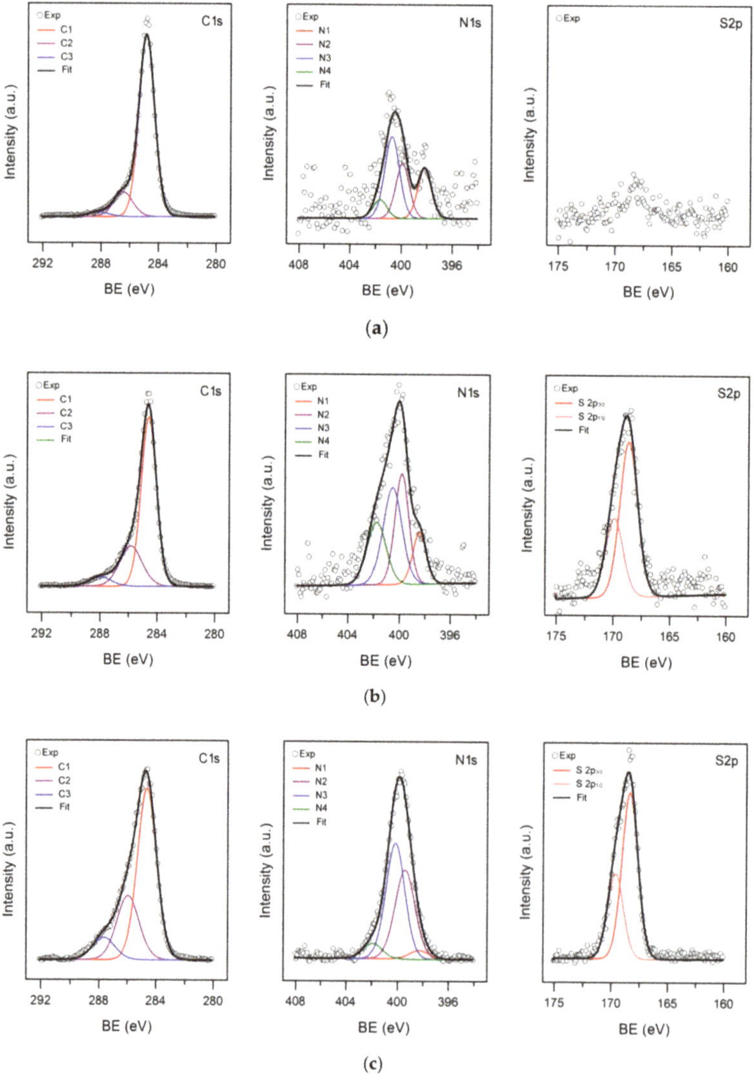

Figure 5. High-resolution C1s, N1s and S2p core-level X-ray photoelectron spectra of (**a**) untreated ACC and hybrid PAni-ACC composites synthesized after (**b**) 40 min and (**c**) 120 min electropolymerization time.

In agreement with the low relative abundance of S, the S2p core-level spectrum of the fresh ACC sample (Figure 5a) shows a weak band centered at ca. 167.9 eV. Because of the low intensity of this band, fitting the possible components under the envelope was not attempted. According to the 10- to 30-fold increase in the amount of S (Table S1), S2p spectra in Figure 5b,c are much more intense. They appear shifted to higher BEs, therefore, pointing to a higher oxidation state of the S atom. The fitting of the S2p line revealed a single atomic environment with a spin-orbit doublet having its $2p_{3/2}$ component located at 168.6 ± 0.1 eV, so confirming the presence of (bi)sulfate anions [44].

The N1s photoemission line (Figure 5) can be decomposed into four distinct components located on average at 398.2 ± 0.1 (N1), 399.7 ± 0.2 (N2), 400.6 ± 0.2 (N3), and 402.0 ± 0.2 eV (N4), (Table S2). As far as activated carbon materials are concerned, the low-BE component was assigned to nitrogen in pyridine-like structures, and the N2 subpeak may be ascribed to aromatic amide or amine moieties, as well as to pyrrolic or pyridone structures [39,45]. The contributions above 400 eV correspond to positively-charged N structures: quaternary nitrogen (N3) and pyridine N-oxides (N4) [39,45]. By contrast, totally different local chemical states for N have been described in the literature dealing with N-containing conducting polymers, like polyaniline and its derivatives [5,41,42]: The N1 peak was attributed to deprotonated imine nitrogen in quinoneimine units, N2 to amine nitrogen in benzenoid rings, and peaks at >400 eV to positively charged N atoms. In particular, a peak component at about 400.5 eV (N3) was assigned to N atoms with delocalized positive charge, while that at the highest BE (N4) was associated with N atoms bearing localized positive charge. These assignments seem more appropriate to interpret the N1s core-level spectrum in Figure 5c, since the high N/C atom ratio suggests that XPS entirely reflects the composition of a layer of conducting polymer. The PAni doping level (S/N ratio) in the hybrid composite formed of Figure 5c is 0.63, and the protonation level (N^+/N ratio) is 0.56. Therefore, sulfur is most likely to be incorporated in the form of (bi) sulfate anions, which compensates for the positive charge residing on nitrogen sites. Under the assumption that cationic N atoms originate solely from protonation of quinoneimine N sites [5], the proportion of neutral imine (=N−, 398.2 eV) plus charged nitrogen (N^+, >400 eV) to total nitrogen gives an intrinsic redox state of 60%. The N1s line corresponding to PAni-ACC composites formed at intermediate electropolymerization time (Figure 5b) can also be resolved into the same four aforementioned components, but their true assignment may be obscured by N photoelectrons from the carbon matrix, which represents nearly 1/3 of the total N (Table S1). Therefore, an estimation of the doping level, protonation level and redox state of polyaniline was not attempted in this case.

(2) Valence-Band Spectra

Figure 6 shows the variation in the Valence-band (VB) spectrum of the ACC substrate after modification with PAni. The spectrum of the bare carbon cloth (Figure 6a) is characterized by two broad peaks of similar relative intensity corresponding to O2s (25 eV) and C2s (19 eV) levels [40]. This latter peak may also involve a possible contribution from N2s orbitals at its low BE energy side (18–16 eV) [40]. The region below 16 eV is characterized by features resulting from the strong overlap of mixed O2p, C2p and N2p components [40]. In Figure 6a, a sharp peak appears in this region (12.2 eV), serving as a distinctive fingerprint for our base carbon fabric material. This characteristic peak is totally suppressed in the VB spectrum of hybrid PAni-ACC composites formed at prolonged electropolymerization time (Figure 6c, 120 min). The VB lineshape in this material shows three weak consecutive peaks laying in the region 10–20 eV and an asymmetric peak of high intensity at 24.5 eV. Moreover, it should be emphasized that the VB edge is far below the Fermi level in the ACC substrate, but the growth of PAni increases the density of states near the Fermi energy to delineate a small shoulder at ~4 eV. A similar VB pattern has been reported earlier for PAni films, although the relative peak intensity and resolution of the various peaks is highly dependent on the particular intrinsic redox state, the doping and protonation levels, and the nature of the dopant ion [46–48]. Nakajima et al. [46] reported a set of peaks in the region 10–20 eV (17.5 eV, 14 eV and 11.5 eV) due to an overlap of molecular orbitals involving benzenoid C and N in leucoemeraldine salt, while the electron density close to the

Fermi level was found to be characteristic of a conjugated structure of double and single bonds. More recently, Bocchini et al. [48] distinguished a C2s peak at 13 eV, a 9–11 eV feature corresponding to a C–N 2p-σ state, and a group of weak photoemission peaks below 8 eV assigned to C2p-σ and C2p-π bands in polyemeraldine doped with camphorsulphonic or aminosulphonic acids. Hybrid PAni-ACC composites formed at intermediate electropolymerization times (Figure 6b) show mixed features from both the underlying carbon matrix and a polymer thin film. Thus, the lineshape, relative intensity and position of peaks above 15 eV change to become similar to that of PAni, whereas the characteristic sharp peak survives, but is shifted to lower BE. Also, the VB edge lies between that of the untreated ACC and that of the fabric coated with thick PAni layers.

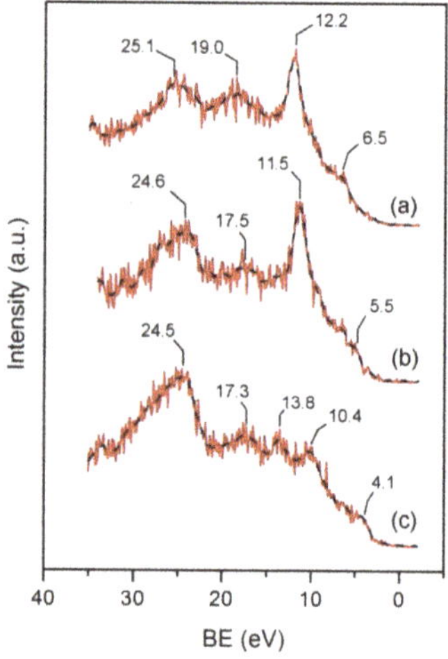

Figure 6. Valence-band X-ray photoelectron spectra of (**a**) untreated ACC and hybrid PAni-ACC composites synthesized after (**b**) 20 min and (**c**) 120 min of galvanostatic electropolymerization.

3.1.3. Porous Texture

Figure 7a shows nitrogen adsorption/desorption isotherms of bare ACC and representative PAni-ACC composite samples. The evolution of the BET specific surface area and that of different pore volumes with the electropolymerization time is plotted in Figure 7b,c respectively. A detailed list of these parameters can be found in Table 1. The shape of the N_2 adsorption curve of the unmodified ACC (Figure 7a) corresponds to type IV isotherms, with a H4 adsorption/desorption hysteresis loop [49]. This form is typical of solids possessing mesopores embedded in a microporous matrix, where the adsorptive uptake proceeds via multilayer adsorption followed by capillary condensation [49]. The shape of the isotherm is similar to those reported for other cellulose-based commercial activated carbon fibers [23,39]. The growth of PAni layers leads to a general decrease in the N_2 uptake, but the isotherm shape remains unchanged for composite textiles obtained at short to intermediate electropolymerization time. However, when the amount of charge passed is sufficiently high to produce thick PAni films, there is a dramatic loss in the N_2 adsorption capacity and the hysteresis does not close at lower pressures. This may be due to diffusional restrictions to gas adsorption/desorption associated to pore narrowing and/or occlusion induced by the grown polymer.

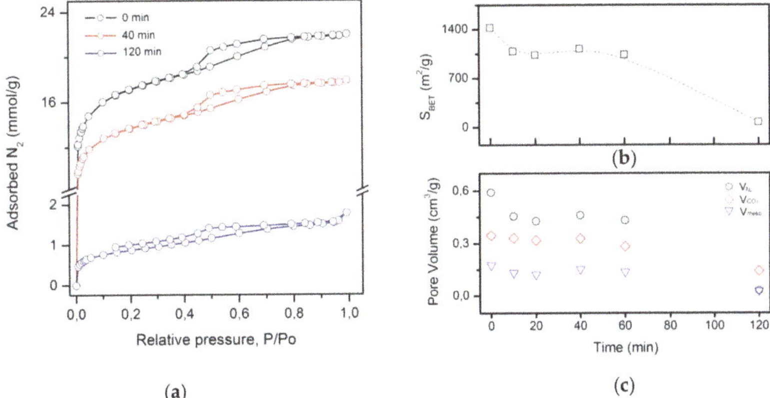

Figure 7. Porous texture characterization of untreated ACC and hybrid Pani-ACC composites synthesized after different electropolymerization times: (**a**) N_2 adsorption–desorption isotherms; (**b**) Brunauer-Emmett-Teller (BET) surface area; (**c**) Pore volume distribution: micropore volume (open circles), ultramicropore volume (open diamonds), mesopore volume (open triangles).

Table 1. Porous texture analysis of as-received ACC and hybrid PAni-ACC composites synthesized after different electropolymerization times.

Time (min)	S_{BET} (m²/g)	V_{tp} (cm³/g)	V_{μ} ¹ (cm³/g)	$V_{ultra\ \mu}$ ² (cm³/g)	V_{meso} (cm³/g)
0	1424	0.77	0.59	0.35	0.18
10	1086	0.59	0.45	0.33	0.14
20	1032	0.55	0.43	0.32	0.12
40	1120	0.61	0.46	0.33	0.15
60	1037	0.57	0.43	0.28	0.14
120	71.1	0.055	0.028	0.14	0.027

¹ Volume of micropores. ² Volume of ultramicropores.

The commercial ACC has a high apparent BET surface area of 1424 m²/g and exhibits both a micropore and mesopore structure (Table 1). In accordance to what is observed in Figure 7a, micro- and mesopore volumes, and therefore the specific surface area, decrease by 20%–25% whenever PAni grows as a uniform thin and dense film at short-to-moderate electropolymerization time (10–60 min, Figure 7b,c). Within this time interval, these parameters remain constant, but they drop pronouncedly in the hybrid composites with thick PAni layers. In contrast, the volume of the narrower micropores (ultramicropores) barely changes with PAni loading, except at long electropolymerization time, when a clear decrease is also observed (Figure 7c).

3.1.4. Thermal Analysis

The thermogravimetric curves of representative PAni-ACC composites obtained at selected electropolymerization times are shown in Figure 8a and compared to the thermal evolution of the untreated ACC. During the first thermogravimetric run (25–120 °C, inset to Figure 8a), the carbon support and the PAni-loaded samples all exhibit a weight loss below 100 °C. This loss is associated with the endothermic release of moisture or weakly adsorbed water solvent molecules, and perhaps with the evaporation of some residual monomers [50–52]. The amount of water lost by the carbon support is lower than the water released from any polymer-modified fabric, probably because PAni imparts a significant hydrophilic character. However, no clear pattern correlating the amount of deposited polymer and the water content was found. During the second thermogravimetric run from 25–1000 °C only negligible residual water is lost from the samples (Figure 8a).

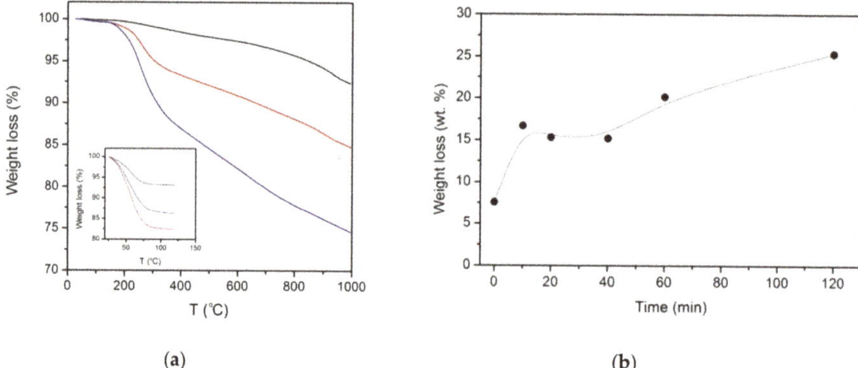

(a)   **(b)**

Figure 8. (**a**) TGA curves of unmodified ACC (black line) and hybrid PAni-ACC composites synthesized after 40 min (red line) and 120 min (blue line) of galvanostatic electropolymerization; the inset shows the thermal pattern during the first thermogravimetric run up to 120 °C; the main panel shows the thermal behavior of the resulting dried samples; (**b**) Evolution of the total weight loss at 1000 °C for heat-dried PAni-ACC composites obtained at different electropolymerization times.

The dry carbon cloth undergoes a first smooth mass loss in the range 200–400 °C, followed by a continuous decomposition and a more pronounced loss between 900–1000 °C (Figure 8a, black line). These features are well connected with the thermal decomposition of oxygen surface complexes in carbonaceous materials, which are known to evolve as CO_2 and CO [30]. The thermogravimetric curves of hybrid PAni-ACC composites show a prominent loss between 150–350 °C and a subsequent steady weight decrease up to 1000 °C. The observed thermal pattern is a characteristic feature of PAni-like powders or films in their emeraldine salt (i.e., doped) state [51–53]. The sharp loss in the temperature range 150–350 °C has been attributed to either chain scission or cross-linking processes (e.g., leading to phenazine-like segments) with dopant removal [50–52], while the loss above 400 °C has been related to structural degradation of the polymer backbone leading to its complete carbonization at the highest temperature [52–54].

Our results show that both the sharp feature within 150–350 °C and the total weight loss are closely related to the amount of CP deposited on the carbon fabric (Figure 8a, red and blue lines). In samples with a moderate PAni loading (red line in Figure 8a), the loss corresponding to the decomposition processes of the underlying carbon cloth (800–1000 °C) is still discerned. However, it is missed in heavily-loaded PAni-ACC composites (Figure 8a, blue line). The total weight loss of dry samples as a function of the electropolymerization time (Figure 8b) reflects the evolution of the polymer amount in the hybrid material. Figure 8b shows an abrupt increase at the early stages of the electropolymerization process, a stabilization region and a second increase at the longest treatment time. Thus, the total weight loss vs. time plot mimics the evolution of the photoelectron N/C vs. time plot (Figure 4).

Thermogravimetric data can be used to estimate the PAni loading of the different hybrid composites. The estimate was done as follows: First, the total weight loss of dry hybrid PAni-ACC at 1000 °C was corrected by the loss corresponding to the carbon support at the same temperature (7.57 wt.%); then, the amount of polymeric material in the composite was estimated after considering that about 45 wt.% residue was left after carbonization at 1000 °C [53,54]. PAni-loadings are expressed either as the mass of polymer (mg) per unit geometric area (cm^2) of carbon fabric or as wt.%. The results are summarized in Table 2 for composites electrosynthesized at 10–120 min time. The data confirm that a massive deposit of PAni occurs at 120 min of electropolymerization that accounts for ca. 50% of the composite material mass.

Table 2. PAni loadings of dried hybrid PAni-ACC composites synthesized after different electropolymerization times.

Time (min)	Initial Dry Mass (mg)	Corrected Weight Loss [1] (%)	PAni Weight Loss (mg)	PAni Total Mass [2] (mg)	PAni Loading	
					(wt.%)	(mg cm^{-2}) [3]
10	8.8250	9.12	0.8047	2.2679	25.70	6.23
20	9.2770	7.75	0.7186	2.0251	21.83	5.03
40	9.9108	7.63	0.7561	2.1307	21.50	4.93
60	8.4070	12.58	1.0577	2.9807	35.46	9.89
120	11.1129	17.79	1.9766	5.5704	50.13	18.1

[1] After subtraction of the total loss from the carbon fabric support (see text). [2] After considering that a 45 wt.% polymer residue remains on the heat-treated sample. [3] Calculated on the basis of cloth aerial density.

3.2. Enhanced Electrical and Adsorptive Properties.

3.2.1. Capacitance and Surface Sheet Resistance.

The electrical properties of the untreated ACC and PAni-coated hybrid composites were obtained from CV and four strip probe conductivity measurements (Figure 9). CVs became stabilized within two cycles and remained unchanged thereafter. No loss of material was seen during CV recording.

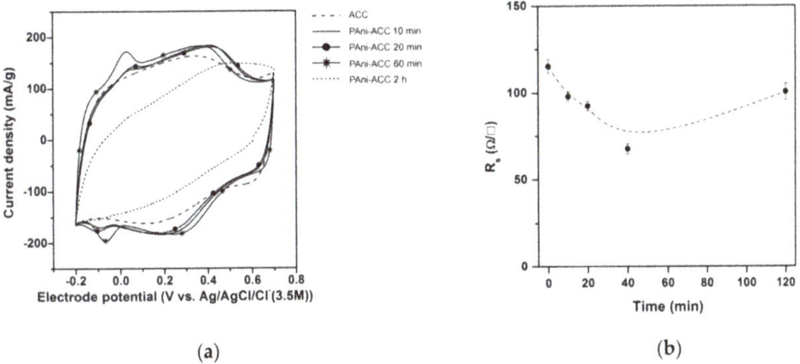

(a)   (b)

Figure 9. (a) Stabilized cyclic voltammograms of untreated ACC and hybrid PANi-ACC composites synthesized at different electropolymerization times. Scan rate: 1 mV·s^{-1}, supporting electrolyte 1.0 M H$_2$SO$_4$; (b) Evolution of the surface sheet resistance upon the time of electropolymerization.

The CV of the untreated ACC (Figure 9a), dashed line) shows a capacitive response with a distorted rectangular shape that indicates a deviation from the ideal double-layer capacitor behavior [11,15]. This is a symptom of a slow charging/discharging response caused by a potential difference across micropores [18,55]. In addition to the main double layer charge/discharge contribution, a redox couple is discerned ($E_{1/2}$ = 0.25 V). This feature is associated with a faradic process involving redox transitions of surface carbon–oxygen groups, thus behaving as a pseudocapacitance [11,30]. At low or intermediate electropolymerization time (0–60 min), a moderate increase in the voltammetric current occurs as a result of PAni film formation. This increase in current has a pseudocapacitive character and is mainly due to reversible redox transitions corresponding to the doping/dedoping processes in PAni. The broad redox couple located within 0.4–0.2 V has been assigned to the reversible transition between leucoemeraldine and emeraldine salt oxidation states of the polymer in several PAni-coated 3D porous carbon materials, like foams [15] and activated carbon fibers or clothes [11,22,55]. The pair of sharp and small peaks at ~0.0 V may be related to the presence of some phenazine- or phenoxazine-like moieties in the PAni backbone [41,56,57]. In accordance with the modest increase in the CV response, the specific capacitance of the PAni-ACC composites rises by up to a 12% (Table 3). Most of this increase occurs within the first 10 min of electropolymerization and then it tends to level off during the first hour of electropolymerization. The CV corresponding to thick PAni coatings formed at the longest treatment (Figure 9a, dotted line) appears noticeably tilted. Furthermore, the pseudocapacitive features of PAni are barely distinguishable and the total specific capacitance falls even below the capacitance of the unmodified ACC (Table 3). The abrupt decrease in the specific capacitance parallels the loss of available BET surface area measured by N_2 adsorption isotherm. The evolution of surface sheet resistance with time (Figure 9b) follows up the reported changes in the CV response and the specific capacitance. Within the first 60 min of electropolymerization, the electrical resistance of the hybrid PAni-ACC composite diminishes, but it rises again at prolonged treatment, i.e., when a thick nanofibrous PAni coating develops.

Table 3. Gravimetric specific capacitance of hybrid PAni-ACC composites as a function of the electropolymerization time.

Time (min)	C_{sp} (Fg^{-1})
0	127
10	138
20	137
40	136
60	142
120	98

3.2.2. Adsorption of Acid Red 27

The experimental equilibrium data for the uptake of Acid Red 27 on unmodified ACC (Figure 10a) at 25 °C were fitted with the Langmuir and Freundlich models (see Appendix A), [58,59]. The isotherm parameters can be obtained from the slope and y-intercept of the C_e/q_e vs. C_e and ln q_e vs. ln C_e plots. The correlation coefficients for the linear regression fittings of Langmuir and Freundlich models to the experimental data were 0.9997 and 0.9397, respectively (Figure S1). Therefore, the equilibrium adsorption of Acid Red 27 on ACC is best described by the Langmuir isotherm, with a maximum adsorption capacity of 146.6 mg g^{-1} and b=0.82 L mg^{-1}. The simulated adsorption equilibrium curve based on Langmuir parameters is also shown in Figure 10a (solid line).

The isothermal kinetic curves for the adsorption of Acid Red 27 on ACC are shown in Figure 10b for concentrations ranging 25–150 mg L^{-1}. The dye uptake is faster at the initial stages of adsorption and then it decreases as the process approaches equilibrium, owing to the increasing occupancy of surface sites and slow diffusion into the internal porous structure [14]. The adsorption rate increases with the increasing initial concentration. The equilibrium uptake shows the same dependence, although it levels off at high initial concentrations because maximum adsorption capacity is approached.

Kinetic data in Figure 10b were modelled according to the well-known pseudo-first order (PFO or Lagergren equation) and Ho's pseudo-second order (PSO) models (Equations (A5) and (A6), Appendix A) [60]. The relevant kinetic parameters can be inferred from the slope and y-intercept of $\ln(q_e - q)$ vs. t (PFO model) and t/q vs. t (PSO model) plots (see Figures S2 and S3, respectively). The calculated rate constants, the theoretically predicted q_e, and the corresponding R^2 are listed in Table 4 for different initial Acid Red 27 concentrations. Also tabulated are the calculated PSO initial adsorption rates, r_0. Both models provide a good fitting of the experimental data over the whole time span according to the square correlation coefficient. Then, the goodness of fit was further evaluated with the so-called average relative error (ARE, Equation (A10), Appendix A). The ARE in Table 4 is lower for PSO-based fittings than for PFO kinetics over the whole range of initial concentrations. Further, the equilibrium uptakes predicted by the PSO model are generally closer to the experimental ones. Then, the PSO model correlates the experimental kinetic data better than the PFO model does.

The second order rate constant shows a decreasing tendency as the initial concentration increases, whereas the initial adsorption rate follows a growing trend. The simulated adsorption curves derived from PSO kinetics are drawn in Figure 10b as solid lines.

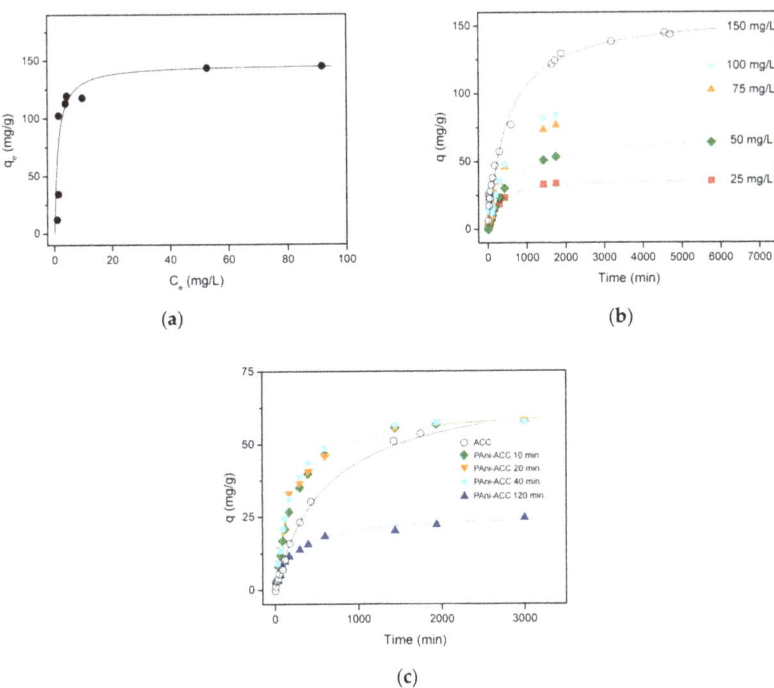

Figure 10. (a) Experimental (solid symbols) and Langmuir-based simulated adsorption isotherm (solid line) for Acid Red 27 on untreated ACC at 25 °C; (b) Experimental (symbols) and PSO modelled (solid lines) kinetic curves for the adsorption of Acid Red 27 on untreated ACC at different initial dye concentrations; (c) Experimental (symbols) and PSO modelled (solid lines) kinetic adsorption curves of 50 mg L^{-1} Acid Red 27 solution onto hybrid PAni-ACC composites formed at different electropolymerization times.

Table 4. Pseudo-first-order and pseudo-second-order kinetic parameters for the adsorption of Acid Red 27 on ACC (C_0: mg·L^{-1}; q_e: mg·g^{-1}; k_1: min^{-1}; k_2: g·mg^{-1}·min^{-1}; r_0: mg·g^{-1}·min^{-1}).

C_0	$q_{e,exp}$	PFO				PSO				
		$k_1 \times 10^3$	$q_{e,cal}$	R^2	ARE	$k_2 \times 10^5$	$q_{e,cal}$	r_0	R^2	ARE
25	35.5	1.85	32.4	0.993	0.35	9.21	37.6	0.130	0.997	0.16
50	63.9	1.05	60.5	0.991	0.27	2.24	71.2	0.113	0.997	0.07
75	102.5	0.79	94.0	0.976	0.37	1.80	109.1	0.215	0.982	0.22
100	119.6	0.67	105.8	0.982	0.69	0.94	134.0	0.169	0.987	0.37
150	143.5	1.05	122.7	0.994	0.52	1.94	153.1	0.454	0.995	0.21
200	144.7	1.14	137.2	0.994	0.26	1.10	163.6	0.294	0.999	0.13

The PSO model is known to show widespread good fit to the adsorption kinetics of most dyes and organic pollutants on activated carbon and other synthetic or natural adsorbents [9,60]. However, it has been claimed [60] that the rate constant, k_2, should not be regarded as being truly representative of intrinsic kinetics, but as an empirical parameter that lumps different controlling mechanisms in a complex manner. In fact, our data invariably show a noticeable deviation from linearity at the initial stage of adsorption (Figure S3). This typical downward curvature was previously realized by Azizian et al. [61], who ascribed it to a mixed rate control by both diffusion and surface reaction. In order to check the significance of mass transfer on the Acid Red 27 adsorption kinetics onto ACC, we used the Vermeulen model (Equation (A9), Appendix A) [60]. According to this model, the Bt vs. t plot (Boyd plot) should be a straight line passing through the origin for a pure intraparticle diffusion mechanism. If the line has a non-zero intercept, the adsorption is also controlled by external film diffusion [31,60]. In our case, the plots are linear for the whole concentration range (see Figure S4), but they do not pass through the origin (Table S3), thus pointing out that film diffusion is also involved in governing the adsorption rate.

The evolution of Acid Red 27 adsorption uptake onto different PAni-ACC hybrid composites is shown in Figure 10c for a single dye initial concentration of 50 mg L^{-1}. The kinetic curve for the bare ACC is also depicted for the sake of comparison. PAni layers deposited at short time (thin and compact films) remarkably enhance the rate of adsorption, but the total amount of dye adsorbed at equilibrium remains basically unchanged. However, higher PAni loading achieved at prolonged electropolymerization (thick and porous nanofibrous coating) leads to both a decrease in the adsorption rate and a considerable loss of the adsorption capacity at equilibrium. As above, PFO and PSO modelling of the experimental data were attempted and the kinetic parameters of interest, as well as proofs for testing the goodness of fit, are summarized in Table 5. Again, the PSO model provides an overall better description of the adsorption kinetics. Solid lines in Figure 10c were then traced by using the calculated kinetic parameters of the PSO fitting. These parameters corroborate that hybrid PAni-ACC composites of low polymer loading and smooth film morphology promote a faster adsorption of the dye, while the equilibrium uptake remains unaffected. On the contrary, the nanofibrous thick PAni coating is detrimental to both the adsorption rate and the dye loading into the adsorbent composite material.

Table 5. Pseudo-first-order and pseudo-second-order kinetic parameters for the adsorption of Acid Red 27 (50 mg L^{-1}) on hybrid PAni-ACC composites synthesized at different electropolymerization times. (t: min; q_e: mg·g^{-1}; k_1: min^{-1}; k_2: g·mg^{-1}·min^{-1}; r_0: mg·g^{-1}·min^{-1}).

t	$q_{e,exp}$	PFO				PSO				
		$k_1 \times 10^3$	$q_{e,cal}$	R^2	ARE	$k_2 \times 10^5$	$q_{e,cal}$	r_0	R^2	ARE
0	63.9	1.05	60.5	0.991	0.27	2.24	71.2	0.113	0.997	0.07
10	57.6	2.20	48.3	0.994	0.17	6.53	63.1	0.260	0.999	0.05
20	58.1	2.29	47.3	0.990	0.18	7.96	62.6	0.312	0.999	0.06
40	57.5	3.29	52.7	0.996	0.13	8.89	61.9	0.340	0.999	0.07
120	24.5	1.05	16.8	0.919	0.23	1.53	25.8	0.102	0.995	0.08

Finally, the involvement of mass transfer on the adsorptive capability of the hybrid composites was evaluated with the aid of Bt vs. t plots (Figure S5 and Table S4). In this instance, only the composites formed at 10 to 60 min of electropolymerization show good linear plots with a non-zero intercept, which again suggests that both external and intraparticle diffusion may play a role in the adsorption process. Instead, the materials obtained at longer treatment time do not show a linear behavior and therefore the involvement of mass transfer is unclear.

4. Discussion

In this work, we show that large sized (4 cm × 5 cm) activated carbon cloths can be easily modified with electrochemically deposited polyaniline by closely attaching the fabric to a stainless steel anode in a filter-press type cell. A simple galvanostatic procedure at a constant current of 100 mA (~14 mA g^{-1} cm^{-2}) allowed the electropolymerization process to be carried out at an anodic potential varying in the range 0.6–0.8 V vs. Ag/AgCl. This electrode potential is sufficient to oxidize aniline monomers, to generate enough PAni nucleation sites and to ensure fast chain growth [16,55,62]. A simple control of the PAni loading could be achieved by passing different amounts of charge, i.e., by adjusting the elapsed electropolymerization time. The highly porous carbon fabric provides the hybrid composite material with high specific surface area, flexibility and mechanical strength to mitigate the stress associated with polymer volume changes accompanying doping/dedoping [11,12,22], while PAni adds some new enhanced characteristics related to its reversible redox properties and electron conductivity [2,12,62].

It is well known that PAni and its composites can be synthesized with a wide variety of micro- and nanoscale structures, with a surprising number of different sizes and shapes, from smooth thin films to nanofibers, nanorods or nanotubes to globular or granular agglomerates [26,27]. The observed supramolecular organization and structure is strongly dependent on the method of synthesis and the experimental conditions [6,26,27]. As for electrochemical methods, the applied current/potential and the way the electrode is polarized (i.e., potentiostatically, galvanostatically, by cyclic potential sweep or pulses), as well as the nature and surface condition of the electrode support, all have a dramatic effect on the morphology of the deposited polymer film [6]. In the present work, the film morphology varied with processing time from dense and uniform thin films covering individual carbon fibers to a loosely adherent and porous, thick coating of nanofibrous or nanowire-shaped structures. Three-dimensional networks of nanofibrilar or nanowire structures seem the most frequently encountered morphology for PAni deposited on high surface area carbon substrates (e.g., carbon fibers or cloths), either by electrochemical [20,21,62,63] or conventional in-situ chemical polymerization [22–24]. However, in agreement with our results, some authors have obtained smooth dense layers [8,63] or a transition from this morphology to irregular or randomly connected nanofibrilar aggregates upon extending the polymerization time [12,62]. It has also been reported that long reaction time [23] or prolonged galvanostatic oxidation [20] lead to incontrollable growth among carbon fibers and result in a detrimental effect via peeling-off from the carbon surface.

Apart from the scattered compact deposits seen in Figure 2, no clear evidence of a PAni film uniformly covering the carbon fibers is provided by SEM imaging. However, the increase in the surface N/C ratio measured by XPS at both sides of the carbon cloth suggests that an (ultra)thin film develops within this timespan. Also, the PAni loadings deduced from thermogravimetric analysis (Table 2) indicate that there is more polymer than what is seen by SEM. We believe that during the pre-conditioning step, anilinium cations are pre-adsorbed on the internal porosity of the carbon cloth [11,55], thus acting as nucleation seeds and reacting with liquid-phase aniline to yield ultra-thin films covering the pore walls. This view agrees with the results by Salinas-Torres and coworkers [17], who used position-resolved micro-Small Angle X-Ray Scattering (SAXS) to demonstrate that a PAni layer of sub-nanometric thickness grows inside the microporosity of activated carbon fibers. The particular non-linear increase in the PAni content revealed by both XPS and thermal analysis (Figures 4 and 8, Table 2) might be explained by a decrease in the polymerization rate caused by the depletion of aniline monomers inside the pores and the induced concentration-gradient between the external bulk solution and the solution filling the pores. Note that dye adsorption studies point out that the bare carbon material possesses a large pore mass transfer resistance (see below). Slow internal pore diffusion provides enough time for self-assembly of intermediate oligomers into thin films. Once PAni reaches the outer carbon surface, external film diffusion facilitates fast access of solution aniline to the growing chains and the polymerization rate increases again to build a 3D network of nanofibrilar structures and nanowire aggregates.

As long as a thin film develops on the carbon fabric surface, XPS shows mixed spectroscopic characteristics from both the support and PAni. When a thick coating is formed, XPS is typical of pure PAni, close to the semioxidized emeraldinde salt state. The temperature-induced changes in the hybrid composites (Figure 8) are also compatible with the thermal behavior of PAni in its doped emeraldine salt state [50–53]. The evolution with the electropolymerization time of the dry composite total weight loss and the calculated PAni loading (Figure 8 and Table 2) are consistent with the trends shown by XPS analysis. The PAni loadings achieved in our hybrid composites ranged from ~5 (within the first 40 min of treatment) to 18 mg cm^{-2}, at the longest polymerization time. These loadings are in the range of those reported by others for carbon fiber substrates [11,23,62,64].

The electrodeposition of PAni is accompanied by a decrease in the apparent BET area and pore volume. Earlier authors also described a remarkable blockage of micro- and mesoporosity in composites of activated carbon fibers [8,11,17] or activated carbon fiber textiles [23,25,64] with PAni films, deposited either by chemical or electrochemical methods. In these reports, the surface area loss ranged from 20% to 40%. In the present investigation, the specific surface area decreases by ca. 20%–25% and remains basically unchanged within the time interval when a uniform and dense thin PAni film covers the carbon surface. The evolution of pore volumes (Table 1 and Figure 7c) indicates that PAni does not significantly grow inside the narrower micropores, in contrast to the results by Salinas-Torres and coworkers [17]. There is no clear evidence why polymerization inside ultramicropores is hampered, but it may arise from too short a contact time in the pre-conditioning step to allow diffusion of aniline into such narrow pores. Also, we can tentatively argue that carbon fibers may possess some kind of hierarchical pore structure, with ultramicropores preferentially distributed near the fiber core. Thus, PAni films blocking the outer micro- and mesopores would hinder efficient monomer supply to sustain polymerization inside the inner pores. In fact, position-resolved micro-SAXS studies revealed a heterogeneous radial distribution of electrosynthesized PAni [17], which preferentially accumulates in the outer shells of carbon fibers. The thick nanofibrous PAni coating grown at prolonged electropolymerization time virtually blocks all the porosity, thereby causing an abrupt decrease in the specific surface area to reach a typical value of pure PAni precipitates [65].

The reversible redox transitions of PAni doping/dedoping processes contribute to moderately enhance the capacitive response of hybrid PAni-ACC composites. The extra pseudocapacitive features observed in the CVs of the hybrid composites (Figure 9) arise from the leucoemeraldine-to-emeraldine state transition (0.2–0.4 V) and a reversible electron-transfer (~0.0 V) probably involving phenazine- or phenoxazine-like segments in the polymer backbone. These latter redox features appear in polymeric materials of the PAni family, like oligomers of o-aminodiphenylamine [41,56], ladder polymers like poly-o-phenylenediamine [57] or poly-o-aminophenol [56], PAni- and poly-o-anisidine-clay nanocomposites [66,67] or reduced graphene oxide-PAni-ACC composites [63]. More recently, it has been claimed that even chemically synthesized PAni contain constitutional phenazine and N-phenylphenazine segments [68] that play a key role in the self-assembly of polymer chains to build different supramolecular structures [27].

The specific capacitance of the hybrid composites increases moderately in hybrid composites with thin dense PAni films. (Table 3). The reported specific capacitance for pure Pani-modified electrodes lies within a broad range from 160 to 815 F/g [69], while the theoretical capacitance for Pani at a 50% doping level is 750 F/g [70]. Upon considering this latter value, the expected capacitance for a 20 wt.% PAni-ACC composite would be about 250 F/g, but if one considers the lowest reported PAni capacitance, an estimated value of about 134 F/g would be obtained, which is close to the capacitances listed in Table 3. Then, the slight increase in composite capacitance could be due to the electrodeposition of PAni with a small specific capacitance. Alternatively, it can also be a consequence of the deleterious effect of decreasing the available surface area, which counteracts both the PAni extra capacitance and the reduced sheet resistance. Composites with the highest PAni loading densities show a strongly distorted CV and a decreased capacitance, as a consequence of the increased sheet resistance and the abrupt diminution of the surface area caused by the thick nanofibrous coating. Note that VB analysis revealed that the nanofibrous PAni layers have a higher DOS near the Fermi level (Figure 6). Then, the increased R_s and lessened electrochemical properties imparted by nanofibrous PAni layers should not be caused by poor intrinsic electron conductivity of PAni, but related to high PAni–PAni and/or PAni-carbon intraparticle resistance to charge transport.

The evolution of the specific capacitance observed in this work is in agreement with the results reported earlier by other researchers [12,16,21,23,62]. It has been generally shown that the capacitance is enhanced by either smooth homogeneous PAni coatings of small thickness [12] or nanofiber-shaped assemblies with a certain degree of 3D order and small diameter [22,23] obtained at low PAni loading densities. Such thin nanostructures are believed to provide fast access of electrolyte and shortened path lengths for ion and electron transport. When the loading is raised at longer polymerization times and/or higher monomer concentration, thicker polymer films build up with a high degree of particle agglomeration [12,23,24]. This results in an increase of the diffusion resistance of electrolyte ions and also in a less efficient electrical contact between the conducting polymer and the textile carbon [22].

Finally, we studied the effect of the modification of the ACC by PAni on the adsorption of Acid Red 27. The dye uptake on both bare ACC and PAni-ACC composites obeys PSO kinetics, with the involvement of mass transport on the adsorption rate control. Our results point out that PAni thin films of moderate loading densities provide accelerated PSO adsorption kinetics of Acid Red 27 from aqueous solution, while keeping the maximum adsorption capacity unchanged. However, the values of k_2 (Table 5) and the apparent intraparticle diffusion coefficients from Boyd plots (Table S4) are still below the range of those commonly reported for dye adsorption on many different adsorbents (in the order of 10^{-2}–10^{-4} g mg^{-1}min^{-1} and ~10^{-17} m^2 s^{-1}, respectively) [14,65,71–73]. These low values signify that the internal mass transfer resistance is very high and is dominated by the carbon porous structure, thereby explaining the very long time taken by the adsorption system to attain equilibrium. The adsorption of dyes on PAni powder [10,74] or its nanocomposites with other conducting polymers [65], carbonaceous materials, metal oxides, and low-cost bioadsorbents [9,14,19] has been extensively studied to date, with a general consensus that the adsorption rate is governed by PSO kinetics. The uptake of dyes on PAni-based adsorbents seems to be driven by different types of binding forces, namely, π–π attraction, hydrogen bonding or electrostatic interaction between charges residing on both the PAni backbone and the dye functional groups [9,14]. In our case, the electrostatic interaction between negatively-charged sulfonate groups in Acid Red 27 and positively-charged N sites in the backbone of the acid-doped PAni films overrides the loss in the specific surface and leads to adsorption improvement. Further, the exchange of some (bi) sulfate ions with dye molecules as dopant anions should not be ruled out as an additional mechanism to promote their incorporation to the hybrid adsorbent. A similar explanation was put forward by other authors for the promoted adsorption of anionic sulfonated dyes on PAni [9,10] or PAni-based nanocomposites [19,65]. Despite the fact that the same interaction forces are operative when a thick nanofibrous network of PAni is formed at the longest treatment time, the strong blockage of carbon pores and the abrupt decrease in the specific surface area resulted in an important adsorption rate slowdown and a dramatic loss in the maximum adsorption capacity.

5. Conclusions

Flexible PAni-ACC composites of relatively large size can be easily fabricated by simple galvanostatic electropolymerization of dissolved aniline in a filter-press electrochemical cell, whose modular and stackable design is particularly well suited for easy scaling-up to pilot plant or even technical scales. The morphology of the deposited PAni is strongly dependent on the polymerization time (i.e., the amount of charge passed), ranging from dense and thin films to porous and thick coatings made of interconnected nanofibrils or nanowires. With the exception of some scattered spots, thin films appear featureless in SEM, and their occurrence is indirectly deduced from the XPS N/C and S/C ratios. Composites with a polymer thin film morphology show mixed XPS and valence band patterns from the carbon fabric and the conducting polymer, whereas those for composites with PAni nanofibrous morphology are typical of pure PAni, in a state close to that of (bi) sulfate-doped emeraldine salt.

PAni loadings range from 5 to 18 mg cm^{-2} (~25–50 wt.%), but they increase non-linearly with the electropolymerization time, as shown by combined XPS and TGA data. We propose that aniline polymerization occurs initially inside the carbon pores, but the polymerization rate levels off once aniline is depleted and supplied by slow internal pore diffusion. This mechanism would also facilitate the self-assembly of aniline oligomers into thin films. The analysis of the N_2 and CO_2 isotherm data indicates that thin film PAni deposition occurs on micro- and mesopores, while the ultramicropore volume remains unaffected. Accordingly, the BET surface area decreases by 20%–25%.

Thick nanofibrous coatings formed in highly-loaded PAni-ACC composites cause a strong blockage of pore entrance and a dramatic loss of the specific surface area to values typical of pure PAni.

The conductivity, electrochemical and liquid-phase adsorptive properties of the hybrid PAni-ACC composites are strongly conditioned by the microstructure of deposited PAni. The cyclic voltammograms of composites with thin film morphology show pseudocapacitive features related to reversible leucoemeraldine-emeraldine transition and electron-transfer in phenazine/phenoxazine-like segments. Accordingly, the specific capacitance is enhanced and the sheet resistance falls to a minimum. In addition, the pseudo-second order kinetics for the adsorption of Acid Red 27 are remarkably promoted. We believe that the extra pseudocapacitance, the improved conductivity and the attractive electrostatic interactions between dye molecules and PAni counterbalance the loss in a specific surface area. On the contrary, in composites with a thick nanofibrous morphology, the pronounced decrease in surface area, and perhaps higher PAni-PAni intraparticle and/or PAni–carbon contact resistances, can be at the origin of the decreased capacitance, conductivity, and adsorption rate and capacity.

Supplementary Materials: The following are available online at http://www.mdpi.com/1996-1944/12/16/2516/s1, Figure S1: Isotherm linearized fittings for Acid Red 27 adsorbed on ACC, Figure S2: Pseudo-first order $\ln(q_e - q)$ vs. t plots for the adsorption of Acid Red 27 on ACC, Figure S3: Pseudo-second order t/q vs. t plots for the adsorption of Acid Red 27 on ACC., Figure S4: Boyd plots for the adsorption of Acid Red 27 on ACC, Figure S5: Boyd plots for the adsorption of Acid Red 27 on hybrid PAni-ACC composites, Table S1: Elemental surface composition (in at.%) of ACC and hybrid PAni-ACC composites, Table S2: C1s and N1s peak positions and relative abundance, Table S3: Boyd plot parameters for the adsorption of Acid Red 27 on ACC, Table S4: Boyd plot parameters for the adsorption of Acid Red 27 on hybrid PAni-ACC composites.

Author Contributions: Conceptualization, C.Q.; Methodology, C.Q. and R.B.; Validation, L.L. and E.B.; Formal Analysis, C.Q. and R.B.; Investigation, L.L. and E.B.; Visualization, L.L., E.B. and C.Q.; Data Curation, L.L., R.B. and C.Q.; Writing-Original Draft Preparation, C.Q.; Writing-Review & Editing, all authors.; Supervision, C.Q.; Funding Acquisition, C.Q. and R.B.

Funding: This research was funded by Spanish Ministerio de Economía y Competitividad and FEDER funds, (grants MAT2016-76595-R and RYC-2017-23618) and Generalitat Valenciana (grant PROMETEO/2018/087).

Appendix A

Appendix A.1 Specific Capacitance, C_{sp}

The gravimetric specific capacitance was obtained from CV according to Equation (A1)

$$C_{sp} = \frac{\int (I/m)dE}{v\,\Delta E} \tag{A1}$$

where I/m (mA g^{-1}) is the voltammetric current normalized by the total mass of the composite material, v is the scan rate (mV s^{-1}), and ΔE (mV) is the potential window spanned in the CV. The reported capacitances are the average of capacitances calculated in both the anodic and cathodic branches of the voltammogram.

Appendix A.2 Surface Sheet Resistance, R_s

To measure R_s, the electrical resistance, R (Ω), was first obtained from the linear slope of V vs. I plots in the range 0–10 mA. Then, R_s was calculated by using Equation (A2):

$$R = \frac{\rho}{t}\frac{L}{W} = R_s\frac{L}{W} \tag{A2}$$

where ρ is the resistivity, t is the thickness of the probed carbon composite sheet, and L and W stand for its length and width, respectively.

Appendix A.3 Adsorption Isotherms and Kinetic Models

Langmuir (Equation (A3)) and Freundlich (Equation (A4)) isotherm models were used in their linearized forms:

$$\frac{C_e}{q_e} = \frac{1}{bq_m} + \frac{C_e}{q_m} \tag{A3}$$

$$\ln q_e = \ln K_F + \frac{1}{n} \ln C_e \tag{A4}$$

where q_e (mg g^{-1}) and C_e (mg L^{-1}) are the adsorbate uptake and solution concentration at equilibrium; and q_m (mg g^{-1}) and b (L mg^{-1}) are the Langmuir parameters, representing the maximum adsorption capacity and the adsorption equilibrium constant respectively [59]. In the Freundlich model, the adsorption coefficient, K_F (mg$^{(1-1/n)}$L$^{1/n}$g^{-1}), is related to the adsorption strength, and the exponent n accounts for the energetic heterogeneity of the adsorbent surface [59].

PFO (Equation (A5)) and PSO (Equation (A6)) kinetic models were used in their respective linear forms:

$$\ln(q_e - q) = \ln q_e - k_1 t \tag{A5}$$

$$\frac{t}{q} = \frac{1}{k_2 q_e^2} + \frac{t}{q_e} \tag{A6}$$

where q and q_e (mg g^{-1}) are the adsorption uptakes at time t (min) and at equilibrium respectively, and k_1 (min^{-1}) and k_2 (g mg^{-1} min^{-1}) stand for the pseudo-first and pseudo-second order rate constants. In the PSO model, the product $k_2 q_e^2$ is the initial adsorption rate, r_0.

The Vermeulen model is an approximate solution of the exact Crank equation for intraparticle diffusion:

$$F(t) = \sqrt{1 - \exp(-Bt)} \tag{A7}$$

where $F(t)$ is the fractional uptake (q/q_e) at time t, and B is the time constant (min^{-1}) that depends on the adsorbent particle radius, r_p, and the effective intraparticle diffusion coefficient, D_e, as follows [72,73]:

$$B = \frac{D_e \pi^2}{r_p^2} \tag{A8}$$

The Vermeulen model is valid over a wide adsorption timespan and is commonly applied in the rearranged form:

$$Bt = -\ln(1 - F^2) \tag{A9}$$

Then, the product Bt is calculated from F at each contact time, and next plotted against time.

Appendix A.4 Average Relative Error

The average relative error (ARE) is defined as [60]:

$$ARE = \frac{1}{N} \sum_{i=1}^{N} \left| \frac{q_{i,exp} - q_{i,cal}}{q_{i,exp}} \right| \tag{A10}$$

where the subscripts *exp* and *cal* refer to the experimental and calculated values, and N is the number of data points.

References

1. Le, T.-H.; Kim, Y.; Yoon, H. Electrical and electrochemical properties of conducting polymers. *Polymers* **2017**, *9*, 150. [CrossRef] [PubMed]

2. Ates, M. Review study of electrochemical impedance spectroscopy and equivalent electrical circuits of conducting polymers on carbon surfaces. *Prog. Org. Coat.* **2011**, *71*, 1–10. [CrossRef]

3. Culebras, M.; Gomez, C.M.; Cantarero, A. Review on polymers for thermoelectric applications. *Materials* **2014**, *7*, 6701–6732. [CrossRef] [PubMed]

4. Choi, H.; Yoon, H. Nanostructured electrode materials for electrochemical capacitor applications. *Nanomaterials* **2015**, *5*, 906–936. [CrossRef] [PubMed]

5. Kang, E.T.; Neoh, K.G.; Tan, K.L. Polyaniline: A polymer with many interesting intrinsic redox states. *Prog. Polym. Sci.* **1998**, *23*, 277–324. [CrossRef]

6. Bhadra, S.; Khastgir, D.; Singha, N.K.; Lee, J.H. Progress in preparation, processing and applications of polyaniline. *Prog. Polym. Sci.* **2009**, *34*, 783–810. [CrossRef]

7. Sowa, I.; Wójciak-Kosior, M.; Strzemski, M.; Sawicki, J.; Staniak, M.; Dresler, S.; Szwerc, W.; Mołdoch, J.; Latalski, M. Silica modified with polyaniline as a potential sorbent for matrix solid phase dispersion (MSPD) and dispersive solid phase extraction (d-SPE) of plant samples. *Materials* **2018**, *11*, 467. [CrossRef]

8. Tian, S.C.; Zhang, Z.H.; Zhang, X.H.; Ostrikov, K. Capacitative deionization using commercial activated carbon fiber decorated with polyaniline. *J. Colloid Interf. Sci.* **2019**, *537*, 247–255. [CrossRef]

9. Nasar, A.; Mashkoor, F. Application of polyaniline-based adsorbents for dye removal from water and wastewatera review. *Environ. Sci. Pollut. R.* **2019**, *26*, 5333–5356. [CrossRef]

10. Mahanta, D.; Madras, G.; Radhakrishnan, S.; Patil, S. Adsorption of sulfonated dyes by polyaniline emeraldine salt and its kinetics. *J. Phys. Chem. B* **2008**, *112*, 10153–10157. [CrossRef]

11. Salinas-Torres, D.; Sieben, J.; Lozano-Castello, D.; Cazorla-Amoros, D.; Morallon, E. Asymmetric hybrid capacitors based on activated carbon and activated carbon fibre-PANI electrodes. *Electrochim. Acta* **2013**, *89*, 326–333. [CrossRef]

12. Fonseca, C.; Almeida, D.; Baldan, M.; Ferreira, N. Influence of the PAni morphology deposited on the carbon fiber: An analysis of the capacitive behavior of this hybrid composite. *Chem. Phys. Lett.* **2011**, *511*, 73–76. [CrossRef]

13. Wang, G.P.; Zhang, L.; Zhang, J.J. A review of electrode materials for electrochemical supercapacitors. *Chem. Soc. Rev.* **2012**, *41*, 797–828. [CrossRef] [PubMed]

14. Kumar, R.; Ansari, M.O.; Barakat, M.A. Adsorption of brilliant green by surfactant doped polyaniline/ MWCNTs composite: Evaluation of the kinetic, thermodynamic, and isotherm. *Ind. Eng. Chem. Res.* **2014**, *53*, 7167–7175. [CrossRef]

15. Sipahi, M.; Parlak, E.A.; Gul, A.; Ekinci, E.; Yardim, M.F.; Sarac, A.S. Electrochemical impedance study of polyaniline electrocoated porous carbon foam. *Prog. Org. Coat.* **2008**, *62*, 96–104. [CrossRef]

16. Mondal, S.K.; Barai, K.; Munichandraiah, N. High capacitance properties of polyaniline by electrochemical deposition on a porous carbon substrate. *Electrochim. Acta* **2007**, *52*, 3258–3264. [CrossRef]

17. Salinas-Torres, D.; Sieben, J.M.; Lozano-Castello, D.; Morallon, E.; Burghammer, M.; Riekel, C.; Cazorla-Amoros, D. Characterization of activated carbon fiber/polyaniline materials by position-resolved microbeam small-angle X-ray scattering. *Carbon* **2012**, *50*, 1051–1056. [CrossRef]

18. Chen, W.C.; Wen, T.C.; Teng, H.S. Polyaniline-deposited porous carbon electrode for supercapacitor. *Electrochim. Acta* **2003**, *48*, 641–649. [CrossRef]

19. Gopal, N.; Asaithambi, M.; Sivakumar, P.; Sivakumar, V. Adsorption studies of a direct dye using polyaniline coated activated carbon prepared from Prosopis juliflora. *J. Water Process Eng.* **2014**, *2*, 87–95. [CrossRef]

20. Horng, Y.Y.; Lu, Y.C.; Hsu, Y.K.; Chen, C.C.; Chen, L.C.; Chen, K.H. Flexible supercapacitor based on polyaniline nanowires/carbon cloth with both high gravimetric and area-normalized capacitance. *J. Power Sources* **2010**, *195*, 4418–4422. [CrossRef]

21. Cheng, Q.; Tang, J.; Ma, J.; Zhang, H.; Shinya, N.; Qin, L.C. Polyaniline-Coated Electro-Etched Carbon Fiber Cloth Electrodes for Supercapacitors. *J. Phys. Chem. C* **2011**, *115*, 23584–23590. [CrossRef]

22. He, X.; Gao, B.; Wang, G.; Wei, J.; Zhao, C. A new nanocomposite: Carbon cloth based polyaniline for an electrochemical supercapacitor. *Electrochim. Acta* **2013**, *111*, 210–215. [CrossRef]

23. Dong, L.; Liang, G.; Xu, C.; Liu, W.; Pan, Z.-Z.; Zhou, E.; Kang, F.; Yang, Q.-H. Multi hierarchical construction-induced superior capacitive performances of flexible electrodes for wearable energy storage. *Nano Energy* **2017**, *34*, 242–248. [CrossRef]

24. Yu, P.; Li, Y.; Yu, X.; Zhao, X.; Wu, L.; Zhang, Q. Polyaniline nanowire arrays aligned on nitrogen-doped carbon fabric for high-performance flexible supercapacitors. *Langmuir* **2013**, *29*, 12051–12058. [CrossRef] [PubMed]

25. Ma, J.; Tang, S.; Syed, J.A.; Meng, X. Asymmetric hybrid capacitors based on novel bearded carbon fiber cloth-pinhole polyaniline electrodes with excellent energy density. *Rsc Adv.* **2016**, *6*, 82995–83002. [CrossRef]

26. Tran, H.D.; D'Arcy, J.M.; Wang, Y.; Beltramo, P.J.; Strong, V.A.; Kaner, R.B. The oxidation of aniline to produce "polyaniline": A process yielding many different nanoscale structures. *J. Mater. Chem.* **2011**, *21*, 3534–3550. [CrossRef]

27. Sapurina, I.; Stejskal, J. The mechanism of the oxidative polymerization of aniline and the formation of supramolecular polyaniline structures. *Polym. Int.* **2008**, *57*, 1295–1325. [CrossRef]

28. Leary, J.D.; Hamouda, F.; Mazé, B.; Pourdeyhimi, B. Preparation of pseudocapacitor electrodes via electrodeposition of polyaniline on nonwoven carbon fiber fabrics. *J. Appl. Polym. Sci.* **2016**, *133*. [CrossRef]

29. Rivera, F.; de Leon, C.; Nava, J.; Wals, F. The filter-press FM01-LC laboratory flow reactor and its applications. *Electrochim. Acta* **2015**, *163*, 338–354. [CrossRef]

30. Tabti, Z.; Ruiz-Rosas, R.; Quijada, C.; Cazorla-Amoros, D.; Morallon, E. Tailoring the surface chemistry of activated carbon cloth by electrochemical methods. *Acs Appl. Mater. Inter.* **2014**, *6*, 11682–11691. [CrossRef]

31. Lopez-Bernabeu, S.; Ruiz-Rosas, R.; Quijada, C.; Montilla, F.; Morallon, E. Enhanced removal of 8-quinolinecarboxylic acid in an activated carbon cloth by electroadsorption in aqueous solution. *Chemosphere* **2016**, *144*, 982–988. [CrossRef]

32. Huang, H.C.; Ye, D.Q.; Huang, B.C. Nitrogen plasma modification of viscose-based activated carbon fibers. *Surf. Coat. Tech.* **2007**, *201*, 9533–9540. [CrossRef]

33. Oh, K.W.; Kim, S.H.; Kim, E.A. Improved surface characteristics and the conductivity of polyaniline-nylon 6 fabrics by plasma treatment. *J. Appl. Polym. Sci.* **2001**, *81*, 684–694. [CrossRef]

34. Banaszczyk, J.; Schwarz, A.; De Mey, G.; Van Langenhove, L. The van der pauw method for sheet resistance measurements of polypyrrole-coated para-aramide woven fabrics. *J. Appl. Polym. Sci.* **2010**, *117*, 2553–2558. [CrossRef]

35. Qu, L.; Tian, M.; Zhang, X.; Guo, X.; Zhu, S.; Han, G.; Li, C. Barium sulfate/regenerated cellulose composite fiber with X-ray radiation resistance. *J. Ind. Text.* **2015**, *45*, 352–367. [CrossRef]

36. Volkov, A.; Tourillon, G.; Lacaze, P.C.; Dubois, J.E. Electrochemical polymerization of aromatic amines-IR, XPS and PMT study of thin-film formation on a Pt electrode. *J. Electroanal. Chem.* **1980**, *115*, 279–291. [CrossRef]

37. Bandosz, T.J.; Ania, C.O. Surface chemistry of activated carbons and its characterization. In *Activated Carbon Surfaces in Environmental Remediation*; Bandosz, T.J., Ed.; Elsevier: Amsterdam, The Netherlands, 2006; Volume 7, pp. 159–229. ISBN 978-0-12-370536-5.

38. Chiang, Y.C.; Lee, C.Y.; Lee, H.C. Surface chemistry of polyacrylonitrile- and rayon-based activated carbon fibers after post-heat treatment. *Mater. Chem. Phys.* **2007**, *101*, 199–210. [CrossRef]

39. Yang, S.; Li, L.; Xiao, T.; Zheng, D.; Zhang, Y. Role of surface chemistry in modified ACF (activated carbon fiber)-catalyzed peroxymonosulfate oxidation. *Appl. Surf. Sci.* **2016**, *383*, 142–150. [CrossRef]

40. Xie, Y.M.; Wang, T.J.; Franklin, O.; Sherwood, P.M.A. X-Ray Photoelectron spectroscopic studies of carbon-fiber surfaces.XV. Core-level and valence-band studies of pitch-based fibers electrochemically treated in ammonium carbonate solution. *Appl. Spectrosc.* **1992**, *46*, 645–651. [CrossRef]

41. Cotarelo, M.; Huerta, F.; Quijada, C.; Mallavia, R.; Vaquez, J. Synthesis and characterization of electroactive films deposited from aniline dimers. *J. Electrochem. Soc.* **2006**, *153*, D114–D122. [CrossRef]

42. Cotarelo, M.A.; Huerta, F.; Quijada, C.; Perez, J.M.; del Valle, M.A.; Vazqueza, J.L. Spectroscopic and electrochemical study of the redox process of poly(2,2'-dithiodianiline). *J. Electrochem. Soc.* **2006**, *153*, A2071–A2076. [CrossRef]

43. Chen, W.C.; Wen, T.C.; Hu, C.C.; Gopalan, A. Identification of inductive behavior for polyaniline via electrochemical impedance spectroscopy. *Electrochim. Acta* **2002**, *47*, 1305–1315. [CrossRef]

44. Beamson, G.; Briggs, D. *High Resolution XPS of Organic Polymers: The Sciencia ESCA 300 Database*; John Wiley & Sons Ltd: Chichester, UK, 1992.

45. Bai, B.C.; Lee, H.U.; Lee, C.W.; Lee, Y.S.; Im, J.S. N_2 plasma treatment on activated carbon fibers for toxic gas removal: Mechanism study by electrochemical investigation. *Chem. Eng. J.* **2016**, *306*, 260–268. [CrossRef]

46. Nakajima, T.; Harada, M.; Osawa, R.; Kawagoe, T.; Furukawa, Y.; Harada, I. Study on the interconversion of unit structures in polyaniline by X-ray photelectron spectroscopy. *Macromolecules* **1989**, *22*, 2644–2648. [CrossRef]

47. Vempati, S.; Ertas, Y.; Babu, V.J.; Uyar, T. Optoelectronic properties of layered titanate nanostructure and polyaniline impregnated devices. *ChemistrySelect* **2016**, *1*, 5885–5891. [CrossRef]

48. Bocchini, S.; Castellino, M.; Della Pina, C.; Rajan, K.; Falletta, E.; Chiolerio, A. Inkjet printed doped polyaniline: Navigating through physics and chemistry for the next generation devices. *Appl. Surf. Sci.* **2018**, *456*, 246–258. [CrossRef]

49. Kruk, M.; Jaroniec, M. Gas adsorption characterization of ordered organic-inorganic nanocomposite materials. *Chem. Mater.* **2001**, *13*, 3169–3183. [CrossRef]

50. Boyle, A.; Penneau, J.F.; Genies, E.; Riekel, C. The effect of heating on polyaniline powders studied by real-time synchroton radiation diffraction, mass-spectrometry and thermal analysis. *J. Polym. Sci. Pol. Phys.* **1992**, *30*, 265–274. [CrossRef]

51. Chen, C.H. Thermal and morphological studies of chemically prepared emeraldine-base-form polyaniline powder. *J. Appl. Polym. Sci.* **2003**, *89*, 2142–2148. [CrossRef]

52. Salavagione, H.J.; Cazorla-Amoros, D.; Tidjane, S.; Belbachir, M.; Benyoucef, A.; Morallon, E. Effect of the intercalated cation on the properties of poly(o-methylaniline)/maghnite clay nanocomposites. *Eur. Polym. J.* **2008**, *44*, 1275–1284. [CrossRef]

53. Trchova, M.; Konyushenko, E.N.; Stejskal, J.; Kovarova, J.; Ciric-Marjanovic, G. The conversion of polyaniline nanotubes to nitrogen-containing carbon nanotubes and their comparison with multi-walled carbon nanotubes. *Polym. Degrad. Stabil.* **2009**, *94*, 929–938. [CrossRef]

54. Kuroki, S.; Hosaka, Y.; Yamauchi, C. A solid-state NMR study of the carbonization of polyaniline. *Carbon* **2013**, *55*, 160–167. [CrossRef]

55. Lin, Y.R.; Teng, H.S. A novel method for carbon modification with minute polyaniline deposition to enhance the capacitance of porous carbon electrodes. *Carbon* **2003**, *41*, 2865–2871. [CrossRef]

56. Quijada, C.; Morallón, E.; Huerta, F.; Montilla, F. Electrochemical strategies for tuning the properties of Polyaniline-based conducting polymers. In *Recent Advances within the Field of Materials Science in Spain*; Berenguer-Murcia, A., Caturla, M.J., Molina, J.M., Morallón, E., Quijada, C., Román, M.C., Sancho, J.C., Vidal, L., Eds.; Universitat d'Alacant: Sant Vicent del Raspeig, Spain, 2015; pp. 429–443. ISBN 978-0-12-370536-5.

57. del Valle, M.; Diaz, F.; Bodini, M.; Alfonso, G.; Soto, G.; Borrego, E. Electrosynthesis and characterization of o-phenylenediamine oligomers. *Polym. Int.* **2005**, *54*, 526–532. [CrossRef]

58. Ruthven, D.M. *Principles of Adsorption and Adsorption Processes*; Wiley & Sons Inc.: New York, NY, USA, 1984; pp. 88–123. ISBN 0-471-86606-7.

59. Worch, E. *Adsorption Technology in Water Treatment. Fundamentals, Processes and Modeling*; De Gruyter GmbH & Co: Göttingen, Germany, 2012; pp. 162–164. ISBN 978-3-11-024022-1.

60. Tan, K.L.; Hameed, B.H. Insight into the adsorption kinetics models for the removal of contaminants from aqueous solutions. *J. Taiwan Inst. Chem. E.* **2017**, *74*, 25–48. [CrossRef]

61. Haerifar, M.; Azizian, S. Mixed surface reaction and diffusion-controlled kinetic model for adsorption at the solid/solution interface. *J. Phys. Chem. C* **2013**, *117*, 8310–8317. [CrossRef]

62. Hu, C.C.; Li, W.Y.; Lin, J.Y. The capacitive characteristics of supercapacitors consisting of activated carbon fabric-polyaniline composites in $NaNO_3$. *J. Power Sources* **2004**, *137*, 152–157. [CrossRef]

63. Zhong, M.; Song, Y.; Li, Y.; Ma, C.; Zhai, X.; Shi, J.; Guo, Q.; Liu, L. Effect of reduced graphene oxide on the properties of an activated carbon cloth/polyaniline flexible electrode for supercapacitor application. *J. Power Sources* **2012**, *217*, 6–12. [CrossRef]

64. Li, Y.; Chen, C. Polyaniline/carbon nanotubes-decorated activated carbon fiber felt as high-performance, free-standing and flexible supercapacitor electrodes. *J. Mater. Sci.* **2017**, *52*, 12348–12357. [CrossRef]

65. Bhaumik, M.; McCrindle, R.; Maity, A. Efficient removal of Congo red from aqueous solutions by adsorption onto interconnected polypyrrole-polyaniline nanofibres. *Chem. Eng. J.* **2013**, *228*, 506–515. [CrossRef]

66. do Nascimento, G.; Constantino, V.; Landers, R.; Temperini, M. Aniline polymerization into montmorillonite clay: A spectroscopic investigation of the intercalated conduct-ling polymer. *Macromolecules* **2004**, *37*, 9373–9385. [CrossRef]

67. Boutaleb, N.; Benyoucef, A.; Salavagione, H.; Belbachir, M.; Morallon, E. Electrochemical behaviour of conducting polymers obtained into clay-catalyst layers. An in situ Raman spectroscopy study. *Eur. Polym. J.* **2006**, *42*, 733–739. [CrossRef]
68. Trchova, M.; Moravkova, Z.; Blaha, M.; Stejskal, J. Raman spectroscopy of polyaniline and oligoaniline thin films. *Electrochim. Acta* **2014**, *122*, 28–38. [CrossRef]
69. Li, H.; Wang, J.; Chu, Q.; Wang, Z.; Zhang, F.; Wang, S. Theoretical and experimental specific capacitance of polyaniline in sulfuric acid. *J. Power Sources* **2009**, *190*, 578–586. [CrossRef]
70. Snook, G.A.; Kao, P.; Best, A.S. Conducting-polymer-based supercapacitor devices and electrodes. *J. Power Sources* **2011**, *196*, 1–12. [CrossRef]
71. Ozcan, A.S.; Ozcan, A. Adsorption of acid dyes from aqueous solutions onto acid-activated bentonite. *J. Colloid Interf. Sci.* **2004**, *276*, 39–46. [CrossRef]
72. Porkodi, K.; Kumar, K.V. Equilibrium, kinetics and mechanism modeling and simulation of basic and acid dyes sorption onto jute fiber carbon: Eosin yellow, malachite green and crystal violet single component systems. *J. Hazard. Mater.* **2007**, *143*, 311–327. [CrossRef]
73. Garcia-Mateos, F.J.; Ruiz-Rosas, R.; Marques, M.D.; Cotoruelo, L.M.; Rodriguez-Mirasol, J.; Cordero, T. Removal of paracetamol on biomass-derived activated carbon: Modeling the fixed bed breakthrough curves using batch adsorption experiments. *Chem. Eng. J.* **2015**, *279*, 18–30. [CrossRef]
74. Ayad, M.M.; Abu El-Nasr, A. Adsorption of cationic dye (methylene blue) from water using polyaniline nanotubes base. *J. Phys. Chem. C* **2010**, *114*, 14377–14383. [CrossRef]

MDPI

St. Alban-Anlage 66

4052 Basel

Switzerland

Tel. +41 61 683 77 34

Fax +41 61 302 89 18

www.mdpi.com

Materials Editorial Office

E-mail: materials@mdpi.com

www.mdpi.com/journal/materials

* 9 7 8 3 0 3 9 3 6 4 9 7 8 *